# Nuclear Dynamite
## The Peaceful Nuclear Explosions Fiasco

Trevor Findlay

Brassey's Australia
A Division of Pergamon Press Australia

Member of Maxwell Macmillan Pergamon Publishing Corporation

Brassey's has imprints in London, Washington and Sydney. The companies' origins lie in *Brassey's Naval Annual*, first published in 1886. Brassey's now specialises in books on international relations, defence, foreign policy, strategy and tactics, military history, military affairs and military technology. Brassey's (UK) also publishes the annuals, papers and journal of the International Institute for Strategic Studies. All three Brassey's companies welcome English language publishing proposals from authors and institutions throughout the world.

Brassey's Australia is a division of Pergamon Press Australia and a member of the Maxwell Macmillan Pergamon Publishing Corporation.

| | |
|---|---|
| AUSTRALIA | Pergamon Press (Australia) Pty Ltd, 19a Boundary Street, Rushcutters Bay, NSW 2011, Australia |
| UK | Pergamon Press plc, Headington Hill Hall, Oxford OX3 0BW, England |
| USA | Pergamon Books, Inc., Maxwell House, Fairview Park, Elmsford, New York 10523, USA |
| PEOPLE'S REPUBLIC OF CHINA | Pergamon Press, Room 4037, Qianmen Hotel, Beijing, People's Republic of China |
| FEDERAL REPUBLIC OF GERMANY | Pergamon Press GmbH, Hammerweg 6, D-6242 Kronberg, Federal Republic of Germany |
| BRAZIL | Pergamon Editora Ltda, Rue Eça de Queiros, 346, CEP 04011, Paraiso, São Paulo, Brazil |
| JAPAN | Pergamon Press, 8th Floor, Matsuoka Central Building, 1-7-1 Nishishinjuku, Shinjuku-ku, Tokyo 160, Japan |
| CANADA | Pergamon Press Canada, Suit 271, 253 College Street, Toronto, Ontario M5T 1R5, Canada |

First published 1990



Cover design by Adrian Young

Typeset in Australia by Post Typesetters
Printed in Singapore by Singapore National Printers Ltd

National Library of Australia Cataloguing in Publication Data:

Findlay, Trevor, 1951–
Nuclear dynamite: the peaceful nuclear explosions fiasco.

Bibliography.
Includes index.
ISBN 0 08 034436 4.

1. Project Plowshare — History. 2 Nuclear explosions. 3. Nuclear excavation. I. Title.

621.48

To my parents,
Len and Linda Findlay

Out of the dust of the desert dawn at Alamagordo in July of 1945, a new age of man began with a suddenness not characterising the beginning of any other epoch in his history. We are not sure what History will call it: the Atomic Age, the Age of Cybernetics, the Age of Science or whatever. Maybe in the circumspection of a millennium it will be called the Age of Gadgetry — the Toyhood Period of Man. But in many respects a uniquely and complexly new set of problems was inherited by all living things, including the species that now holds dominance over the planet's biota.

*Report* of the Committee on Enviromental Studies
for Project Chariot, Plowshare Program, 1966

# Contents

# Acknowledgements

I would like to thank the following individuals for their assistance, in various ways, in the writing of this book: Desmond Ball, Peter Basham, Nenne Bodell, Iris Borg, Susan Bower, Nick Carrera, Pearce Corden, Kathy Crittenburger, David Fischer, Ed Fleming, Jozef Goldblat, John Gunderson, Warren Heckrotte, Eleanor Hodges, Mary-Lou Hickey, Fred Holzer, Gerald Johnson, Pauline Kerr, Andrew Mack, John Miskel, Keith Mitchell, Milo Nordyke, Christopher Reberger, Vivien Read, Lawrence Scheinman, Tonia Shand, Carlene Stockton, Edward Teller, Alan Waite, Alan Wilson, Hugh de Witt, Herbert York, Gerard Van Bohemen, Peter and Margaret Varghese.

I would also like to acknowledge the assistance of the following organisations: the Australian National Library, Canberra; the Australian Nuclear Science and Technology Organisation, Sydney (formerly the Australian Atomic Energy Commission); the Australian National University, Canberra; the Australian Seismological Centre, Canberra; the Canadian Centre for Disarmament and Arms Control, Ottawa; the Cartography Department, Research School of Pacific Studies, Australian National University, Canberra; the Department of Foreign Affairs and Trade, Canberra; the Lawrence Livermore National Laboratory, California; the Peace Research Centre, Canberra; the Stockholm International Peace Research Institute, Stockholm; the US Arms Control and Disarmament Agency, Washington DC; and the US State Department, Washington DC.

*Nuclear Dynamite* was written under the auspices of a Disarmament Fellowship at the Peace Research Centre, Australian National University, funded by the Australian Department of Foreign Affairs and Trade. I am grateful to the Department and the Peace Research Centre for their support. I am especially indebted to Andrew Mack, Head of the Centre, for his unwavering enthusiasm and encouragement, to Mary-Lou Hickey, the best proof-reader in the business, and to Carlene Stockton for her multifarious assistance.

Unless otherwise attributed, responsibility for errors of fact or judgement in *Nuclear Dynamite* resides, of course, with me.

# Preface

Serious examination of the possibilities of 'nuclear dynamite', the use of nuclear explosives for non-military purposes, began in the mid-1950s with American and Soviet theoretical studies using data from thermonuclear weapon tests. Subsequently both countries established substantial peaceful nuclear explosion (PNE) programs, conducting experiments at nuclear weapon test sites and in the field. The uses envisaged for PNEs included: earthmoving, earthbreaking, harbour, canal and highway construction, oil and gas stimulation, underground cavity formation, seismic surveys, scientific experiments (such as producing transuranic elements), steam-power generation and the extinguishing of oil and gas well fires.

Despite their early promise, peaceful nuclear explosives did not prove to be the salvation of humankind, as early enthusiasts hoped. Instead they proved to be uneconomic, environmentally hazardous, an impediment to disarmament and arms control efforts and irrevocably tainted by their consanguinity with nuclear weapons. By the early 1970s interest in their use had declined dramatically.

The story of the declining fortunes of nuclear dynamite is partly the tale of the collapse of the American PNE program, Project Plowshare, and partly the saga of international efforts to cope with the arms control implications of PNEs. For Plowshare it was a case of questionable management, doubtful economics, poor industrial response, environmental opposition and declining governmental support due to budgetary pressures. Internationally, the PNE story has been one of genuine idealism, self-serving nationalism and underhandedness. While in technical terms the benefits of PNEs have not been proven or disproven, in political terms nuclear dynamite has largely been discredited. Only the Soviets continue a PNE program, claiming, with decreasing certainty and regularity, that for them it is both economic and effective.

While a complicated net of international laws now governs the use of nuclear dynamite, several countries besides those possessing nuclear weapons are still legally entitled to develop and use PNEs. The PNE story will therefore only end with a complete ban on the development and use of nuclear dynamite by all states, preferably as a result of a comprehensive nuclear test ban treaty.

# Abbreviations

| | |
|---|---|
| AAEC | Australian Atomic Energy Commission |
| AEC | US Atomic Energy Commission |
| ACDA | US Arms Control and Disarmament Agency |
| BOG | Board of Governors of the IAEA |
| CCD | Conference of the Committee on Disarmament |
| CD | Committee on Disarmament, later Conference on Disarmament |
| CSC | Interoceanic Canal Study Commission |
| CTBT | Comprehensive Nuclear Test Ban Treaty |
| ENDC | Eighteen-Nation Disarmament Committee |
| GURC | Gulf Universities Research Council |
| FRG | Federal Republic of Germany |
| IAEA | International Atomic Energy Agency |
| ICC | International Control Commission |
| JVE | Joint Verification Experiment |
| MHF | Massive Hydraulic Fracturing |
| NATO | North Atlantic Treaty Organisation |
| NEPA | National Environmental Protection Act |
| NEVOO | Nevada Operations Office |
| NPT | Nuclear Nonproliferation Treaty |
| NTB | Nuclear Test Ban |
| OPANAL | Agency for the Prohibition of Nuclear Weapons in Latin America |
| PNE | Peaceful Nuclear Explosion or Explosive |
| PNET | Peaceful Nuclear Explosions Treaty |
| PTBT | Partial or Limited Test Ban Treaty |
| SALT | Strategic Arms Limitation Talks |
| SIPRI | Stockholm International Peace Research Institute |
| SPNFZ | South Pacific Nuclear-Free Zone Treaty |
| TTBT | Threshold Test Ban Treaty |
| UN | United Nations |
| USAEC | US Atomic Energy Commission |

# 1

# Swords Into Plowshares: The Invention of Nuclear Dynamite

*They shall beat their swords into plowshares, and their spears into pruning hooks: nation shall not lift up sword against nation, neither shall they learn war any more.*

Isaiah 2:4

THE realisation that the splitting of the atom would release enormous amounts of explosive energy led inexorably to the idea that such energy could be harnessed for peaceful purposes. Just as chemical explosives had proved useful in both military and peaceful applications, so it was predicted that nuclear weapons would prove useful as 'peaceful nuclear explosives' (PNEs)[1] or 'nuclear dynamite'. According to Glenn Seaborg, former chairman of the US Atomic Energy Commission (AEC), the constructive possibilities of nuclear explosives were apparent even to Enrico Fermi and his group during the Manhattan Project.[2] Of these nuclear pioneers, it was the eminent American mathematician, Dr John Von Neumann, who was apparently the first to specifically suggest that nuclear explosives could be used for 'peaceful' purposes. During the early days of the nuclear era, however, such proposals were not given much credence. Nuclear materials were scarce and required for weapons, explosive yields were low, costs were high, and atomic bombs were assumed to be too radioactively 'dirty' to use for non-military purposes. Scientists were in any case preoccupied with producing nuclear weapons, with little time to consider the peaceful uses of nuclear energy.

The catalyst for serious consideration of the possibilities of nuclear dynamite by American scientists came in 1949, ironically, from the Soviet Union.

Following the first Soviet atomic bomb test, Foreign Minister Vishinsky declared that the Soviet Union was not using atomic energy for weapons but rather:[3]

> We are razing mountains, we are irrigating deserts, we are cutting through the jungle, we are spreading life, happiness, prosperity and welfare in places where the human footstep has not been seen for thousands of years.

Even though they knew the Soviets were dissembling, scientists at the Los Alamos Laboratory in New Mexico looked into the feasibility of 'moving mountains' with nuclear explosions. They concluded that 'it did not look like a very practical idea'.[4] In June 1950, in the *Bulletin of the Atomic Scientists*, a Los Alamos staff member, Fred Reines, summarised the sceptical views of his colleagues. Taking literally the Soviet claim that mountains could be razed, Reines concluded that while an atomic bomb of the type dropped on Nagasaki[5] could move a hill a quarter of a mile high and a quarter of a mile across, mountains could not so easily be moved.[6] Atomic bombs were simply not powerful enough. As for the bizarre suggestion that atomic blasts could be used to melt the Arctic icecap to produce water, Reines commented that such a method would melt much less ice than it would pulverise. He concluded that the use of atomic blasts in engineering applications appeared 'at best to be extremely limited in scope, owing to the high concentration of energy and the radioactivity associated with atomic explosions'.[7] In a somewhat unscientific, and in retrospect ironic, jibe at the Soviets, Reines added:[8]

> Of course, if the radioactivity is overlooked, more things become possible, but it is obvious that the Russians have chosen to ignore known facts and to take all possible propaganda advantage of their newly acquired weapon.

In 1952 two events were to set the scene for overturning the scepticism of the Los Alamos scientists. The first was the successful detonation, at Eniwetok Atoll in the South Pacific, of the world's first full-scale thermonuclear or fusion device, otherwise known as the hydrogen bomb, H-bomb or 'superbomb'. The explosion, on 1 November 1952, obliterated the island of Elugelab, dramatically demonstrating the possibilities of nuclear excavation. This and subsequent H-bomb explosions in the Pacific, according to physicist Gerald Johnson, 'set the technical stage' for peaceful applications such as excavation. They did so by demonstrating major advantages that fusion explosives have over atomic (fission) explosives and conventional chemical explosives.[9]

First, unlike fission explosives, there is in theory no limit to the power or yield of fusion explosives. Fission bombs, being dependent on a 'critical size' of uranium to explode, are inherently limited in yield: if a fission bomb exceeds the critical, it explodes at once; if it is smaller it cannot be exploded at all.[10] A fusion bomb cannot be smaller than a fission bomb since it must use a fission bomb as a detonator. But it can be many times bigger. Its size depends on the amount of tritium and deuterium added to fuel the fusion process begun by the fission trigger. Thermonuclear bombs can therefore be made powerful enough for any 'planetary engineering' job imaginable.[11]

Second, thermonuclear devices are cheaper, since deuterium and tritium are isotopes of comparatively abundant hydrogen, while atomic devices use expensive uranium and plutonium. Moreover, because of their very high energy density, on either a volume or weight basis, fusion devices are cheaper than conventional chemical explosives of the same power, by a factor of ten or more.[12] A 2 megaton[13] thermonuclear device, for example, could fit into a 40 inch (100cm) drill hole, while TNT of the same power would be in the shape of a cube almost 300 feet (90m) on each side.[14] Moreover, the economic gain from thermonuclear explosives theoretically increases with the yield. For example, increasing the yield from 10 to 2000 kilotons — a 200-fold increase — would not even double the cost. Hence the bigger the bang, the cheaper the nuclear dynamite.

Third, fusion explosions are radioactively 'cleaner' than fission explosions of the same yield, creating fewer and less dangerous radioactive by-products. While fusion does create *some* harmful radioactive isotopes, notably tritium, these are relatively minor compared with fission's roughly 200 radioactive by-products.[15] Since the 'dirtiest' part of a thermonuclear explosion is the fission explosion used to trigger it, the key to producing even 'cleaner' thermonuclear explosions is to minimise the size of the fission trigger.

After the successful detonation of the H-bomb, the second event that hastened the development of nuclear dynamite was the establishment of the Livermore Laboratory in California.[16] Designed as an alternative centre of nuclear weapons research and development to Los Alamos, Livermore was established largely at the insistence of Edward Teller. Teller, the Hungarian-born physicist commonly considered the 'father of the H-bomb', disagreed sharply with both the pace and direction of work on the bomb at Los Alamos. He saw the H-bomb as a vital element in America's defence against the Soviet threat and therefore deserving of almost unlimited financial and scientific support from the nation.

Years later Teller would deny that he knew at the time that thermonuclear explosives would lead to 'clean' bombs that could be used for peaceful purposes.[17] The development of PNEs *per se* or investigation of their specific uses was thus not among his principal reasons for advocating work on the H-bomb or the establishment of the Livermore Laboratory. Nonetheless, almost from the time that the possibility of a 'super' had been conceived, Teller had harboured vague notions about its peaceful applications. These undoubtedly constituted a further justification for proceeding with its development. As early as October 1945 Teller intimated to Enrico Fermi that among his reasons for pursuing a thermonuclear explosive was '. . . a certainty that the super bomb will allow us to extend our power over natural phenomena far beyond anything we can at present imagine'.[18] Teller confirmed in 1976 that:[19]

> ... I did believe in science and progress and in finding out what can be done. And I did have the confidence that there was at least a good possibility that what we developed would be properly used rather than misused.

Whatever his intention, the establishment of the new laboratory not only gave Teller the freedom to work on the H-bomb in his own way, but allowed him to pursue his vision of using nuclear technology, including nuclear explosives, for peaceful purposes. He was supported by physicist Ernest Lawrence, mentor of the laboratory and, like Teller, a strong proponent of the H-bomb.[20]

Livermore would become the 'chief idea factory' of the United States' peaceful nuclear explosions program.[21] In Glenn Seaborg's words, the 'far-sighted American PNE program was crystalised by Livermore scientists, and this laboratory ... had scientific responsibility for this project from the beginning'.[22] This was partly because of the enthusiasm of Teller and Lawrence, and partly because of the bright young scientists recruited to the laboratory.[23] As Seaborg explains:[24]

> ... the late Ernest O. Lawrence and Edward Teller assembled a corporal's guard of amazingly young men, most of whom had only recently received their PhDs ... Their job was to help improve and diversify the nuclear weapons that are so important to American security and to explore some prospects for applying nuclear energy for peaceful purposes. They were infected with the 'gung ho' spirit of Ernest Lawrence, a spirit that has prevailed at Livermore ever since. Their performance has been brilliant, imaginative, and invaluable to the nation in both general areas of their responsibility.

Theodore Taylor, former director of the US Defense Nuclear Agency, describes Livermore (and Los Alamos) in the 1950s as:[25]

> ... nurtured by massive government funding, (with) a high degree of local authority ... to pursue exciting new possibilities and unqualified government enthusiasm for our activities. Perhaps most stimulating of all was the intense exhilaration that comes from exploring radically new technical concepts ... For many of us ... the climax of this exhilaration ... was nuclear tests.

Another factor in Livermore's pursuit of 'peaceful' uses for nuclear explosives was undoubtedly the sense of competition between Livermore and Los Alamos. Herbert York, the first director of Livermore, writes:[26]

> I have little doubt that the Livermore Laboratory was seen as competition by Los Alamos, and I know for a fact that we at Livermore saw ourselves as being in competition with them.

This, in York's view, resulted in the proliferation of a 'very great variety' of nuclear weapon designs for the US arsenal.[27] It also encouraged the search for exotic new uses of nuclear energy, such as nuclear dynamite.

In 1956, four years after the Eniwetok explosion and six years after Los Alamos scientists had dismissed the idea of PNEs, Harold Brown, one of the young 'corporal's guard' recruited to Livermore in its formative months, initiated a 'brainstorming' session to look afresh into the possibilities of nuclear dynamite.[28] His 'brainstorming' group, comprising Arthur Biehl, Edward Teller, Gerald Johnson and Herbert York, concluded that the idea of using nuclear devices for 'peaceful purposes' was promising enough to merit study.[29] It was at this meeting, according to York, that the Livermore Laboratory 're-invented' PNEs.[30] Teller, Brown and Johnson subsequently wrote to the Atomic Energy Commission with a number of imaginative proposals for using nuclear dynamite.[31]

One of the ideas considered by the Livermore group arose from the 1956 Suez Crisis. In November 1956, with Egypt blockading the Suez Canal, Brown proposed the use of nuclear explosives to cut a new 300-mile (480-kilometre) canal through Israel, from Gaza on the shores of the Mediterranean to the Gulf of Aqaba.[32] Congressman Craig Hosmer, a Californian Republican, later described this as one of the better ideas for circumventing the blockade: '... if you can't get through the existing canal, dig a new one! And do it with nuclear explosives'.[33] While Teller's biographers claim that the Suez Crisis was the 'political trigger' for the American PNE program,[34] Gerald Johnson describes the canal proposal as simply an 'exercise' that was not taken seriously.[35] Johnson recalls that when the Suez was closed, 'we simply pulled out a map, developed a profile from the Gulf of Aqaba to the Mediterranean and rapidly decided in principle it was technically possible'. But no proposal was submitted to the AEC or raised with the Israelis. When the Suez Canal was reopened, discussion evaporated.[36] Nevertheless Johnson testified to the Congressional Joint Committee on Atomic Energy (JCAE) in 1964 that while the Suez Canal project 'turned out to be not too feasible... the study did plant the seed of the possibility of nuclear excavations that is still growing'.[37]

In February 1957, 'under the inspired leadership of Dr Harold Brown and with the enthusiastic support of Professor Ernest O. Lawrence and Dr Edward Teller',[38] a more formal, but still secret conference was convened at Livermore, to 'sift fact from fancy in the new field of using nuclear explosives for peaceful and beneficial pruposes'.[39] This time, in addition to Livermore scientists, including Brown, Johnson and Teller,[40] the meeting was attended by representatives of the AEC, Los Alamos and the Sandia Corporation.[41] The basic concepts that emerged at this meeting are said to have guided the subsequent peaceful nuclear explosions program of the United States.[42] It focused on nuclear excavation, 'the most natural extension of the conventional use of high explosives' and, in Glenn Seaborg's words, the application that ultimately 'most stirred the world's imagination'.[43] According to an official report:[44]

> Egypt's seizure of the Suez Canal in 1956 and the long-expressed interest of the Panama Canal Company in excavating a sea-level canal across the American Isthmus placed the initial focus... on large scale earth excavation.

Other ideas 'kicked around' included the use of underground explosions to facilitate open-cut mining and stimulate oil production.

As a direct result of the conference, the Atomic Energy Commission decided in June 1957 to formalise the Livermore Laboratory's hitherto informal investigation of PNEs by establishing Project Plowshare.[45] Harold Brown seized on the name, derived from the Biblical exhortation to nations to beat their swords into plowshares, after mentioning his interest in PNEs to physicist Dr I.I. Rabi. Rabi responded: 'So you want to beat your old atomic bombs into plowshares!'[46]

'Plowshare' was, however, clearly a misnomer. The program would not heed the Biblical exhortation by turning nuclear weapons into 'harmless' peaceful nuclear explosives. New types of nuclear devices would be developed and additional nuclear devices produced specially for the program. Moreover, even so-called 'peaceful nuclear explosives' could still be used, if necessary, as weapons. Plowshare would consequently not result in fewer nuclear weapons being available for military purposes. On the contrary, it would impede and complicate efforts aimed at halting the nuclear arms race by means of negotiated disarmament and arms control agreements. Its existence would also pave the way for India's claim that its 1974 nuclear weapon test was a 'peaceful nuclear explosion'. As Pringle and Spigelman note, if those who coined the name Plowshare had read further, 'they might have been given pause'.[47] In the book of Joel the plowshares are readily turned back into swords.

In 1957 none of these considerations weighed on the minds of Plowshare's founders. On the contrary, the establishment of Plowshare was accompanied by 'enthusiasm and optimism' at the Livermore Laboratory.[48] It also received the 'active and enthusiastic support' of the AEC and the Joint Committee on Atomic Energy.[49] Enthusiasm was tempered, according to Livermore geologist Iris Borg, only by the knowledge that nuclear explosive technology was still in its infancy.[50]

No sooner was Plowshare conceived, however, than it was threatened by growing international controversy over radioactive fallout from nuclear testing in the atmosphere by the United States and the Soviet Union. In June 1957, the month that Plowshare was formally established, special Presidential Assistant for Disarmament, Harold Stassen, was preparing to recommend to President Eisenhower that the United States join a moratorium on all nuclear tests. Eisenhower had himself already expressed interest in such an idea. On 14 June the Soviet Union pre-empted Stassen with its own moratorium proposal. Ten days later Edward Teller, Ernest Lawrence, AEC Chairman Strauss, and Mark Mills, one of Teller's principal assistants, were summoned to a meeting with the President on the issue.[51] Teller opposed the moratorium. His proferred solution to the fallout controversy was the 'clean' H-bomb. He predicted that his scientists could reduce the fission (or radioactively 'dirty') component of such explosives to 5 per cent within seven years.[52] Such 'clean' bombs would not only ease the fallout problem but would be useful for peaceful purposes such as excavation. Teller recalls:[53]

> One point was raised in the discussion which was and is of great importance. We can perfect 'clean' nuclear explosives. These can be used in war to destroy an intended target without releasing radioactivity to be carried by the winds to do damage indiscriminately where no damage was intended. These 'clean' explosives can also be used in peace as powerful workhorses in mammoth construction jobs.

Eisenhower seemed highly interested in Teller's presentation.[54] In Gerald Johnson's view however, he was mostly interested in the use of 'clean' bombs in war, rather than in their potential peaceful applications.[55] Eisenhower's reported 'great interest' in PNEs[56] was, Johnson says, simply politeness. Nevertheless, at a press conference the following day, Eisenhower argued against ending nuclear tests on the grounds that it 'might impede progress on the production of a fallout-free nuclear bomb *and* [emphasis added] on the development of nuclear energy for peaceful purposes'.[57] This saved Plowshare — for the time being.

In September 1957, only three months after Plowshare had formally begun, another event of crucial importance to its future occurred. This was the detonation of the world's first fully-contained underground nuclear explosion. First proposed early in 1956 by Livermore geophysicist David Griggs,[58] the idea of testing underground, says Gerald Johnson, 'fell on fertile soil with us at Livermore' as a way of avoiding the problems associated with atmospheric testing: primarily the fallout, but also the 'very bright flash' and airblast.[59] It also promised to reduce costly delays caused by unfavourable weather conditions at the Nevada Test Site.[60] Johnson, later director of Plowshare at Livermore, realised that underground testing could have also 'an important bearing' on the laboratory's plans for nuclear dynamite.[61]

In mid-1956, at about the time that Plowshare was being conceived, Livermore's Test Division, headed by Johnson, began planning the first underground test explosion.[62] Codenamed Operation Rainier, it involved the detonation of a 1.7 kiloton device inside a mesa at the Nevada Test Site. It was designed to discover whether radioactivity could be contained underground, whether there would be any seismic consequences, such as earth tremors, and whether the costs would be reasonable.[63] The experiment was opposed by scientists at Los Alamos, including director Norris Bradbury, who believed it was unlikely to work.[64]

Rainier was detonated on 19 September 1957. As Johnson and Brown describe it:[65]

> In its first instants the explosion melted and vapourized some 700 tons of rock. The small chamber inflated to a hollow cavity 110 feet in diameter enclosed in a four-inch thick shell of liquefied rock, which dripped from the ceiling in stalactites and condensed in glassy 'raindrops' as it fell... As an assurance of the safety of future underground explosions, it was found that almost all of the radioactivity was concentrated in this hard, insoluble material. About one minute after the explosion the ceiling of the cavity

> collapsed under the burden of the tons of crushed and broken rock above it; successive caving produced a 'chimney' which ran upward several hundred feet through the rock toward the top of the mesa.

According to Edward Teller, all the radioactivity from the explosion was trapped in the cavity where it '... was watched for years. It moved inches, feet. Long before it could reach any living thing, it would decay'.[66] For Teller:[67]

> All this was highly interesting, and it was most important to Project Plowshare. Our experiment had confirmed that we could break up large quantities of rock and make it permeable to water. And other nuclear explosions had taught us that we could greatly reduce the total radioactive output of a bomb. In time, we could make our Plowshare tools 'clean'.

The Rainier experiment inspired several 'underground engineering' ideas under the Plowshare program, including the recovery of natural gas, oil, copper and iron-ore. Three months after Rainier, the AEC General Advisory Committee recommended that a study group investigate the use of underground explosions for the production of radioactive isotopes and large earth moving projects.[68] Rainier also stimulated the provision of specific funding for Plowshare:[69] $300 000 in 1958.[70] Plowshare was off and running.

In the birth of Plowshare the role of Edward Teller is often perceived to be seminal. Teller's biographers, for instance, credit him with 'initiating' Plowshare. He also originated several ideas for using nuclear dynamite, some of which he admitted, 'Most people responded to... with a shrug and an incredulous smile'.[71] One involved the use of nuclear explosions to generate thermal electricity. Another was the creation of diamonds underground. Teller warned, however, that these would not be large crystals ('so ladies need not sell their jewellery right away'!), but in the form of diamond powder, useful for machine tools.[72] Teller also looked to the day when the weather might be controlled by nuclear energy.[73]

While not formally in charge of the Livermore Laboratory's programs, including Plowshare, Teller wielded great influence there as a result of his scientific eminence and the veto power he had been given over decisions of the laboratory's Scientific Advisory Committee.[74] In March 1958 he was made Director of Livermore. His biographers note that during his two-year tenure:[75]

> ... much of the ground was plowed for several decades of work in nuclear weapons. He also played an equally important role in the pursuit of peaceful uses of nuclear energy (Plowshare)...

Was Teller then the 'father of Plowshare' as well as 'father of the H-bomb'? Clearly he was not the first American to conceive of PNEs. Nor was Project Plowshare his idea. He himself attributes it to Harold Brown.[76] Herbert York also credits Brown with producing most of the new ideas that emanated from Plowshare.[77] Others record that a triumvirate of Teller,

Brown and Gerald Johnson were Plowshare's 'chief movers and articulate spokesmen'.[78] The official Livermore version is that Plowshare was formed 'under the leadership of Dr Harold Brown and with the support of Professor Ernest O. Lawrence and Dr Edward Teller'. [79] An examination of the early history of Plowshare confirms that it was indeed a group effort, certainly encouraged by Teller, but not attributable to him or any other individual.

Glenn Seaborg describes Teller as not so much founding as 'urging onward' the Plowshare program.[80] After Teller suggested to the Senate Foreign Relations Committee that Plowshare might be useful in colonising the moon, the bemused Committee chairman and Seaborg had the following exchange:[81]

> *Dr Seaborg:* ... Edward is very much sold on the Plowshare program.
> *Chairman:* On the moon?
> *Dr Seaborg:* Anywhere.
> *Chairman:* This is on the moon.
> *Dr Seaborg:* Anywhere.
> *Chairman:* Oh, anywhere?
> *Dr Seaborg:* He feels very strongly about it. I do not know whether he is quite the originator of the Plowshare program, but he certainly has been one of the strongest advocates from the very beginning.

Gerald Johnson recalls that, as 'professor' of the laboratory, Teller was extremely interested in Plowshare, kept in touch with the Plowshare team and contributed actively to the ideas produced. Essentially though, Teller was looking 'out rather than in'.[82] He was, above all, Plowshare's high-level salesman, viewing promotion as vital to the program's survival. As Pringle and Spigelman put it, Teller was convinced, 'like most men propelled by a messianic zeal', that if only everyone knew as much as he did, they would agree with him about the value of the 'clean' bomb and of Plowshare.[83] Johnson indeed credits Teller with 'breaking the dam at the Washington level' in terms of acceptance and funding for Plowshare.[84] If Teller was not the father of Plowshare, then he was certainly its midwife and chief guardian.

Pringle and Spigelman, in *The Nuclear Barons*, imply that the idea of nuclear dynamite was seized upon by American weapon scientists to improve a public image tarnished by concern about atmospheric fallout.[85] They describe Plowshare as 'a good public relations ploy' with a 'catchy title'.[86] William Epstein goes further, claiming that:[87]

> Many members of the American arms control community believed that both the notion of clean bombs and that of underground peaceful nuclear explosions were invented by these scientists because of their fears that nuclear tests might be halted and permanently ended.

Wolfgang Panofsky, Director of the Stanford Linear Accelerator, notes that 'it is very difficult indeed to distinguish between sincere, but possibly misguided, advocacy of PNEs and cynical promotion of PNEs to further advances in nuclear weapons technology'.[88]

According to Arnold Kramish the confluence of the nuclear test moratorium proposal and the advent of Plowshare resulted in the 'feeling on the part of many that Project Plowshare was initially designed merely as a political mechanism'.[89] Even Edward Teller's technical book on PNEs, *The Constructive Uses of Nuclear Explosives*,[90] was greeted with the comment that 'Plowshare has often seemed to have a larger component of politics than of economics in it . . .'[91]

Was Plowshare then a purely political creation designed to head off a moratorium and later a comprehensive test ban treaty? The circumstantial evidence certainly points strongly in that direction. It seems more than coincidental, for instance, that PNEs should be 're-invented' in 1956, when the nuclear fallout debate was dominating the presidential election campaign. It is also at the very least amazingly coincidental that Project Plowshare should be formally initiated in June 1957, the very month that President Eisenhower was seriously considering a moratorium.

Also suspicious was the vigour with which Plowshare, from its inception, was touted both nationally and internationally by the Livermore Laboratory and the AEC. US government agencies, private industry, foreign governments and international organisations were all courted, even before a single Plowshare experiment had been conducted.[92] For instance, in an extraordinary move, the Livermore Laboratory in 1958 published the results of the Rainier shot and a declassified record of the February 1957 conference, now dubbed the First Plowshare Symposium. For its part, the AEC declassified the whole subject of Plowshare except details of the nuclear devices likely to be employed.[93] Edward Teller reacted excitedly: 'It is wonderful to see that the Atomic Energy Commission has declassified everything about the action of nuclear explosives in Plowshare'.[94] Teller wanted the AEC to go further, urging 'very careful investigation of the degree to which certain tools to be used in Plowshare can be declassified'.[95] The purpose was obviously public promotion of the program.

Plowshare's first international exposure came at the Second International Conference on the Peaceful Uses of Atomic Energy held in Geneva in 1958. For Glenn Seaborg, 'enthusiasm for PNEs became international' after the United States delegation painted the potential economic benefits 'in glowing terms'.[96] In the only paper on PNEs, Gerald Johnson described such potential applications as harbour construction and oil stimulation.[97] Unexpectedly, it provoked hostile reactions from Romania and the Soviet Union. The Soviets, despite having invented the concept of a 'peaceful nuclear explosion', accused the United States of using Plowshare as a subterfuge for nuclear weapons testing.[98]

The campaign to promote Plowshare was made more suspicious by the prominent role played by Edward Teller, who also happened to be the leading apologist for atmospheric nuclear testing. Teller had a vested interest in both Plowshare and testing. He had also largely forfeited any claim to being a dispassionate scientific authority by his political activism and tendentious public pronouncements. According to Robert Jungk:[99]

> Teller's interviews and public lectures, even his purely scientific papers, which are considered by his colleagues to be for the most part insufficiently thought out and consequently imprecise, seem often to be contrived with a view to obtaining as many newspaper headlines in as thick type as possible.

Teller adopted a high public profile on the fallout issue, including participating in a televised debate with Nobel Laureate Linus Pauling in 1958. Teller's view of nuclear fallout was simply that it was 'not worth worrying about' and dangerous only 'in the imagination'.[100] Peter Metzger has commented:[101]

> These opinions, presented as facts from a scientist, were not accompanied by any supporting evidence and depended entirely upon the reputation of Dr Teller for their credibility.

Another commentator noted that 'there are those who might more conservatively estimate the radiation hazards'.[102]

But does all this justify a conspiracy theory concerning the origins of Plowshare? Clearly Teller recognised that nuclear weapons had a poor public image, which reflected on their inventors. He often expressed impatience with 'the tendency of most Americans to see only evil in the atom, to equate atomic power with the suffering and devastation of Hiroshima'.[103] He undoubtedly hoped that the opprobrium suffered by the creators of nuclear weapons could be diluted if the bomb could be used for peaceful purposes such as PNEs. As Herbert York less charitably puts it, Teller was 'eager to push any program or idea anywhere that made nuclear weapons appear to be good and useful'.[104] Physicist Ernest Sternglass recalls:[105]

> I think a lot of people who worked on the nuclear weapons system — especially Edward Teller — felt desperate about these weapons. Nobody wants to build a gadget that is never used. They were always looking for ways to put the atom to 'peaceful' uses.

Glenn Seaborg acknowledges that:[106]

> The opportunity to turn the menacing technology of nuclear explosions into something so beneficial was irresistible, especially for those of us who had worked in the wartime atomic bomb project.

Clearly, Teller used Plowshare to help head off a moratorium and, later, a test ban. But Teller could hardly have been under the illusion that Plowshare would be a decisive factor in either case. After all, PNEs would not solve the fallout problem. Indeed they implied more atmospheric nuclear explosions, not fewer. Plowshare was in fact a relatively minor weapon in Teller's arsenal and only marginally useful in enhancing the image of nuclear testing. Moreover, far from dreaming up Plowshare simply as a political ploy, Teller was patently a 'true believer', supporting it vociferously even into the 1970s when it had become a political liability rather than a political asset. Even Herbert York, a strong critic of Teller, concedes that while Teller seized on the idea of Plowshare and used it for his own purposes, he did not invent or promote Plowshare simply because of its political value.[107]

Teller's main weapons against a test ban were first, the argument that testing was needed to ensure America's continued nuclear superiority over the Soviet Union and second, the development of 'clean' nuclear devices, the very same explosives that would ensure Plowshare's utility. In the meantime, Teller proposed that the Russians and Americans should agree that in a given period they would not reintroduce into the atmosphere more than half the radioactive material that was disappearing due to natural decay. Teller believed that this restriction would goad nuclear scientists on both sides of the Iron Curtain into developing 'clean' nuclear devices.[108]

Seen in the wider context of their time, the 'invention' of PNEs and the establishment of Plowshare need no Machiavellian explanation. They are, on the contrary, quite easily explicable as yet one more product of the 'gung ho' spirit of the Livermore Laboratory and the enthusiasm of the US Atomic Energy Commission for harnessing the power of the atom for the 'good' of mankind. As the US Arms Control and Disarmament Agency has noted, Plowshare:[109]

> ... accorded well with a widespread desire to see the atom cast in a peaceful role. Consequently, the AEC was acting in the spirit of the time when... it formally established a PNE program under the compelling name of 'Plowshare', symbolic of great transformations.

In contrast, Gerald Johnson notes drily that the origin of Plowshare at the 'working level' was simply its 'technical feasibility'.[110]

Plowshare's true place in history then is not simply as a political gimmick, but alongside such other atomic age 'gadgets' as the nuclear-powered aeroplane, the plutonium pacemaker and wristwatch, and the nuclear-powered rocket. As Glenn Seaborg put it: 'Where science fiction goes, can the atom be far behind? My only fear is that I may be underestimating the possibilities'.[111]

# Notes

1 The acronym PNE or PNEs refers to either or both peaceful nuclear *explosions* and *explosives*. The intended meaning in each case should be clear from its context.
2 G. T. Seaborg and W. R. Corliss, *Man and Atom: Building A New World Through Nuclear Technology*, E. P. Dutton and Co., New York, 1971, pp. 99–100; 'Peaceful Nuclear Explosions and Arms Control', *Backgrounder*, US Information Service, Canberra, 16 August 1988, p. 1.
3 H. F. York (ed.), *Arms Control*, W. H. Freeman and Co., San Francisco, 1973, p. 38.
4 Ibid., p. 346.
5 The Nagasaki bomb had a yield of approximately 20 kilotons. A kiloton is a measure of the yield or power of a nuclear device equivalent to 1000 metric tons of TNT. The Hiroshima bomb had a yield of 12–15 kilotons.
6 F. Reines, 'Are There Peaceful Engineering Uses of Atomic Explosives?', *Bulletin of the Atomic Scientists*, June 1950, pp. 171–72.
7 Ibid., p. 172.
8 Ibid.
9 From correspondence with Gerald Johnson, Palos Verdes, California, 17 January 1987.
10 L. N. Ridenour, 'The Hydrogen Bomb' in York (ed.), p. 53.
11 G. T. Seaborg *et al.*, *Plowshare*, Division of Technical Information, US Atomic Energy Commission, Oak Ridge, Tennessee, 1966, p. 22.
12 *PNE Activity Projections for Arms Control Planning* (The *GURC Report*), Gulf Universities Research Consortium, Galveston, Texas, US Arms Control and Disarmament Agency Document, ACDA/PAB-253, Washington DC, 1975, p. 2; US Congress, Senate Foreign Relations Committee, Hearings on the Threshold Test Ban Treaty and Peaceful Nuclear Explosions Treaty (TTBT and PNET), 28 July, 3 August, 8 and 15 September 1977, US Government Printing Office, Washington DC, 1977, p. 132.
13 A megaton is equivalent to one million metric tons of TNT.
14 H. P. Metzger, *The Atomic Establishment*, Simon and Schuster, New York, 1972, p. 232.
15 T. S. Lough, 'Peaceful Nuclear Explosions and Disarmament', *Peace Research Reviews*, Canadian Peace Research Institute, Ottawa, 1968, p. 34.
16 During its existence the laboratory has had a number of name changes, most notably after the death of physicist Ernest Lawrence in mid-1958, when it became the Lawrence Livermore Laboratory. Today it is known as the Lawrence Livermore National Laboratory. To avoid confusion, the laboratory will mostly be referred to throughout the text as the Livermore Laboratory or simply Livermore.
17 S. A. Blumberg and G. Owens, *Energy and Conflict: The Life and Times of Edward Teller*, G. P. Putnam's Sons, New York, 1976, p. 216.
18 R. Rhodes, *The Making of the Atomic Bomb*, Simon and Schuster, New York, 1986, pp. 757–8.
19 Blumberg and Owens, p. 216.
20 Ibid., p. 400.
21 R. Sanders, *Project Plowshare*, Public Affairs Press, Washington DC, 1962, p. 158.
22 G. T. Seaborg, *Nuclear Milestones*, W. H. Freeman and Co., San Francisco, 1972, p. 214.
23 Sanders, p. 158.
24 Seaborg, *Nuclear Milestones*, p. 211.
25 T. B. Taylor, 'Endless Generations of Nuclear Weapons', *Bulletin of the Atomic Scientists*, November 1986, p. 12.
26 H. F. York, 'The Origins of the Lawrence Livermore Laboratory', *Bulletin of the Atomic Scientists*, September 1975, p. 14.
27 H. F. York, *The Advisors: Oppenheimer, Teller and the Superbomb*, W. H. Freeman and Co., San Francisco, 1976, p. 136.
28 From correspondence with Gerald Johnson, Palos Verdes, California, 17 January 1987. Twenty years later Brown would become President Carter's Secretary of Defense.
29 York (ed.), p. 346.
30 From correspondence with Herbert York, La Jolla, California, 25 September 1986. See also B. A. Bolt, *Nuclear Explosions and Earthquakes: The Parted Veil*, W. H. Freeman and Co., San Francisco, 1976, p. 181.

31 Sanders, p. 43; D. B. Brooks and J. V. Krutilla, *Peaceful Use of Nuclear Explosives: Some Economic Aspects*, Resources for the Future, Inc., Washington DC, 1969, p. 1.
32 Bolt, p. 181; E. Teller *et al.*, *The Constructive Uses of Nuclear Explosives*, McGraw-Hill, New York, 1968, p. vi; J. Koop, 'Plowshare and the Nonproliferation Treaty', *Orbis*, XII, 3, Fall 1968, p. 795.
33 US Congressional Record, 14 April 1969, US Government Printing Office, Washington DC, 1969, p. 8945.
34 Blumberg and Owens, p. 400.
35 From correspondence with Gerald Johnson, Palos Verdes, California, 17 January 1987.
36 Blumberg and Owens, p. 400.
37 US Congress, Joint Committee on Atomic Energy (JCAE), Hearings on AEC Authorizing Legislation for Fiscal Year 1965, 20 February 1964, US Government Printing Office, Washington DC, 1964, p. 1220.
38 G. W. Johnson, 'The Plowshare Program — History and Goals', US Congress, JCAE, Hearings on Peaceful Applications of Nuclear Explosives — Plowshare, 5 January 1965, US Government Printing Office, Washington DC, 1965, p. 62.
39 Bolt, p. 182; E. Teller, *The Legacy of Hiroshima*, Doubleday and Co., New York, 1982, p. 81. The meeting later became known as the First Plowshare Symposium.
40 York (ed.), p. 346.
41 F. J. Clark Jr., 'Plowshare Program Development Activities' in L. E. Weaver (ed.), *Education for Peaceful Uses of Nuclear Explosives*, University of Arizona Press, Tucson, Arizona, 1970, p. 234; Blumberg and Owens, p. 400.
42 F. Kreith and C. B. Wrenn, *The Nuclear Impact*, Westview Press, Boulder, Colorado, 1976, p. 7.
43 T. Ginsburg, *The Question of Peaceful Explosions for the Benefit of Non-Nuclear Weapon States*, Document A/CONF. 35/DOC. 2, Conference of Non-Nuclear Weapon States, Geneva, July 1968, p. 4; G. T. Seaborg and B. S. Loeb, *Stemming the Tide: Arms Control in the Johnson Years*, Lexington Books, Lexington, Mass., 1987, p. 310.
44 *Energy and Technology Review*, Lawrence Livermore National Laboratory, September 1982, p. 14.
45 I.Y. Borg, *Nuclear Explosions for Peaceful Purposes*, Lawrence Livermore National Laboratory Paper No. UCRL–95026, Rev. II, November 1986, p. 2.
46 Teller, *The Legacy of Hiroshima*, p. 82.
47 P. Pringle and J. Spigelman, *The Nuclear Barons*, Avon Books, New York, 1981, p. 256.
48 Borg, p. 2.
49 Johnson, p. 62.
50 Borg, p. 2.
51 Blumberg and Owens, p. 397; G.T. Seaborg and B.S. Loeb, *Kennedy, Khrushchev and the Test Ban*, University of California Press, Berkeley, 1981, pp. 8–9.
52 According to Lawrence Livermore scientist Hugh de Witt, Teller and Lawrence also had in mind a pure fusion device, with the fusion set off by conventional high explosives, thereby avoiding fission altogether. The idea of a pure fusion device was, according to de Witt, 'an empty dream'. He notes that, 'In spite of thirty years' work on this concept there is no possibility we can see of obtaining a fusion reaction on the earth with high explosives. That whole effort has had to be abandoned'. See *Interdisciplinary Seminar on Nuclear Technology, Society & Health*, University of California, Berkeley, Transcript of seventh meeting (unpublished), 30 October 1986, p. 20.
53 Blumberg and Owens, p. 398.
54 Ibid., p. 397.
55 From correspondence with Gerald Johnson, Palos Verdes, California, 17 January 1987.
56 Blumberg and Owens, pp. 397–98.
57 Pringle and Spigelman, p. 252.
58 T.C. Findlay, 'Twenty Minutes with Edward Teller', *Newsletter*, Peace Research Centre, Australian National University, Canberra, 2, 1, 1987, p. 2.
59 From correspondence with Gerald Johnson, Palos Verdes, California, 17 January 1987.
60 Johnson, p. 62.
61 Teller, *The Legacy of Hiroshima*, p. 82.
62 From correspondence with Gerald Johnson, Palos Verdes, California, 17 January 1987.
63 Ibid.

64 From an interview with Gerald Johnson, Washington DC, 10 December 1986; see also Seaborg and Corliss, pp. 409–10.
65 G.W. Johnson and H. Brown, 'Non-military Uses of Nuclear Explosives' in York (ed.), pp. 410–11.
66 Teller, *The Legacy of Hiroshima*, p. 83.
67 Ibid.
68 The *GURC Report*, p. 3.
69 All monetary values quoted are in US dollars unless otherwise stated.
70 From an interview with Gerald Johnson, Washington DC, 10 December 1986.
71 Blumberg and Owens, pp. 82 and 89.
72 Ibid., p. 89.
73 Physicist Hans Bethe has remarked that '... nine out of ten of Teller's ideas are useless. He needs men with more judgement, even if they be less gifted to select the tenth idea which is often a stroke of genius' (H. Bethe, 'Comments on the History of the H-Bomb, 1954' in R.C. Williams and P.L. Cantelon (eds), *The American Atom: A Documentary History of Nuclear Policies From the Discovery of Fission to the Present, 1939–1984*, University of Pennsylvania Press, Philadelphia, 1984, p. 137).
74 York, 'The Origins of the Lawrence Livermore Laboratory', p. 13.
75 Blumberg and Owens, p. 405.
76 Teller, *The Legacy of Hiroshima*, pp. 81–82. Teller confirmed this in an interview with the author at the Livermore Laboratory on 9 December 1986.
77 From an interview with Herbert York, La Jolla, California, 1 December 1986.
78 Sanders, p. 158.
79 *Energy and Technology Review*, p. 14.
80 Seaborg and Corliss, p. 100.
81 US Congress, Senate Foreign Relations Committee, Hearings on the Nuclear Test Ban Treaty, 12–15 August, 19–23 August, 26–27 August 1963, US Government Printing Office, Washington DC, 1963, p. 268.
82 From an interview with Gerald Johnson, Washington DC, 10 December 1986.
83 Pringle and Spigelman, p. 257.
84 From an interview with Gerald Johnson, Washington DC, 10 December 1986.
85 Pringle and Spigelman, p. 220.
86 Ibid., p. 256.
87 W. Epstein, *The Last Chance*, The Free Press, New York, 1976, p. 172.
88 Testimony of Wolfgang Panofsky to US Senate Foreign Relations Committee, 12 September 1974, reprinted in US Congress, Senate Foreign Relations Committee, Hearings on TTBT/PNET, 13 and 15 January 1987, US Government Printing Office, Washington DC, 1987, p. 271.
89 A. Kramish, *The Peaceful Atom in Foreign Policy*, Harper and Row, New York, 1963, p. 117.
90 Teller *et al.*
91 P. Morrison, 'Constructive Uses of Nuclear Explosives' in York (ed.), p. 417.
92 While it could be argued that a great deal of data had been accumulated from nuclear weapon tests that could be of relevance to Plowshare, Plowshare Director John Kelly seemed to imply in his testimony to the JCAE in 1966 that such data could not necessarily be simply extrapolated to peaceful uses: '...we were beginning to recognise that the simple empirical formula may not be adequate...' Dr Spofford English also testified that: 'It turns out in most technology development that after you do a few experiments, you really think you know an awful lot. You are very enthusiastic. As you do more to try to identify some of the problems, you get shaken up in many cases. You find out you did not really understand the basic phenomena. Then you have to settle down and you climb slowly back up by a series of experiments. We think we have gone through the minimum in the Plowshare program, and we now understand much more than we actually thought we did several years ago' (see US Congress, JCAE, Hearings on AEC Authorizing Legislation for Fiscal Year 1967, US Government Printing Office, Washington DC, 1966, pp. 1367–68).
93 Sanders, p. 43.
94 US Congress, JCAE, Hearings on Frontiers in Atomic Energy Research, 22–26 March 1960, US Government Printing Office, Washington DC, 1960, p. 51.
95 Ibid., p. 52.

96 Seaborg and Loeb, *Stemming the Tide*, pp. 309–10.
97 *Report of the US Delegation to the International Conference on the Peaceful Uses of Atomic Energy, 1–13 September 1958*, USAEC, US Government Printing Office, Washington DC, 1958, p. 20.
98 From an interview with Gerald Johnson, Washington DC, 10 December 1986; see Chapter 5 for further details.
99 R. Jungk, *Brighter Than A Thousand Suns*, Penguin, Harmondsworth, 1960, p. 299.
100 Metzger, p. 108.
101 Ibid.
102 Morrison, p. 417.
103 Teller, *The Legacy of Hiroshima*, p. 92.
104 From correspondence with Herbert York, La Jolla, California, 25 September 1986.
105 L. J. Freeman, *Nuclear Witnesses: Insiders Speak Out*, W. W. Norton and Co., New York, 1981, p. 67.
106 Seaborg and Loeb, *Stemming the Tide*, p. 347.
107 From an interview with Herbert York, La Jolla, California, 1 December 1986.
108 Blumberg and Owens, p. 396.
109 *The American Experience with Peaceful Nuclear Explosions*, US Arms Control and Disarmament Agency, US Government Printing Office, Washington DC, 1978, p. 1.
110 From correspondence with Gerald Johnson, Palos Verdes, California, 17 January 1987.
111 D. Ford, *The Cult of the Atom*, Simon and Schuster, New York, 1982, p. 24.

# 2

# Project Plowshare: The Moratorium and Beyond

*Plowshare confronts the difficult task of gaining acceptance in an anxious world, convulsed by social and political upheavals and constantly threatened by atomic calamity. The program has faced, and most probably will continue to face scepticism and political opposition. Too much of the world stands either in awe or in dread of atomic power, and sometimes exhibits a curious mixture of both. Negative attitudes, moreover, are not confined to the general public. A few vocal scientists have greeted the innovation of this program with misgivings and doubts, basing their caveats on fear, scientific reservations and political inclinations. The sensible way to settle differences of opinion about the feasibility and safety of Plowshare is to experiment — to test the scientific judgements of Plowshare's proponents against results in the field.*[1]

Ralph Sanders, 1962

DESPITE his previous misgivings and the opposition of the Livermore Laboratory, President Eisenhower, influenced by pro-test ban members of his administration, announced on 22 August 1958 that the United States would be willing to join a moratorium on nuclear testing. He proposed however that PNEs be exempted. Such experiments would be open to observation by invited foreign representatives and the results fully disclosed in the scientific press.[2] The President also proposed that any test ban treaty 'deal with the problem of detonations for peaceful purposes, as distinct from weapons tests'.[3]

The moratorium began on 31 October 1958. As it turned out, the United States conducted no 'peaceful' detonations during this period, despite the fact that the AEC was eager to commence Plowshare experiments and announced plans to do so on several occasions. AEC Chairman McCone announced in January 1959, for instance, that the Commission was considering conducting tests later in the year relating to the 'peaceful uses of nuclear energy'.[4] Several American scientists protested, advocating a postponement so as not to disrupt the test ban treaty negotiations due to commence in Geneva in October 1959. Even after the test ban talks began however, the AEC persisted. On 16 March 1960 the Commission announced, with the express approval of President Eisenhower, that the United States was preparing to detonate its first Plowshare shot in early 1961.[5] The Soviet Union was even invited to send observers. Edward Teller greeted the announcement enthusiastically, saying 'It is wonderful that we are planning to proceed with a very important shot in New Mexico. We plan to proceed with international observation. All this is fine'.[6] The Administration also announced plans for nuclear detonations for seismic study purposes. The Soviet Union protested that such explosions would violate the moratorium.[7]

Also in March 1960 a row erupted in Canada over a statement by a Mr Floberg of the AEC, to the effect that the test ban negotiations would have no effect on the Commission's plans for detonating a 9 kiloton PNE in Canada's Athabasca tar sands later that year.[8] The experiment, codenamed Project Oilsand, was to be jointly sponsored by the AEC and the Richfield Oil Company of Los Angeles and its partners. It would test the theory that an underground nuclear explosion could stimulate the flow of oil from such tar sands. The experiment would take place fifty-four miles south-east of Fort McMurray in Alberta.

The Canadian Prime Minister, Mr Diefenbaker, was asked in Parliament whether the explosion would be conducted without the agreement of the Canadian government. AEC Chairman McCone quickly gave assurances that it would not proceed without the permission of both the Canadian government and the US President. The Canadian government subsequently announced that planning for the experiment was being postponed 'pending clarification of certain political considerations'.[9] The only physical preparation for the experiment that had in fact been undertaken was the planting of a stake in the ground at the proposed ground zero.[10] The Canadian Department of Mines and Technical Surveys, which had been studying the proposal,[11] concluded in the meantime that there was a lack of data on the potential radioactivity hazard. It also noted the possibility that thousands of nuclear devices might be needed to exploit the tar sands fully.[12] The Canadian government decided not to proceed with a detailed site analysis,[13] the Prime Minister having apparently personally ruled that no nuclear device should be detonated in Canada.[14]

The Eisenhower Administration's approach, or at least that of its Atomic Energy Commission, to the question of PNEs during the moratorium was clearly naïve. For most observers, and for the Soviet Union, a Plowshare test, whether conducted in Canada or New Mexico, could be interpreted as a nuclear weapon test in disguise. This could jeopardize not only the moratorium and the test ban negotiations, but the new atmosphere in US/Soviet relations engendered by Khrushchev's visit to the United States in September 1959.[15] Fortunately, the United States refrained from carrying out any PNE detonations during the moratorium.

As a result, the moratorium, which was to last almost three years, from 31 October 1958 to 15 September 1961, was a setback for Plowshare. Not only were no Plowshare detonations conducted, but no new experimental data from nuclear weapon tests could be obtained. The moratorium also discouraged potential industrial partners which the AEC was attempting to coax into joint 'commercial' Plowshare ventures. Gerald Johnson lamented to the Joint Committee on Atomic Energy (JCAE) in 1964:[16]

> Plowshare has had great difficulties, because about the time Plowshare was being born, the United States entered into a moratorium on testing for three years. When the moratorium ended the main effort properly went first into weapons improvement... We would have been much further down the road in terms of developing [Plowshare] technology.

Today Johnson believes that, on balance, the moratorium helped rather than hindered Plowshare, by providing a 'breathing space' for consolidation of knowledge and plans.[17] Indeed, during the moratorium Plowshare scientists conducted as many theoretical studies as personnel resources and funding permitted. They also performed conventional explosive cratering experiments and studied data from past nuclear weapon tests. Feasibility studies for several potential Plowshare projects were also begun, most notably in regard to the construction of a new Panama Canal.[18]

Plowshare also geared up organisationally. In December 1958 a Peaceful Nuclear Explosions Branch was established in the AEC's Division of Military Applications to administer the Plowshare Program.[19] To advise the AEC on the selection and carrying out of specific projects, a Plowshare Advisory Committee was established in September 1959, headed by Dr Spofford English, special assistant to the AEC General Manager.[20] In August 1961 the PNE Branch became the Division of Peaceful Nuclear Explosives.[21] Other agencies were enlisted to support Plowshare. The US Geological Survey agreed to conduct support studies.[22] The Sandia Corporation, which was involved in the production of nuclear weapons, commenced a Plowshare research and development effort.[23] In January 1959 the Bureau of Mines and the AEC held a joint symposium in Dallas, Texas, on the use of PNEs to recover shale oil.[24]

Apparently confident that the moratorium was only a temporary setback, Livermore scientists could hardly contain their enthusiasm for Plowshare. Writing in *Scientific American* in December 1958, Johnson and Brown proclaimed: 'It is not too much to say that we are on the verge of a new period of "geographical engineering"'.[25] They suggested using PNEs for such purposes as geothermal electric power generation, mining and excavation. They cautioned however that 'surface explosions for these desirable ends...would have as undesirable side effects all the destructive consequences that make nuclear bombs such formidable weapons'. These were listed as: the incendiary effects of the fireball; the so-called 'prompt' radiation at the instant of the explosion; the 'residual' radiation due to fallout and the creation of isotopes in the surrounding soil; and the air-blast and ground-shock effects. Johnson and Brown were 'confident' nonetheless that appropriate precautions would reduce each of these disadvantages to 'acceptable levels'.[26]

The AEC also continued its Plowshare publicity campaign during the moratorium. A Second Plowshare Symposium was held in San Francisco in May, 1959. Unlike the first, it was open to the public and attracted public criticism.[27] A letter to the *Bulletin of the Atomic Scientists*, commenting on the official report of the symposium, remarked:[28]

> Your reviewer can't decide whether it is the heady air of California that has caused the contributors to throw caution to the winds, or whether the AEC chose to document what was in essence a scientific salesmanship session in order to arouse public interest.

Sensitive to criticism of Plowshare, the AEC allegedly made efforts to head it off where possible. Dr Gary Higgins, a Division Director at Livermore, was scheduled to give a speech before the American Society of Mechanical Engineering in Minneapolis in mid-1959. His paper, on nuclear excavation, included mortality estimates for the radiation likely to be released during nuclear excavation of a new Panama Canal. The AEC found this offensive, informing Higgins so on the evening before the delivery of his paper. Higgins was obliged to delete the mortality estimates from his text, which was published 'thus bowdlerised' in *Mechanical Engineering* in January 1960.[29]

One important source of support for Plowshare that emerged at an early stage was the Congressional Joint Committee on Atomic Energy. The JCAE supported a 'vigorous, imaginative, and aggressive atomic energy program... [stressing] boldness and risk-taking rather than caution and economy'.[30] Established in 1946 as a watchdog over the US nuclear industry, the committee had 'unusually great powers',[31] including the power to authorise the spending of appropriated funds.[32] By 1958, in the absence of strong presidential control or interest in atomic energy matters, the committee, which had at first been a healthy critic of the AEC, had transformed itself into the virtual manager of the agency it was supposed to be 'watchdogging'.[33]

This situation was to benefit Plowshare greatly, first by allowing it to escape the critical scrutiny it would otherwise have attracted under a Congressional committee's steely gaze. Second, in line with its strong support for all 'innovations' in nuclear technology, the JCAE promoted Plowshare zealously, at times more dogmatically than the AEC itself. One committee member, Representative Craig Hosmer, (Republican, California), rivalled Edward Teller in his enthusiasm. Hosmer advocated transforming Plowshare into a 'commercial' service offering a wide variety of 'off the shelf' nuclear explosives for industrial purposes. He was supported by fellow Californian

Representative, Chet Holifield, a Democrat, who was to become the longest-serving and most formidable chairman of the JCAE.

In March 1960, while the moratorium was still in force, the committee organised hearings on the topic: 'Frontiers in Atomic Energy Research'. Plowshare figured prominently. Edward Teller testified in its favour, asserting, with somewhat less certainty than when he briefed President Eisenhower:[34]

> I am sure today that we can make nuclear explosives much cheaper, more flexible, and much cleaner. I can say, not with certainty, but *with quite a bit of hope* [emphasis added], that we can make nuclear explosives for peaceful purposes so clean that the worry about radioactivity in its peaceful applications *may* [emphasis added] disappear completely.

Teller also tried to convince the committee that Plowshare had no military implications:[35]

> The tools which we want to develop are not military explosives. They are related to military explosives like a passenger car is related to a tank. The principles, or some of the principles, are similar. The execution is quite different, and the technical knowledge obtained from the one is only indirectly relevant to the other. The specific reasons for the differences are these: A weapon must be light, easily deliverable. It can, on the other hand, be moderately expensive, because the overwhelming importance of the specific job that it performs justifies the expense. A peaceful nuclear explosive can be clumsy, big, heavy, and still it is an excellent experimental tool... Therefore, we seek to accomplish quite different aims. We accomplish them along different roads, and the two approaches, while related, are really quite distant from each other.

The only critic of Plowshare to testify was the Chairman of the University Research Board at Princeton University, Professor Henry Smyth.[36] Declaring that Plowshare seemed to have in it 'the seeds of controversy', Smyth urged that the debate on PNEs 'should take into account many factors besides technological feasibility and costs':[37]

> Surely this committee and other committees of the Congress should consider very carefully whether the use of nuclear explosions for peaceful purposes will jeopardize our efforts at international control. Virtuous assurances that such explosions are not being used for weapon development will, I am afraid, simply not be believed. Guarantees that there are no present or future hazards to water supply will be viewed with scepticism. Since safety control will almost certainly require such explosions to be carried out by the Government, the economics of any results achieved will be hard to ascertain...
>
> I am sure I shall find the presentations this afternoon interesting. Perhaps I shall be convinced that this is a cheap way to dig harbors, to release oil from shale, and salt from sea water. I hope the witnesses will forgive me for considering these points only of peripheral importance. Even if our use of nuclear explosions for peaceful purposes promises to contribute significantly to our material welfare, I still question its political advisability in terms of international understanding and peace.

Smyth recommended a retardation or indefinite suspension of PNE experiments. His warnings fell on deaf ears. The JCAE never did carry out an examination of the social and political implications of Plowshare. And all Professor Smyth's fears would prove justified.

In early 1961 Glenn Seaborg, who with Edwin McMillan had shared the Nobel Prize in Chemistry for the discovery of plutonium, succeeded John McCone as chairman of the AEC. McCone had been a Plowshare supporter. He had testified to the JCAE in 1960 that:[38]

> The nuclear explosive is a new and inexpensive source of energy which can be put to work to perform herculean feats now beyond man's ability or economic means... Through [Plowshare] we hope to obtain the data necessary to convert a nuclear explosive into an energy and research tool so that the atom will then indeed be the servant of man.

Glenn Seaborg however was in a different league to McCone. According to Daniel Ford, Seaborg was obsessed by the historic importance of his discovery and had decided to devote the rest of his life to promoting plutonium as the 'salvation of mankind'.[39] Seaborg believed that:[40]

> ...nuclear energy was a magician's potion that could free industrial society permanently from all practical bounds. Millions of homes could be heated and lighted [sic] by a single large nuclear reactor. Factories could be powered forever by the cheap and inexhaustible electricity generated by nuclear fission. The deserts could be made to bloom, sea water could be made potable, mountains could be moved, rivers diverted — all as a result, he prophesised, of 'planetary engineering' made possible by the miraculous element he had discovered.

Barring Teller and Craig Hosmer, Plowshare could not have had a more receptive chairman at the helm of the Commission. To the already bizarre list of suggested Plowshare applications, Seaborg would add weather and earthquake control and defending the earth against cosmic projectiles.[41]

His personal interest in Plowshare centred however on the search for elements which do not occur in nature — so-called 'transuranic' or 'heavy' elements. Seaborg hoped to see his discovery of plutonium extended by the discovery of even heavier new elements in the intense 'neutron fluxes' created in nuclear explosions.[42] Seaborg described this as being 'scientific work of the highest order, and with the potential of opening large new fields of technology'.[43] The uses envisaged for such isotopes were in diagnostic and therapeutic medicine and as 'miniature sources of energy'.[44]

## *Gnome*

In September 1961 the Soviet Union resumed nuclear testing, thereby ending the moratorium. This permitted Plowshare to begin in earnest. The world's first ever 'peaceful' nuclear explosion was scheduled for detonation within just three months, on 10 December 1961, at a desert site twenty-five miles south-east of Carlsbad, New Mexico.[45] The President had authorised preparation of the site during the moratorium.[46] The 5 kiloton underground explosion, code named Gnome, was designed to:[47]

(a) determine whether heat from an underground explosion could produce steam to power electric generators
(b) produce radioactive isotopes
(c) measure the characteristics of nuclear detonations in salt formations
(d) test several theories in the field of nuclear physics, and
(e) provide data for the design of future PNE devices.

President Kennedy described the experiment as 'a further example of this country's desire to turn the power of the atom to man's welfare rather than his destruction'.[48] To underscore these sentiments the White House invited 400 observers from United Nations member states, the press and the scientific community. According to an eyewitness account, an almost festive air prevailed among the spectators gathered at the site. A report compiled for the US Arms Control and Disarmament Agency some years later recorded that:[49]

> The local people were reportedly pleased by all the activity, 'clustering around each time Edward Teller, the widely celebrated father of the H-Bomb, who is also father of Plowshare', was filmed by television crews. A prominent citizen of the area was quoted as saying, 'This bomb is for the good of mankind, and we're for it.' Perhaps not all the foreign visitors shared precisely this sentiment; but most of them must have felt, at the very least, that this was an important event in the unfolding of a technology which the United States would carry to a high degree of refinement.

The test was not an unqualified success. The explosion, in an eight by ten foot chamber cut into a strata of rock salt, caused clouds of intensely radioactive steam to escape from a water-filled shaft connected to the detonation chamber to test whether usable steam would be produced. Radioactivity at the shaft entrance reached 10 000 roentgens an hour (the maximum safe exposure for humans being half a roentgen annually).[50] Gerald Johnson recalls hovering over the site in a helicopter with Edward Teller and members of the press at the time of the explosion. When a reporter asked Teller what was the unanticipated vapour coming out of the ground, he replied casually, 'White vapour'![51] The AEC was not nearly so relaxed. It closed nearby highways and took other precautions. The Public Health Service recorded, however, that 'no persons off-site were harmfully exposed to radioactivity from the steam'.[52]

Ground motion was also greater than expected. The official explanation was that 'elastic waves generated by the explosion travelled through the surrounding rock with unexpectedly high amplitude and velocity, causing concern for the effects of ground motion on nearby structures'.[53] Much of the expected data from the detonation was lost, including film footage. Later the cavity created by the explosion collapsed. Saturated brine liberated by the detonation hindered attempts to recover thermal energy.[54]

The AEC nevertheless hailed the experiment as 'a human scientific and engineering achievement of enormous complexity'.[55] The *New York Times* enthused: 'This is a good and cheerful occasion... yesterday's explosion may point a better way, to the use of the atom for the good of all mankind'.[56] According to Gerald Johnson there was some embarrassment about the Gnome shot, but the failure did not seem to have any long-term effect on Plowshare. It was not, however, an auspicious start to the program.

## *The Panama Canal Project*

A major impetus for Project Plowshare from its inception was the dream of building a new sea-level transisthmian, interoceanic canal. According to Mason Willrich, the canal idea was a 'major incentive' for Plowshare, providing it with most of its 'political momentum'.[57] Edward Teller contends that the Panama Canal proposal was 'central in guiding the research and development program for nuclear cratering technology'.[58] For Ed Fleming, formerly Assistant Director for PNEs at Livermore, the project was the 'chief driving force behind Plowshare'.[59]

The idea of a new canal had been mooted even before the existing Panama Canal, which uses a series of locks, had been constructed. The lock system was slow, vulnerable to attack or sabotage and increasingly unable to accommodate either the size or volume of interoceanic traffic expected of it. The idea of a new canal also had political utility for the United States, which controlled the Panama Canal Zone. Periodically, the United States used the threat of building a new canal, particularly one located outside Panama, to keep the rumbustious Panamanians in line. It even had a 1916 treaty with Nicaragua providing for the construction of a new canal within Nicaraguan territory.[60] As Farnsworth and McKenney explain it:[61]

> If a route were found outside Panama, or even... in the sparsely populated eastern portion of Panama, the present canal and other facilities located near the terminal cities of Panama and Colon would be obsolete. These terminal cities had become the major cities of Panama because they were on the Canal and would suffer serious economic decline if a sea-level canal were built some distance away.

The idea of using PNEs to build such a canal featured in the First Plowshare Symposium in 1957, having originated as one of the early 'desk studies' at Livermore.[62] When the study was published in 1958 it naturally

*Figure 1 Cavity produced by the Gnome underground nuclear explosion, Plowshare's first nuclear experiment. Note the man standing in the lower centre of the photograph. (Courtesy of Lawrence Livermore National Laboratory.)*

aroused the strong interest of the Panama Canal Company, the US government agency that operated the canal. The company studied the matter and submitted its encouraging findings to President Eisenhower in 1960.[63] Sometime thereafter Edward Teller and Gerald Johnson briefed the President on the idea, providing him with (necessarily) rough cost estimates.[64] They impressed upon the President that the project could not be carried out without a large experimental program to perfect nuclear excavation techniques. Eisenhower, noting that a canal through Columbia would cost $500 million, quipped: 'It would be worth $500 million just to get out of Panama'. Vice-President Nixon, who was also present, asked several questions and seemed genuinely intrigued by the possibilities of nuclear dynamite. According to Johnson, Eisenhower's interest was, on the other hand, due almost entirely to his perception of the project's 'political' value in the context of US/Panama relations.[65]

In 1960 Eisenhower authorised a preliminary study, to be carried out by the AEC's new Plowshare Division, of the possibility of using PNEs to build a new canal. Of the 30 possible routes for a new canal identified by the Panama Canal Company in 1947, only two, Route 17 in Panama and Route 25 in Colombia, were deemed suitable for nuclear excavation.[66] Both had sparse populations, adequate distance from large urban centres and suitable topography. The AEC, using rough engineering estimates, put the cost of using nuclear excavation for Route 17 at $750 million, compared with approximately $2,300 million for conventional earth-moving methods, a saving of approximately one-and-a-half billion dollars.[67] However no further action on the Panama Canal project was made during the Eisenhower Administration.

In 1961 Panamanian President Chiari visited Washington to pressure President Kennedy to renegotiate the 1905 Panama Canal Treaty. Kennedy refused, using as one of his excuses the fact that negotiations on the existing canal would be unnecessary if a new canal was built.[68] In May 1962 Kennedy directed the AEC and the US Army Corps of Engineers to undertake a five-year joint assessment of nuclear excavation techniques for building such a canal.[69] In June 1962 the corps established the US Army Engineer Cratering Group for this purpose,[70] with an annual budget of approximately $1 million.[71]

In December 1962, after presenting Edward Teller with the annual Fermi Award for his contribution to nuclear physics, Kennedy asked Teller about the feasibility of using nuclear dynamite to build a new Panama Canal. The President wanted to know how long it would take to complete the necessary geological research, survey work, soil sampling, and the economic and nuclear calculations. 'While he was asking me all these questions', Teller reported later, 'I interrupted. I said, "Excuse me, Mr President. All *that* will take less time than it will take you to make the right political decision"'.[72] According to Teller's biographers, he had come close to insulting Kennedy, who gave what Teller described as 'a sour grin' and walked off.[73]

## *Sedan*

The first experiment in Plowshare's excavation program, designed to provide data relevant to the Panama Canal proposal and other nuclear excavation ideas, was code-named Sedan. The 100 kiloton device, the first nuclear explosive specifically designed for Plowshare, was detonated on 6 July 1962 at the Nevada Test Site.[74] It was the largest man-made explosion detonated in the continental United States to that time. According to the AEC the blast threw dirt and rocks 7000 feet (2 kilometres) into the air, leaving a crater 300 feet (90 m) deep and 1200 feet (300 m) in diameter. In the words of seismologist Bruce Bolt, the explosion removed alluvium more than three football fields wide and one football field deep.[75] A visit to the Sedan crater even today, according to Bolt, 'leaves fixed in the mind an image of massive power'.[76] A group of four Soviet technicians who visited the crater described it as 'of outstanding interest'.[77]

AEC spokesmen asserted that 95 per cent of Sedan's radioactivity had either been trapped in the earth or had quickly returned to earth in the heavier portion of the debris thrown up by the blast.[78] Glenn Seaborg explained that the explosion had deposited a small quantity of radioactivity, most of which would have decayed rapidly, in an irregular fan downwind from ground zero. He claimed that people living fifty miles downwind would have received on average only one-seventh the radiation dose they would receive each year from natural background radioactivity.[79] Seaborg informed the JCAE in 1964, however, that Sedan had probably deposited detectable amounts of radioactivity beyond US borders.[80]

After the Sedan shot Plowshare activity accelerated. Between 1961 and 1973 at least twenty-seven dedicated Plowshare explosions were conducted, most of them at the Nevada Test Site.[81] At least another twenty-two non-Plowshare nuclear explosions, mostly nuclear weapon tests, had some Plowshare component. (For a full list of Plowshare and Plowshare-related experiments see Annex I.) The interplay between Plowshare and the nuclear weapon test program was close, with much of the data from weapon tests having direct relevance to Plowshare. For example, scaling laws, which determine how deep an explosion must be conducted to be contained, were based on data from both weapon and Plowshare tests.[82] Plowshare experiments had four broad goals.[83]

1. to test 'underground engineering' or excavation techniques
2. to discover new heavy elements
3. to test new or improved explosive devices
4. to reduce the radioactivity hazard of nuclear explosives.

The last of these goals was the most crucial. Plowshare enthusiasts realised at an early stage that radioactivity would be the single most important factor in determining the usefulness of nuclear dynamite, especially in excavation projects.

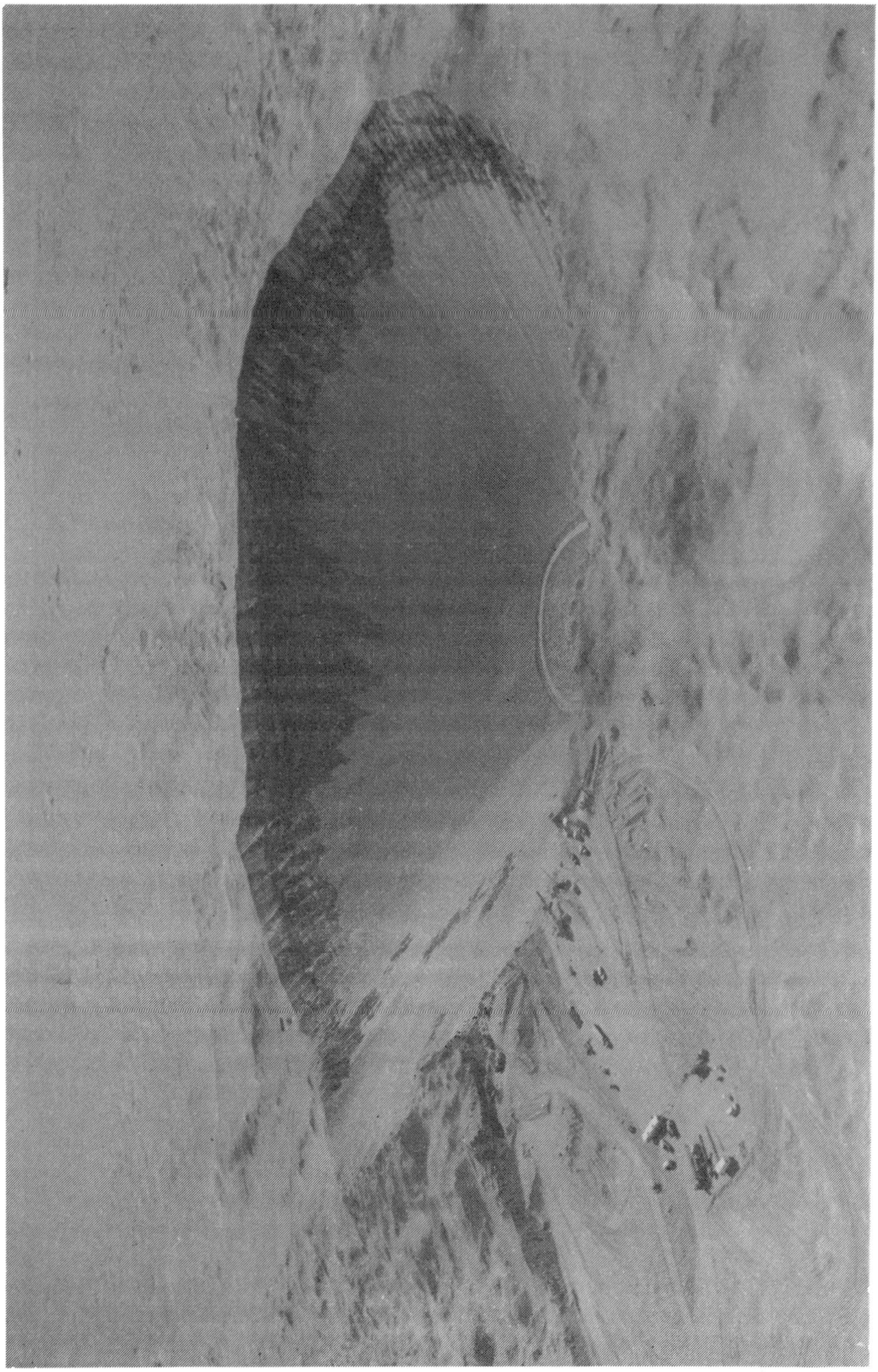

Three methods of reducing the radioactivity hazard were studied. First, attempts were made to reduce the fission component of PNE explosives to reduce the amount of radioactivity produced by the devices themselves. According to Gerald Johnson, by the end of 1963, after two years of work, this program had made a successful start. The projected fallout from excavation projects was by then estimated to be a hundred times less than forecast at the close of testing in 1958.[84] Second, different 'emplacement' techniques for burying nuclear devices were studied to discover how best to trap radioactivity in the material that falls back into a crater immediately after an excavation detonation. Basically this involved discovering the optimum burial depth to secure maximum 'fallback' of radioactivity, without constraining the excavation effect of the blast. When questioned about the mysteries of this technique in late 1986, Edward Teller said he could not reveal them because: '(a) they are secret, and (b) I have forgotten!'[85] A third problem tackled was the so-called 'neutron activation' of naturally-occurring materials, such as lithium, in the vicinity of a nuclear explosion, which increases the overall level of radioactive contamination resulting from the blast. Attempts were made to limit such 'activation' by wrapping the nuclear device in a neutron-absorbing 'blanket' of boron.[86]

Research was also directed towards developing and testing specialised nuclear devices for Plowshare. In the early days it had been assumed that many PNE applications would employ nuclear explosives from the weapons program, as had been the case with the Gnome shot. However, as various possibilities were examined, the desirability of tailor-made explosives for at least some Plowshare purposes became apparent.

## *Project Chariot*

Plowshare's first 'planetary engineering' experiment was to be a harbour construction project codenamed Project Chariot. Planning for Chariot was authorised by the AEC in early 1958, soon after Plowshare was inaugurated. While Chariot was intended primarily as a nuclear excavation *experiment*, it was hoped that the harbour created would be of some practical use if the experiment worked.[87] A site at the mouth of the Ogotoruk Creek near Cape Thompson on the north-west coast of Alaska was chosen, largely because of its proximity to potentially huge oil deposits.

The experiment envisaged the firing of four 20 kiloton explosives in a row to make a shipping channel, and one 200 kiloton explosive at the landward end of the channel to create a turning basin.[88] Because there

*Figure 2 Crater produced by Sedan, Plowshare's first nuclear excavation experiment. The vehicles and equipment in the lower left of the photograph give an idea of the scale. (Courtesy of Lawrence Livermore National Laboratory.)*

was little scientific data on the area and no experience with nuclear testing in an arctic environment, a major environmental study of the region was begun in the late northern spring of 1959. The study was under the auspices of a committee led by Dr John Wolfe of the AEC's Division of Biology and Medicine,[89] whom Gerald Johnson describes as 'an outstanding scientist and philosopher'.[90]

Curiously, the AEC appeared to pre-empt the findings of the study by issuing a report in June 1960 based on shots conducted at Nevada. It asserted, for example, that the effect of the Cape Thompson project on caribou 'should not be noticeable', a conclusion reached by studying the effects of nuclear detonations on a herd of cattle grazing near the Nevada Test Site.[91] More disturbing was the reported 'blackballing' of a biologist, Dr William Pruitt Jr, who was working under an AEC grant at the University of Alaska as part of the Cape Thompson study. According to Peter Metzger, Pruitt's research had led him to the conclusion that the detonation of several nuclear devices at the Cape would have detrimental effects, particularly on the Eskimos. The AEC allegedly pressured the university to edit Pruitt's findings and have him depart his post. The Commission then allegedly conspired to prevent Pruitt's appointment to a position at the University of Montana.[92]

In 1962 Chariot was ignominiously abandoned before the first nuclear device could be fired. As Gerald Johnson puts it, 'Plowshare had a habit of getting into trouble with the best intentions'.[93] In view of Chariot's importance for the future of nuclear excavation, particularly projects such as the proposed new Panama Canal, its cancellation came as a surprise.[94] The official explanation given by the AEC was that the project had been 'overtaken by events'. The Commission claimed it could obtain all the excavation data it required from experiments at the Nevada Test Site. In that case it is difficult to see why Chariot was proposed in the first place. The AEC's explanation is also plainly at odds with its later enthusiasm for a harbour building project in Australia, which it claimed would be an important part of the experimental program for the Panama Canal project.[95] In 1969 Gerald Johnson, who felt the cancellation of Chariot was a mistake, told a symposium on PNEs at the University of Arizona that Chariot was shelved 'not because of the possible biological impact but primarily because':[96]

> From a technical standpoint, with the exception of the documentation of the explosion effects on the ecology, much of the experimental data has already been obtained in Nevada or will be available soon following experiments in Nevada and the Australian harbor project.

In 1962, however, the Australian project was only a vague possibility and cannot have been a factor in Chariot's demise.

It is clear that the real 'event' which had 'overtaken' Chariot was public concern about the environmental impact of exploding nuclear devices in the Arctic tundra. Plowshare enthusiast Arnold Kramish asserts that the project was 'deferred' because of 'local opposition'.[97] Edward Teller charges that the project was halted by 'environmentalists'.[98] Congressman Holifield told Gerald Johnson:[99]

> We had a lot of letters, you know, that this was going to destroy the reindeer and that sort of thing in Alaska. Our correspondents warned that the moss reindeer ate would be impregnated with radioactivity [and] that they would shed their horns in the wrong season.

In his 1969 address at the University of Arizona, Johnson confirmed that among the 'major problems in trying to get authorisation of the Chariot project' was what he termed 'public information and also failure to get independent support for the need for the harbour'.[100] According to Johnson, 'fears were generated in the minds of the local population that were difficult to alleviate, particularly in view of the lack of support for the harbour'. He concluded: 'This experience emphasises what I believe to be the greatest problem in the Plowshare program — that of establishing public confidence and acceptance'.[101]

This lack of public 'confidence and acceptance' is not surprising in view of the fact that the planned explosions would have unavoidably vented radioactivity into the atmosphere. While it is true that both the United States and the Soviet Union were at that time conducting nuclear tests of up to 60 megatons in the atmosphere, this did not alleviate concerns about local fallout from Chariot. In 1961 the St Louis-based Committee for Nuclear Information (CNI), a group of scientists, doctors and citizens concerned about nuclear issues, published an analysis of Chariot which purported to show the 'peculiar biological risks' from fallout in the arctic region.[102] The article was tape recorded in the language of the Eskimos, who have no written language, and distributed to the scattered communities along the Alaskan coast. CNI claims to have been a 'large factor' in disseminating the information about Chariot which eventually led to its cancellation.

The AEC was also unable to find support for Project Chariot on economic or commercial grounds. Glenn Seaborg has noted that the project was dropped 'because the economic incentives were not good enough'.[103] As Ralph Sanders, a staunch apologist for Plowshare pointed out, apparently oblivious to the implications of his remarks:[104]

> The AEC has no ambitions for economic development of the area. The harbor will be small, incapable of handling a great amount of traffic; it also would probably be iced during eight months of the year. The hinterland consists of frozen or soggy tundra with the ground most likely all in permafrost. While the northwestern coast of Alaska reportedly possesses rich mineral deposits — including coal, oil, tin, lead or zinc — the ground, frozen solid below the surface for a few feet, makes mining difficult. No commercial companies have shown interest in exploiting possible mineral deposits there. Therefore, the AEC views the Cape Thompson project as an experiment.

That the AEC could even have considered detonating a nuclear device in such a fragile environment for whatever purpose seemed incredible, not just to some Alaskans, but to the scientists involved in the environmental study. One of the most beneficial outcomes of the entire Plowshare program, the environmental study of the Cape Thompson region cost $3 million and involved almost a hundred scientists. Completed a year after Project Chariot was cancelled, it has been described as 'the most integrated investigation of any piece of real estate on earth'.[105]

The preface to the report has a poetic and emotional style quite unlike most scientific tomes. It makes clear the scientists' commitment to the preservation of the Cape Thompson environment and their barely disguised disdain for the very idea of detonating nuclear devices there:[106]

> The circumpolar arctic tundra is a unique environment on this planet, having no counterpart in the southern hemisphere... it was in [this] primeval arctic tundra, last on earth of undisturbed areas of vast extent, that it was proposed by the US Atomic Energy Commission's Lawrence Radiation Laboratory to excavate an inlet as an experiment in what has been appropriately called 'geographical engineering'.

The study concluded that if the detonations had been carried out, 'the chance of biological cost at the ecological level, including jeopardy to the Eskimo or to the plants and animals from which he derives his livelihood, appears exceedingly remote'. This was followed however by the warning that 'risk is always involved' and that 'there are necessarily some uncertainties involved in some predictions... that can be resolved only by experimentation'. More tellingly, the report concluded that:[107]

> The problems in such massive nuclear undertakings involve far more than radiation, blast, seismic shock, and heat — they involve also the total biota and physical environment and impinge abundantly on man himself.

## *The Arms Control Implications of Nuclear Dynamite*

Nuclear dynamite had other implications besides the environmental. The term 'peaceful nuclear explosive', originated by the Soviets for political purposes and taken up with alacrity by the AEC, introduced a new and dangerous ambiguity into the nuclear arms race. As Gerald Johnson, in contrast to Edward Teller, points out, there can be no such thing as a 'peaceful' nuclear explosive, just as there cannot be a 'peaceful' conventional explosive. It is the application of the explosive, whether nuclear or conventional, that can be military or constructive. To emphasise the point, Johnson used to quip about the 'military uses of peaceful explosives'.[108]

It is certainly possible to develop specialised nuclear devices for non-military purposes. These may not be suitable for military purposes in the sense that they have not been 'configured' as weapons or 'weaponised'. Bulky and awkward to handle, they may not be easily mounted on an intercontinental ballistic missile or dropped as a bomb from an aircraft.[109] It is equally true that some types of nuclear weapons are ill-suited for peaceful purposes such as stimulating natural gas or building harbours. Nevertheless, so-called peaceful nuclear explosives and nuclear weapons share several fundamental characteristics which make them to all intents and purposes the same animal: they are transportable, capable of releasing enormous amounts of explosive energy and, in their various applications, encompass the same range of yields.[110] They can be used to block a river or destroy Geneva, Melbourne or Caracas. As Professor Franklin Long of Cornell University has noted:[111]

> ... the nuclear explosive that is developed, labelled PNE or not, still is essentially the explosive that one could use in a bomb.

In the early 1960s Plowshare enthusiast Arnold Kramish, formerly a physicist with the Manhattan Project, envisaged the development of a distinctive PNE device using scarce uranium-235 and deliberately designed to be too heavy and large to use as a weapon. He proposed that while blueprints of such devices could be made freely available, states would be required to submit their devices to an international commission for certification as to their 'safety and non-military' character. Such ideas however ignored the increasing availability of nuclear materials, including uranium-235, the fact that many PNE applications require small, lightweight nuclear devices, and the danger that once a non-nuclear weapon state acquired the means for making large and bulky PNEs, it would very soon be able to 'weaponise' such devices.[112] In 1974 Wolfgang Panofsky told the US Senate Foreign Relations Committee that:[113]

> For reasons I cannot discuss here I doubt the feasibility [of] whether one can indeed make a nuclear explosion sufficiently 'awkward' to prevent military applications.

Just as troubling as the concept of a 'peaceful nuclear *explosive*' is the notion of a 'peaceful nuclear *explosion*'. As a Canadian Ambassador has colourfully put it, in the manner of Gertrude Stein: 'A nuclear explosion is a nuclear explosion is a nuclear weapon'.[114] Indeed, any so-called 'peaceful nuclear explosion', whether conducted for a constructive purpose or to assist in developing devices for future constructive purposes, is ambiguous in arms control terms, since any nuclear test explosion invariably provides information useful for the development of nuclear weapons. Ulf Ericsson has concluded:[115]

> ... it would be very difficult or impossible to provide for effective development and use of [peaceful] nuclear explosions and to avoid, at the same time, all opportunities for intentional or unintentional and limited nuclear weapons testing. Some opportunities for limited misuse could be suppressed by safeguards, but that would be at the expense of crippling especially the further development of methods for peaceful uses of nuclear explosions, developments essential to efficient exploitation.

The Stockholm International Peace Research Institute concurs:[116]

> The risk that peaceful nuclear explosions may be used for clandestine military purposes can be reduced by expert observation of their preparation, conduct and effects, but even the most intrusive inspection would not deny all weapon-related benefits to a state carrying out such explosions.

Short of their actual use as weapons, the main threat posed by the development of PNEs is therefore their potential for proliferating knowledge about how to construct and perfect nuclear devices for military purposes.

Prior to the 'invention' of PNEs, the explosion of a nuclear device was an unmistakable sign that a country was testing nuclear weapons. Now, any country interested in clandestinely developing nuclear weapons could claim, either to avoid alarming its neighbours and the international community or to circumvent arms control agreements, that it was developing 'peaceful nuclear explosives'. As Fred Ikle, former director of the US Arms Control and Disarmament Agency (ACDA) would lament:[117]

> Peaceful nuclear explosions pretend to make a distinction where there can't be one and make it easier for governments to move in the direction that could lead to weapons production.

As Australian PNE expert, Alan Wilson, notes:[118]

> Even the first step along the PNE development road can contribute significantly to a state's weapon development capability, since the testing of even a primitive device can be used to confirm the underlying design concepts and calculations.

Another danger is that an existing nuclear weapon state, either during a moratorium on nuclear tests or under a test ban, could use PNEs to secretly improve or check the reliability of its nuclear arsenal and maintain its laboratories' nuclear weapon testing skills.

Such subterfuge could occur in two ways. Nuclear weapon tests could be conducted as though they were part of a PNE project, the explosive power of the devices being used for an ostensibly 'peaceful' purpose such as damming a river. Alternatively, the state could simply claim, as India would do in 1974, that a nuclear test conducted at a proving ground was a 'test' of a PNE device. In either case there would be little objective evidence to dispute the claim.

Seismic signals recorded by a verification network would show no difference between a weapon test explosion and a PNE. On-site inspection after the event would reveal only that an explosion had occurred. Similarly, analysis of the fallout from the explosion, if any, would not reveal whether the device was a weapon or a PNE. Neither would examination of the device itself before detonation. It would also be impossible to detect any difference between a nuclear weapon production facility and one for PNEs. As Ulf Ericsson has pointed out:[119]

> The main difference would be the absence, in the peaceful case, of related production of military delivery vehicles and related planning, organization and instructing of the armed forces.

The existence of weapon systems capable of delivering both conventional and nuclear weapons, and the secretiveness of much military training and organisation could lessen the significance of even these differences.

One solution to the PNE problem that has been suggested from time to time is the establishment of an international organisation employing scrupulously 'internationalised' civil servants to manage a PNE service, including the detonation of PNE devices.[120] Such an idea is flawed. If, for instance, the organisation produced its own nuclear devices, and even if it recruited nuclear scientists from the nuclear weapon states for the work, there would still be an unavoidable proliferation of knowledge of nuclear weapons, not to a nation state but to an international entity. Since such bodies have great difficulty handling secret information, no nuclear weapon state would, in any case, be likely to offer it sensitive information on the design of nuclear explosives. If only ready-made nuclear devices were provided by the nuclear weapon states, the international organisation would still need sufficient familiarity with nuclear weapon designs to be able successfully to detonate them and ensure that devices appropriate to particular peaceful applications were used. If the nuclear weapon states were called upon to detonate such devices (either their own or those supplied by other nuclear weapon states) on behalf of the international body, such states would then be able to obtain weapons-related information from the event. As a 'solution' to the PNE problem, 'internationalisation' would clearly result in the very proliferation of knowledge about nuclear weapons which arms control aims to prevent. The idea has consequently never been seriously entertained.

Such schemes are also irrelevant to the psychological threat to arms control inherent in the development and use of nuclear dynamite. Some observers argue that the use of PNEs would increase public acceptance of the military use of nuclear explosives. As Ernest Sternglass has put it:[121]

> As long as you can convey the idea that setting off bombs in people's backyards is no threat, you can use these bombs to fight nuclear wars. If it's okay to dig a canal this way, and the radioactivity from that fallout isn't going to hurt anybody, then setting off nuclear weapons in Europe to defend ourselves isn't going to hurt anybody either.

Others argue that widespread use of PNEs would help weaken moral and psychological taboos against the acquisition of nuclear weapons by non-nuclear weapon states and against expansion of the nuclear arsenals of the existing nuclear powers.[122]

Such considerations were to propel the issue of nuclear dynamite into international politics. Unhappily for Plowshare this would have a major impact on the future of the program. More importantly, it would profoundly affect international efforts to halt the spread of nuclear weapons, both horizontally, to more and more states, and vertically, in greater numbers in the arsenals of states already possessing them.

# Notes

1 R. Sanders, *Project Plowshare*, Public Affairs Press, Washington DC, 1962, p. 4.
2 G.W. Johnson, 'The Plowshare Program — History and Goals', US Congress, JCAE, Hearings on Peaceful Applications of Nuclear Explosives — Plowshare, 5 January 1965, US Government Printing Office, Washington DC, 1965, p. 63.
3 H.K. Jacobsen and E. Stein, *Diplomats, Scientists and Politicians: The US and the Nuclear Test Ban Negotiations*, University of Michigan Press, Ann Arbor, Michigan, 1976, p. 90.
4 Ibid., p. 159.
5 L.A. Sobel (ed.), *Disarmament and Nuclear Tests 1960–63*, Facts on File Inc., New York, 1964, p. 22. This was the 10 kiloton Gnome shot eventually detonated after the moratorium ended.
6 US Congress, JCAE, Hearings on Frontiers in Atomic Energy Research, 22–26 March 1960, US Government Printing Office, Washington, DC, 1960, p. 51.
7 See Chapter 3.
8 US Congress, JCAE, Hearings on Frontiers in Atomic Energy Research, p. 9.
9 Sanders, p. 85.
10 From correspondence with Gerald Johnson, La Jolla, California, 28 September 1987.
11 US Congress, JCAE, Hearings on Frontiers in Atomic Energy Research, p. 34.
12 *Major Activities in the Atomic Energy Programs, January–December 1959*, US Atomic Energy Commission, US Government Printing Office, Washington DC, January, 1960, p. 94.
13 J. Koop, 'Plowshare and the Nonproliferation Treaty', *Orbis*, XII, 3, Fall 1968, p. 805.
14 From correspondence with Gerald Johnson, Palos Verdes, California, 17 January 1987. The Diefenbaker government also refused to permit nuclear weapons in Canada (for Bomarc anti-aircraft missiles and air-launched missiles); see K. McNaught, *The Pelican History of Canada*, Penguin, Harmondsworth, 1975, p. 302.
15 A. Kramish, *The Peaceful Atom in Foreign Policy*, Harper and Row, New York, 1963, p. 120.
16 US Congress, JCAE, Hearings on AEC Authorizing Legislation for Fiscal Year 1965, US Government Printing Office, Washington DC, 1964, p. 1241.
17 From correspondence with Gerald Johnson, Palos Verdes, California, 17 January 1987.
18 Johnson, p. 63.
19 *PNE Activity Projections for Arms Control Planning (*the *GURC Report)*, Gulf Universities Research Consortium, Galveston, Texas, US Arms Control and Disarmament Agency Document, ACDA/PAB-253, Washington DC, 1975, p. 3.
20 *Major Acitivites in the Atomic Energy Programs, January–December 1959*, p. 392.
21 F. Kreith and C.B. Wrenn, *The Nuclear Impact*, Westview Press, Boulder, Colorado, 1976, pp. 7–8.
22 The *GURC Report*, p. 3.
23 Ibid.
24 Ibid.
25 G.W. Johnson and H. Brown, 'Non-military Uses of Nuclear Explosives in H.F. York (ed.), *Arms Control*, W.H. Freeman and Co., San Francisco, 1973, p. 410.
26 Ibid., p. 415.
27 B.A. Bolt, *Nuclear Explosions and Earthquakes: The Parted Veil,* W.H. Freeman and Co., San Francisco, 1976, p. 182.
28 Sanders, p. 175.
29 H.P. Metzger, *The Atomic Establishment*, Simon and Schuster, New York, 1972, p. 276.
30 H. P. Green and A. Rosenthal, *Government of the Atom*, Atherton Press, New York, 1963, p. 105.
31 Metzger, p. 19.
32 R. L. Beckman, *Nuclear Non-Proliferation: Congress and the Control of Peaceful Nuclear Activities*, Westview Press, Boulder, Colorado, 1985, p. 96.
33 Ibid., p. 95.
34 US Congress, JCAE, Hearings on Frontiers in Atomic Energy Research, p. 51.
35 Ibid., p. 52.
36 Smyth was a former AEC Commissioner and author of the seminal *Report on Atomic Energy for Military Purposes*, written for the US Government in 1945.
37 US Congress, JCAE, Hearings on Frontiers in Atomic Energy Research, p. 15.

38 Ibid., p. 8.
39 D. Ford, *The Cult of the Atom*, Simon and Schuster, New York, 1982, p. 22.
40 Ibid., p. 23.
41 G. T. Seaborg and W. R. Corliss, *Man and Atom: Building a New World Through Nuclear Technology*, E. P. Dutton and Co., New York, 1971, pp. 191–97.
42 The search was for elements beyond atomic number 100 (fermium) in the periodic table of chemical elements. Uranium-235 and plutonium-239 are among the 'heaviest' elements known to man.
43 US Congress, JCAE, Hearings on AEC Authorizing Legislation for Fiscal Year 1965, p. 1197.
44 *Energy and Technology Review*, Lawrence Livermore National Laboratory, September 1982, p. 16.
45 O. Dahlman and H. Israelson, *Monitoring Underground Nuclear Explosions*, Elsevier Scientific Publishing Co., Amsterdam, 1977, p. 296.
46 *Energy and Technology Review*, p. 15.
47 Sobel, p. 59.
48 Sanders, p. 63.
49 *The American Experience with Peaceful Nuclear Explosions*, US Arms Control and Disarmament Agency, US Government Printing Office, Washington DC, 1978, p. 2.
50 Sobel, p. 59.
51 From an interview with Gerald Johnson, Washington DC, 10 December 1986.
52 D. B. Brooks and J. V. Krutilla, *Peaceful Use of Nuclear Explosives: Some Economic Aspects*, Resources for the Future, Inc., Washington DC, 1969, p. 13.
53 *Energy and Technology Review*, p. 15.
54 Ibid.
55 G.T. Seaborg *et al., Plowshare*, Division of Technical Information, US Atomic Energy Commission, Oak Ridge, Tennessee, 1966, p. 27.
56 Sanders, p. 71.
57 M. Willrich, *The Non-Proliferation Treaty: Framework for Nuclear Arms Control*, The Mitchie Company, Charlottesville, Virginia, 1969, p. 139; M. Willrich, *Global Politics of Nuclear Energy*, Praeger, New York, 1971, p. 152.
58 E. Teller, *The Legacy of Hiroshima*, Doubleday and Co., New York, 1982, p. 216.
59 From an interview with Ed Fleming, Livermore Laboratory, California, 3 December 1986.
60 D. N. Farnsworth and J. W. McKenney, *US/Panama Relations 1903–1978: A Study In Linkage Politics*, Westview Press, Boulder, Colorado, 1983, p. 112.
61 Ibid., p. 105.
62 G. T. Seaborg and B.S. Loeb, *Stemming the Tide: Arms Control in the Johnson Years*, Lexington Books, Lexington, Mass., 1987, p. 311.
63 Ibid.
64 From an interview with Gerald Johnson, Washington DC, 10 December 1986. Johnson is not certain whether the meeting with Eisenhower took place in 1959 or 1960.
65 From correspondence with Gerald Johnson, Palos Verdes, California, 17 January 1987.
66 Bolt, p. 193.
67 Ibid.
68 Farnsworth and McKenney, p. 37.
69 The *GURC Report*, p. 3; Seaborg and Loeb, p. 311.
70 US Congress, JCAE, Hearings on AEC Authorizing Legislation for Fiscal Year 1970, US Government Printing Office, Washington DC, 1969, p. 306.
71 Johnson, p. 64.
72 Blumberg and Owens, p. 423.
73 Ibid.
74 Bolt, p. 183.
75 Ibid., p. 184.
76 Ibid., p. 183.
77 *The American Experience*, p. 3.
78 Sobel, p. 78.
79 Seaborg and Corliss, p. 103.
80 US Congress, JCAE, Hearings on AEC Authorizing Legislation for Fiscal Year 1965, p. 1200.

81 US Congress, *Nuclear Non-Proliferation Fact Book*, US Government Printing Office, Washington DC, 1977, pp. 375–76.
82 *An Analysis of the Economic Feasibility, Technical Significance and Time Scale for Application of Peaceful Nuclear Explosions in the US, with Special Reference to the GURC Report Thereon (The Long Report)*, Program on Science, Technology and Society, Cornell University, US Arms Control and Disarmament Agency, Washington DC, 1975, p. 8.
83 Koop, p. 798.
84 Johnson, p. 63.
85 T. C. Findlay, 'Twenty Minutes with Edward Teller', *Newsletter*, Peace Research Centre, Australian National University, Canberra, 2, 1, 1987, p. 2.
86 I. Y. Borg, *Nuclear Explosions for Peaceful Purposes*, Lawrence Livermore National Laboratory Paper No. UCRL-95026, Rev. II, November 1986, p. 5.
87 Ibid.
88 G. W. Johnson, 'Technological Development of Nuclear Explosives Engineering' in L. E. Weaver (ed.), *Education for Peaceful Uses of Nuclear Explosives*, University of Arizona Press, Tucson, Arizona, 1970, p. 15.
89 *Environment of the Cape Thompson Region, Alaska*, USAEC, US Government Printing Office, Washington, DC, 1966, p. iii.
90 From an interview with Gerald Johnson, Washington DC, 10 December 1986.
91 Sanders, p. 78.
92 Metzger, pp. 257–58.
93 From an interview with Gerald Johnson, Washington DC, 10 December 1986.
94 Ibid.
95 See Chapter 9.
96 Johnson in Weaver (ed.), p. 15.
97 Kramish, p. 119.
98 From an interview with Edward Teller, Livermore Laboratory, California, 9 December 1986.
99 US Congress, JCAE, Hearings on AEC Authorizing Legislation for Fiscal Year 1965, p. 1242.
100 Johnson in Weaver (ed.), p. 15.
101 Ibid.
102 S. Novick, *The Careless Atom*, Delta, New York, 1969, p. 200.
103 Seaborg and Corliss, p. 189.
104 Sanders, p. 74.
105 Ibid., p. 79.
106 *Environment of the Cape Thompson Region, Alaska*, p. v.
107 Ibid., p. vii.
108 From correspondence with Gerald Johnson, La Jolla, California, 28 September 1987.
109 The first full-scale H-bomb, detonated in November 1952, required a massive refrigeration plant, larger than a two-storey house and weighing 65 tons, to keep its hydrogen fuel in a liquid state prior to detonation. (P. Pringle and J. Spigelman, *The Nuclear Barons*, Avon Books, New York, 1981, p. 243.)
110 A.R.W. Wilson, 'Technical Aspects of Peaceful Nuclear Explosions Relevant to their Possible Role in the Future Proliferation of Weapon-Usable Nuclear Materials' in *Nuclear Energy and Nuclear Weapon Proliferation*, SIPRI, Taylor and Francis, London, 1979, p. 315.
111 US Congress, Senate Foreign Relations Committee, Hearings on the TTBT/PNET, 28 July, 3 August, 8 and 15 September 1977, US Government Printing Office, Washington DC, 1977, p. 130.
112 Kramish, pp. 126–27.
113 Testimony of Wolfgang Panofsky to US Senate Foreign Relations Committee, 12 September 1974, reprinted in US Congress, Senate Foreign Relations Committee, Hearings on the TTBT/PNET, 1987, US Government Printing Office, Washington DC, 1987, p. 270.
114 Eighteen-Nation Disarmament Committee (ENDC) Document, ENDC/PV. 299.
115 U. Ericsson, *The Question of Nuclear Explosions for Peaceful Purposes by Non-Nuclear Weapon States and the Possibility of Misuse of Such Technology for the Production*

*of Nuclear Weapons*, Document A/CONF. 35/DOC. 3, Conference of the Non-Nuclear Weapon States, Geneva, 1968, p. 26.

116 *SIPRI Yearbook 1977*, Almqvist and Wiksell, Stockholm, 1977, p. 356.

117 Beckman, p. 116.

118 Wilson, 'Technical Aspects of Peaceful Nuclear Explosions', p. 316.

119 Ericsson, p. 16.

120 T. S. Lough, 'Peaceful Nuclear Explosions and Disarmament', *Peace Research Reviews*, Canadian Peace Research Institute, Ottawa, 1968, p. 46.

121 L. J. Freeman, *Nuclear Witnesses: Insiders Speak Out*, W. W. Norton and Co., New York, 1981, p. 67.

122 Arnold Kramish, on the contrary, portrayed the occasional PNE as being beneficial to arms control in reminding the world of the potentially catastrophic effects of nuclear war; see Kramish, pp. 126–27.

# 3

# Enter the Arms Controllers

*Probably the most exasperating obstacles to progress in this area have been and still are those so-called 'liberals' whose conscience pangs cause them to view any peaceful application of atomic energy in terms of a mushroom cloud... to hear them tell it, Plowshare, by itself, is the single major obstacle to total and complete world disarmament.*[1]

Craig Hosmer, 1969

THE issue of peaceful nuclear explosions would have an impact on nuclear disarmament and arms control efforts at several levels. First, it would corrupt the arms control objectives of the United States and create dissension within successive US Administrations. Second, it would exacerbate differences between the United States and the Soviet Union. Third, it would create ill-feeling between nuclear weapon states and non-nuclear weapon states over the latter's access to nuclear technology. Fourth, it would provide an added justification for ambitious developing states to frustrate or opt out of disarmament and arms control agreements. Finally, it would contort several arms control agreements eventually concluded. Without nuclear dynamite, arms control efforts would have been considerably less fraught, the issues less confused and the negotiations less divisive. As Fred Ikle, former director of the US Arms Control and Disarmament Agency, exclaims: 'From an arms control point of view, peaceful nuclear explosions are a diabolical invention'.[2]

The cause of the problem was not just the invention of nuclear dynamite and the existence of Plowshare and its later Soviet equivalent. It was also the over-zealous promotion of PNE technology, first by the United States and then by the Soviet Union, and their attempts, at varying times, to preserve their right to conduct and develop PNEs, even while urging other states to give up such rights for the sake of arms control. Herbert York appears to lay most of the blame at the feet of Plowshare:[3]

> Internationally, the problems created by Project Plowshare have proved to be very serious. Several of the nuclear arms control treaties have had to be distorted because influential advocates insisted that the treaties allow for the possibility of future uses of peaceful nuclear explosives.

## *The Antarctic Treaty*

Whether by design or serendipity, the first arms control agreement to deal with PNEs was the 1959 Antarctic Treaty. In what could have been a model for future agreements, Article V of the treaty banned 'any nuclear explosions in Antarctica'.[4] However, it also provided that if international agreements were negotiated concerning the use of nuclear energy, 'including nuclear explosions', the rules established under such agreements would apply to Antarctica, provided that all Antarctic Treaty parties were also parties to such agreements. Even at this early stage then, before any PNE program had begun, and even in relation to the fragile environment of Antarctica, the nuclear weapon states were keeping their PNE options open.

## *The Test Ban Treaty Negotiations, 1958–63*

It was, however, during the protracted negotiations on a nuclear test ban in the late 1950s and early 1960s that the arms control implications of PNEs were first seriously confronted. The negotiations, between the United States, the United Kingdom and the Soviet Union, began in Geneva in October 1958 at the Conference on the Discontinuance of Nuclear Weapon Tests. While the PNE question did not loom large in the talks, it was a continuing irritant. The problem facing the negotiators was how to accommodate the US Plowshare program under an otherwise total nuclear test ban, without allowing the United States to obtain weapons-related benefits from that program. Little consideration was given to the problem of non-nuclear weapon states acquiring and using PNEs. It was evidently assumed that such states would simply be banned from conducting any nuclear explosions, including PNEs.

From the outset of the talks the US delegation, at the urging of the AEC, insisted that the use of PNEs by nuclear weapon states should be permitted and that ways could be found to prevent military benefits being derived from such applications.[5] Ambassador Wadsworth lauded the possibilities of nuclear dynamite thus:[6]

> Nuclear explosive devices may be used, we believe, for such purposes as opening up vast resources of underground petroleum products not presently available for use on an economic basis... to create a reservoir of heat in the earth's crust and to tap it at will... creation of underground reservoirs for the storage of rainfall water which now runs off into rivers and the oceans, and in making

navigable harbors in areas where such facilities are needed but do not now exist.

In contrast, the Soviet Union, with no Plowshare program to defend, was adamant that PNEs be totally prohibited. Although its draft Agreement on the Cessation of Atomic and Hydrogen Weapon Tests of 31 October 1958 contained no specific reference to PNEs, it was assumed that they were covered by the reference to a ban on 'tests of atomic and hydrogen weapons of any type'.[7] On 25 December 1958 Soviet Foreign Minister Gromyko, in a speech to the Supreme Soviet, appeared to signal some movement in the Soviet position by declaring that it might be possible for a test ban treaty to permit PNEs. However the rigorous conditions he set — parity between East and West in the number of PNEs permitted, as well as complete internal and external examination of each other's PNE devices to ensure that weapons-related information could not be obtained — were probably deliberately designed to be unacceptable to the United States.[8]

On 30 January 1959 the United States proposed its own solution to the PNE problem. It tabled a draft article which provided that each treaty party deposit packaged and sealed PNE devices at a designated location within its own territory on or before the date the treaty entered into force. Thereafter, representatives of an International Control Commission (ICC), presumably staffed by the nuclear weapon states, would provide continuous surveillance to ensure that the devices were not tampered with and that no additional devices were added.[9] The devices would not be subject to internal inspection, nor would the depositing state be required to reveal blueprints. While the devices could be serviced by their owners, parts could only be replaced under ICC surveillance. Should a party wish to add a nuclear device to its stockpile after the treaty entered into force, it would have to permit internal and external inspection as well as providing blueprints to the Commission.

This proposal was intended to satisfy concurrently several domestic and international requirements of the Eisenhower Administration.[10] Domestically, it was intended to avoid opposition in Congress, whose Atomic Energy Act prohibited disclosure of the details of nuclear devices. Internationally, it would ensure that devices in any Soviet PNE stockpile would be based on current or old weapons technology, thereby, in theory, preventing improvements in weapon designs being obtained from PNE detonations. The provision for full inspection of devices added to the stockpile after ratification was probably intended to deter the Soviets from making such additions. The Americans had already declared that they could not accept device inspection themselves.

Curiously perhaps, the United States did not insist on the right to test PNE devices. It appeared to assume from the outset that a test ban should prohibit such testing along with testing of nuclear weapons. This would mean restricting Plowshare to using either previously tested nuclear weapon designs or untested new PNE designs — a severe limitation. The problem with permitting the *testing* of PNEs under a test ban was that it would present enormous opportunities for obtaining weapons-related information, far greater than would the use of PNEs in actual PNE projects.

A key problem was diagnostic data. Such data is obtained from nuclear tests, using a variety of on-site diagnostic instruments, to permit scientists to determine the yield, radioactive products, efficiency and effects of nuclear explosions. This data is used for improving the design of nuclear devices — and herein lies the problem — whether for use as weapons or as nuclear dynamite.

In the case of a PNE project, such as building a harbour or stimulating natural gas, it is possible to ban certain types of diagnostic instruments altogether from the site without jeopardising the project. Even then it is impossible to completely avoid the likelihood of military-related information being obtained. As Ulf Ericsson has noted:[11]

> Even if all peaceful explosions were closely observed by experts, to ensure (as well as possible) that no diagnostic measurements are made, the measurement of yield, radioactive debris and peaceful explosions effects necessary for the pursuit of the peaceful purpose itself would provide the nuclear-weapon State making the explosion with some limited opportunities to improve its nuclear weapons.

In the case of a developmental test of a PNE however, the recording of diagnostic data is fundamental to the aim of the explosion, which is to improve the type of nuclear device involved. Banning diagnostic instruments altogether from PNE test sites would render such tests useless.

The Soviets responded to the US proposals in typical 'carrot and stick' fashion. On the one hand they charged that the scheme was simply a cover for continuing US nuclear weapon tests. Tsarapkin, the Soviet delegate, complained that since the proposal did not provide for internal inspection of all stockpiled devices, the United States would be able to surreptitiously include nuclear weapons, either those already tested or new types invented prior to the treaty entering into force. In addition, Tsarapkin charged, the right to replace device parts would permit clandestine modernisation. Moreover, since there was no proposed limit on the number of PNE detonations, the 'doors were wide open' to unlimited nuclear tests. Despite these criticisms, Tsarapkin expressed a willingness, 'without particular enthusiasm', to 'study' the problem of permitting PNEs under a test ban, in order to prevent the issue from derailing the negotiations altogether.[12] He denied that the USSR was itself interested in using nuclear dynamite.

The US stockpile proposal was indeed flawed. Without severe restrictions on diagnostic instruments at PNE sites, the use of even old, obsolete, stockpiled nuclear devices for peaceful purposes could provide a nuclear weapon state with the very type of weapons-related information that a test ban was designed to preclude. Without a ban on diagnostic instruments, such measurements as the yield and the extent and nature of radioactive debris could, for instance, provide evidence of the extent of deterioration of nuclear weapons. Such explosions could also provide data on the effects of such weapons.[13] Perhaps the major drawback, one that would become increasingly clear as Plowshare advanced, was that PNE applications would require nuclear devices as sophisticated as the latest nuclear weapons, not the obsolete varieties contained in a stockpile. As Ulf Ericsson notes:[14]

> The explosive properties desirable for peaceful applications ... call for rather advanced explosive designs. Some of the requirements are not very different from those for advanced military nuclear explosives, even if the environmental requirements and the handling and firing arrangements are somewhat different.

The Soviets tabled a counterproposal on 23 February.[15] It stipulated that treaty parties provide complete descriptions and blueprints of all stockpiled devices, permit internal and external inspection of devices and international control of detonation sites, and disseminate information obtained from detonations to all parties. Finally, the Soviets reiterated that the United States and the United Kingdom, on the one hand, and the Soviet Union on the other, should be permitted to detonate PNEs only on a one-to-one basis.

Like the US proposal, the Soviets' was problematical. No state would be willing to hand over blueprints for, or permit inspection of, PNE devices which were as sophisticated as its latest nuclear weapons.[16] Hence the Soviet proposal would be even more likely than the US proposal to force the nuclear weapon states to use unsophisticated, obsolete nuclear devices unlikely to be suitable for PNE applications. Another drawback was that device inspection might not necessarily reveal whether any military uses could be obtained from a particular PNE detonation. Weapon designers, according to Thomas Lough, could incorporate into nuclear devices particular experimental materials that could not easily be detected. While he admits that this may be far-fetched, it does illustrate, he says, the difficulty of 'knowledge prevention' which is at the heart of the PNE problem.[17] Finally, while the Soviets had mentioned international 'control' of PNE sites and detonations, they had not specified any restrictions on diagnostic instrumentation.

On 23 February 1960 the United States tabled a revised version of its original proposal which was simpler and less costly to administer, but which contained no substantive change other than a provision permitting each party to assist other states in carrying out PNE projects.[18] This would allow the United States to provide PNE 'services' to other countries either for a price or as part of its foreign aid program.[19]

The Soviets did not respond substantively to this new draft and the PNE problem was not discussed again in detail, at least in formal sessions at Geneva, until the advent of the Kennedy Administration in 1961. The Eisenhower Administration was, however, apparently sanguine about reaching an agreement on the issue. On 23 March 1960 AEC Chairman John McCone told the JCAE that:[20]

> There has been an agreement in principle between the Soviets and our negotiators that the use of nuclear explosions for peaceful purposes will be permitted. When I say this has been agreed to in principle, I mean just that, because the specific language of that article has not been discussed.

This was clearly a highly optimistic reading of the negotiations to that point. When a JCAE member sought clarification with the remark: 'Then it must be understood that you will not use nuclear weapons in such a project as Plowshare, but a nuclear device', McCone replied with the half-truth: 'A nuclear device; that is correct'.[21]

In the meantime an issue arose which illustrated the difficulties involved in preventing military benefits being obtained from PNEs. The issue was that of using nuclear explosions during the moratorium to determine the feasibility of seismic verification of a ban on underground nuclear tests.[22] On 3 May 1960, after months of heated debate, the Soviets reluctantly agreed to join the United States and the United Kingdom in such a research program.[23] Rejecting Western suggestions that the explosions be carried out nationally under three-power observation, the Soviet Union said it would agree to a 'strictly limited number of joint underground nuclear detonations' conducted by Western and Soviet scientists and strictly confined to seismic research.

On 7 May 1960 President Eisenhower announced that the United States, on the recommendation of a scientific committee on seismic verification, the Berkner Panel, would itself conduct 'an experimental program of underground detonations encompassing... high explosives and where necessary, nuclear explosions' to improve US ability to detect and identify underground nuclear explosions.[24] Administration officials naïvely explained that the program, codenamed Project Vela, would not involve nuclear weapons. On 9 May the Soviets protested that the program would violate the moratorium.[25] On 11 May Eisenhower told reporters there had been a 'misunderstanding' of his announcement.[26] Compounding the confusion, the President said the proposed detonations would 'not have anything to do either with weapons development or the Plowshare project'. The United States, he said, was not currently 'involved in any nuclear explosions' except the planned seismic experiments with Britain and the Soviet Union.

On 31 May 1960 the panel of American, British and Soviet scientists charged with preparing the joint seismic program reported to the Geneva Conference that the Soviets had objected to seven of the twelve nuclear explosions the United States had proposed, two of them on the grounds that the tests would have military significance.[27] The United States on 2 June offered the Soviets several safeguards to reassure them that seismic experiments could not be used to obtain military benefits.[28] These were:

1. to use only previously tested types of devices
2. to deposit them under international supervision before detonation
3. to permit Soviet observers to carry out instrumentation of the detonations, and

4. to refrain themselves from instrumentation of the tests.

These safeguards were justifiably rejected by Tsarapkin, who described them as 'fictitious'.[29] He warned that any US detonations under these conditions would be grounds for the Soviets to resume nuclear testing.[30] The proposed safeguards would clearly not prevent the United States from using nuclear weapons with new designs, thereby obtaining military benefits. Even with an effective ban on on-site instrumentation, the fact that the AEC could satisfy itself that such devices worked, whether they were new or old, would be sufficient 'military benefit'. In any case, remote detection methods, such as seismic monitoring, collection of fallout samples, and satellite monitoring could provide militarily useful data.

The Soviets insisted that the United States allow Soviet observers to inspect the blueprints and interior of seismic research devices — the same safeguards the Soviets were demanding for PNEs under a test ban treaty.[31] While these safeguards would add to the problems of a state wishing to obtain military benefits from nuclear explosions conducted for seismic study purposes, even they would not entirely remove the possibility.

On 14 June the Americans invited the Soviets and British to witness an underground *non-nuclear* Plowshare explosion.[32] Tsarapkin rejected the invitation on the grounds that the test was not 'relevant' to the Geneva negotiations. Tsarapkin added that since the Soviet Union was not planning any such tests, Western scientists would not be invited to visit Soviet test sites.[33]

On 12 July the United States offered to join Britain and the Soviet Union in contributing nuclear devices to a joint 'pool' for use in the proposed three-power seismic research program.[34] Each country would contribute equal numbers of outmoded nuclear devices and would be free to inspect any of them. The pool would be placed under international guard. Each of the three nations could withdraw any device, under international supervision, to conduct seismic experiments. The proposal was rejected by Tsarapkin on 19 July. He remarked that if the United States and the United Kingdom deemed it necessary to 're-test' seismic detection theories, the Soviet Union would be prepared to inspect their devices beforehand, but it would not permit Western inspection of Soviet devices.[35] Soviet demands for device inspection were obviously not intended to be reciprocal. This Soviet rebuff was followed by a long period of silence on the subject by both sides.

With the advent of the Kennedy Administration, a meeting of US officials was convened on 2 March 1961 in Washington to discuss the US position.[36] Presidential Special Assistant for Disarmament, John McCloy, favoured acceding to Soviet demands for inspection of seismic research devices. He proposed that only obsolete explosives be used. The new AEC Chairman, Glenn Seaborg, responded that the secrecy provisions of the Atomic Energy Act would have to be amended to permit device inspection. Instead Seaborg disingenuously suggested that the devices be 'altered so that they could not be called weapons'.[37] In addition, to prevent the Soviets claiming afterwards that seismic experiments had been weapon tests, Seaborg suggested that a 'competent neutral' or United Nations representative be present to certify to the contrary. McCloy felt it might be considered strange that the United States could not show a nuclear device to a NATO ally but could show it to a neutral. Secretary of State Dean Rusk asked McCloy to give further thought to the neutral observer suggestion, but noted that otherwise the proposal for opening seismic research devices to the Soviets could be considered approved.

Seaborg also reluctantly agreed to inspection of devices for peaceful purposes other than seismic research, but objected to a State Department suggestion that a limit be placed on the number of PNE explosions permitted. Seaborg records that the AEC's concession on PNE device inspection was difficult to make. Opening Plowshare devices to Soviet inspection would restrict the Commission to using obsolete devices which, while generally satisfactory for seismic research, were much less so for Plowshare because of their bulkiness and heavy fallout.[38] (Such a drawback also applied of course to the original US stockpile proposal.) Seaborg says he compromised only because he sympathised with President Kennedy's 'zealous pursuit' of a test ban treaty. He notes that some members of the AEC 'community', presumably including the Plowshare staff at the Livermore Laboratory, disagreed strongly with his position.[39] Nevertheless, Seaborg recalled some years later, he felt that important as Plowshare was, it 'should not be allowed to be a source of discord impeding progress toward a test ban'.[40] In fact it was largely as a result of Seaborg's efforts that Plowshare would become an impediment to a test ban, albeit not the only one.

On 21 March 1961, at the Geneva talks, the United States offered, 'if agreement on other treaty provisions was in sight', to seek to amend the Atomic Energy Act to permit the Soviet Union to examine the internal mechanisms of nuclear devices employed in the US seismic research and Plowshare programs. British and Soviet representatives would be allowed to examine, but not copy, blueprints of US devices and inspect but not handle the parts. Finally, they could be present when devices were reassembled and fired, and they would be permitted to install instruments and receive data on the explosions.[41] The purpose of the latter two provisions was apparently that if there were weapons-related benefits to be obtained, then both sides would benefit. However the side which designed and manufactured the device would obviously still have more familiarity with it, including the blueprints, thereby enabling it to carry out weapons improvements based on the explosions.

In announcing its proposals, the United States made clear that it would expect reciprocal rights should the Soviets begin their own seismic or PNE programs.[42] On 4 April Ambassador Tsarapkin declared that the Soviet Union 'agreed in principle' with the US proposals, while still maintaining that PNEs were 'superfluous and even dangerous'.[43] He suggested that the

Conference 'record agreement' on this and several other issues. Tsarapkin added, however, that the Soviet Union would have to study the 'technical aspects' of these subjects. The Western delegations rejected this idea because of the 'incompleteness' of the Soviet statement and continuing disagreement over the treaty budget.[44]

On 10 April Tsarapkin reiterated Soviet acceptance of the Western proposals. The Soviet Union, he said, understood the proposed safeguards to mean that Soviet representatives would have 'the possibility of control over all the processes connected with preparing the device for detonation'; would be allowed to set up instruments in the immediate vicinity of the device, in tunnels leading to the detonation chamber and at other points on the surface; and would be permitted to inspect the detonation system and 'the environment surrounding the detonation chamber' both 'before and after the explosion'.[45] Tsarapkin continued to insist, however, on a veto over the number and type of US seismic experiments, on the conclusions to be drawn from such experiments, and on numerical parity for Western and Soviet PNEs.

On 18 April 1961 the United States and United Kingdom introduced a complete draft test ban treaty incorporating their PNE proposals.[46] (See Annex II.) Chief US delegate Arthur Dean said it represented a considerable sacrifice on the part of the United States, forcing it to use primitive, inefficient PNE devices which, with their large yields, would impose prohibitively high costs.[47] According to Dean, this would seriously inhibit, for some time at least, some of the more ambitious Plowshare projects envisaged. To overcome this he suggested that the proposed ICC be authorised to prescribe alternative safeguards to permit the use of more advanced PNE devices. He also suggested that the Soviet Union now withdraw its parity proposal since it amounted to giving one side a veto over the other's PNE program.[48]

The Western draft treaty was rejected by the Soviets. While this was not because of its PNE provisions, the extent of the agreement reached on PNEs remained ambiguous. Tsarapkin declared that those aspects of the PNE issue the Soviets still considered unresolved were '... the number of explosions for peaceful purposes to be carried out by the sides, and the question of the proportion in which such explosions are to be carried out...'[49] Ambassador Dean charged that the Soviet Union still wished to 'require a one for one correspondence between Soviet and Western experiments'. This presumably meant, he said, that 'after the first United States explosion we will be unable to conduct a second until the USSR has carried out its first blast'.[50]

Strangely, despite having accepted the Western PNE proposals, the Soviets were to give as one of their reasons for breaking the moratorium in August 1961, the claim that PNEs were a 'cover for military developments':[51]

> ... it is obvious to every informed person that the carrying out of such explosions, even if it is claimed that they are conducted for peaceful purposes,

> is nothing else but a hidden form of perfecting the existing nuclear weapons or putting the finishing touches to their new types; if a nuclear explosive device is effective, for example, for 'moving ground' — and the Western Powers want to assure for themselves the right of carrying out such explosions — the same explosive devices will also be effective for military purposes. Thus while coming out in words for the discontinuance of nuclear tests, the United States and Britain in fact show concern for quite a different thing — for inserting in the treaty a loophole for further perfection of thermonuclear weapons by conducting underground explosions or explosions for so-called peaceful purposes.

Following the breaking of the moratorium, President Kennedy proposed, on 2 September 1961, an immediate ban on all nuclear tests in the atmosphere, with no exception for PNEs.[52] The consensus in the Administration was that atmospheric PNEs could be sacrificed because the arms control imperatives were so strong. As Seaborg notes:[53]

> The effect on the peaceful-uses program [Plowshare] was also considered, but it was agreed that the stakes were so high that this could be disregarded.

This proposal also proved unacceptable to the Soviets. On 28 November 1961 they tabled in Geneva a draft partial test ban treaty that would prohibit all nuclear tests, including PNEs, in the atmosphere, under water and in outer space.[54] Since it was combined with a moratorium proposal, it proved unacceptable to the Americans. Shortly thereafter, the Conference on the Discontinuance of Nuclear Weapon Tests was itself discontinued, never to resume.

On 14 March 1962 the newly-established Eighteen-Nation Disarmament Committee (ENDC), comprising a mixture of Western, Soviet bloc and nonaligned countries, convened in Geneva for the first time. On 21 March it established a sub-committee consisting of the Soviet Union, the United States and the United Kingdom, to continue consideration of a test ban treaty.[55] On 27 August the United States and United Kingdom jointly tabled two draft treaties: a comprehensive treaty banning tests in all environments, and a limited test ban treaty banning tests in the atmosphere, under water and in outer space.[56] Now that negotiations were being conducted under the auspices of a body that included non-nuclear weapon states, the two drafts dealt for the first time with the implications of PNEs for such states.

The treaties implied the right of all states to carry out PNEs, but only with the 'unanimous consent of the original parties' (the United States, the United Kingdom and the Soviet Union) and in conformity with provisions to be specified in an annex. The annex was not tabled since there was no consensus within the US Administration as to its contents, an omission Glenn Seaborg describes as 'awkward and embarrassing'.[57] The draft treaties were, however, the first indication that the nuclear weapon states intended strictly controlling the use of PNEs by non-nuclear states. The Soviets rejected both treaties, but not because of the PNE provisions. The three-power sub-committee adjourned on 20 December 1962 and, like the Geneva Conference before it, never met again.[58] However the Anglo-American draft limited test ban text eventually formed the basis of the 1963 Partial Test Ban Treaty.

Even in the absence of negotiations with the Soviets, debate continued within the US Administration over American PNE policy. This was in part due to the emergence of a new bureaucratic constituency for arms control. Since its establishment in September 1961, the Arms Control and Disarmament Agency (ACDA) had assumed a central role in interdepartmental test ban deliberations.[59] Its inclination was to support a complete ban on PNEs. Failing that, it advocated strong verification measures to ensure that PNEs did not provide a loophole for clandestine nuclear weapon tests. Plowshare's guardians, the AEC and the JCAE, were naturally suspicious of ACDA. Describing ACDA staff as 'nutball people', the JCAE's Craig Hosmer declared that: 'I voted for ACDA . . . I believe it is a good idea to collect all the people who are of that mind in one place where you can keep an eye on them'.[60] The AEC, for its part, withheld information on PNEs from the Agency.[61]

In February 1963 ACDA drafted a PNE annex for the Western draft treaties tabled in Geneva.[62] It envisaged unprecedented verification provisions, although still not necessarily completely watertight, for the use of nuclear dynamite. Any treaty party wishing to conduct a PNE would be required to:

1. give 4 months' notice, including a detailed description of the project, to the ICC
2. give full consideration to comments received from ICC members
3. invite all treaty parties to inspect the nuclear device(s) involved, both externally and internally, and observe the preparation, instrumentation and detonation of each explosion (an exception could be made if all treaty parties agreed).

By this stage AEC Chairman Seaborg was having second thoughts about international inspection of Plowshare devices. Despite his claim that he felt Plowshare was not worth the sacrifice of a test ban, Seaborg now 'felt it necessary' to argue strongly against the ACDA draft.[63] In March 1963 he fired the first salvo in what he later called 'My Effort To Save The Plowshare Program'.[64] Seaborg claims that the AEC was not just concerned about the fate of Plowshare, but about 'the fate in Congress of any test ban treaty that imperilled the program', because 'we knew that Plowshare had some very ardent support in the Congress, particularly in the Joint Committee on Atomic Energy'.[65] This was partly a rationalisation of the AEC's natural desire to preserve Plowshare from the effects of a test ban. It was also, however, a recognition of the JCAE's power. According to Freeman Dyson, the Kennedy Administration's entire negotiating position on PNEs was 'designed to win votes for ratification from senators who strongly supported Plowshare'.[66]

Writing to ACDA Deputy Director Adrian Fisher, Seaborg pointed out that Plowshare was incompatible with Soviet inspection of nuclear devices because, as was becoming increasingly clear, 'for both safety and economic reasons, Plowshare experiments required devices employing advanced weapon design principles that should not be revealed'.[67] According to Seaborg 'the necessity to reveal devices would virtually bring the Plowshare program to a halt'. As an alternative he suggested, despite having previously opposed such an idea, that each side be limited to an annual quota of five or six PNEs.

Seaborg also pointed to the 'considerable progress' made in the past eighteen months in developing nuclear excavation technology, which promised to provide 'large economic benefits in certain types of major construction activities'.[68] This, Seaborg claimed, was a 'significant factor in our current relations with the Government of Panama', a reference to proposals for a new sea-level Panama Canal. The 'progress' mentioned by Seaborg was, however, largely illusory. Plowshare scientists had certainly made some progress in reducing the radioactivity that would result from nuclear excavations. However other aspects of the technology, such as the stability of crater slopes and the possibility of joining craters to form canal trenches, were still largely unexplored. Indeed, by that stage there had been just one Plowshare excavation experiment using an actual nuclear explosion, the 1962 Sedan shot. Seaborg was clearly being over-optimistic in proclaiming the virtues of nuclear excavation.

ACDA Director William Foster replied to Seaborg on 28 March 1963, acknowledging the desirability of Plowshare, but claiming that the Soviets could misuse an annual quota to improve their weapons technology. He was also concerned that such a proposal could subject the United States to a 'barrage of propaganda that we are not sincere in our desire for a test ban'.[69] Despite this rebuff, Seaborg persisted in promoting alternatives to device inspection. At a White House meeting on 17 April 1963 he suggested that, in addition to an annual quota, there might be a yield limit on PNEs, such as 50 kilotons.[70] This seems rather low if it were designed to permit the use of PNEs in major excavation projects. (Blasts of up to 15 megatons would later be recommended for the Panama Canal project.) Seaborg subsequently admitted that his proposal would slow, but not halt, Plowshare. He was confident however that if a major excavation project were undertaken, the treaty could be amended. In a letter to Foster on 11 June, Seaborg suggested two additional safeguards to replace device inspection: a ban on all diagnostic instruments and the submission to the ICC of debris samples.[71]

At a meeting on 14 June 1963 called to discuss the US position, these proposals received very little support.[72] According to Seaborg:[73]

> It was becoming evident that the AEC's enthusiasm for Plowshare was not widely shared, whereas there was considerable concern about the adverse

diplomatic consequences that would arise if PNEs seemed to violate a test ban treaty.

Dean Rusk said he did not see how one could have PNEs and a significant test ban treaty at the same time and 'wondered, semi-facetiously, whether it would be practical to have the Soviets conduct our Plowshare experiments while we conducted theirs'.[74] In Rusk's view, if Plowshare was the only obstacle to a treaty, then the cost of forgoing it would 'probably be negligible compared with the billions that would otherwise go into a competitive arms race'.[75] Seaborg replied that if Plowshare alone were preventing a test ban he might agree. He also reiterated his claim that nuclear excavation was cheaper than conventional methods. The meeting failed to reach consensus on the PNE annex, which in any case soon proved to be redundant.

## *The 1963 Partial Test Ban Treaty*

In July 1963 the United States, the United Kingdom and the Soviet Union commenced negotiations in Moscow on a ban on nuclear test explosions in outer space, under water and in the atmosphere. On 3 July, a week before the talks began, the AEC briefed Chief Negotiator Averell Harriman, Under-Secretary of State George Ball and others on the PNE issue.[76] The purpose of the meeting and a subsequent memorandum, was to suggest some 'liberal criteria', presumably barring device inspection, under which Plowshare excavation detonations might be allowed under a limited test ban treaty.

The draft treaty tabled by the Soviets in Moscow at the outset of the talks made no exception for PNEs. The United States, on the other hand, tabled its August 1963 draft limited test ban treaty, which permitted PNEs, subject to the safeguards still to be finalised in ACDA's annex. The Soviets demurred from this draft on only two points: PNEs and the withdrawal provisions.[77] According to Freeman Dyson, an ACDA officer at the time, the PNE issue was 'one of the stickiest points in the negotiations'.[78] This may be an exaggeration. A report commissioned by ACDA on its experience with arms control over the past twenty-five years fails to mention PNEs in its account of the test ban negotiations.[79]

Seaborg says that Averell Harriman had no chance to engage the Soviets on the subject. He found them resolutely opposed to making any provision whatever for PNEs.[80] Soviet Foreign Minister Gromyko argued that allowing an exception for PNEs, in addition to permitting underground tests, might arouse international suspicion and reduce the appeal of the treaty. He claimed that some progress on PNEs could be achieved through underground explosions.[81] Arthur Dean says that the Soviets were in fact 'terribly suspicious that under the guise of peaceful uses we really were going to conduct some advanced weapons tests'.[82]

Incredibly, Seaborg claims the Soviet position came as 'something of a surprise since they were thought to have plans at least as extensive as our own for such explosions'.[83] This was despite the fact that the Soviet draft treaty plainly made no allowance for PNEs and that the Soviets' preferred position since 1958 had been one of opposition to PNEs. The Soviet attitude was clarified some days later when Harriman reported a conversation with Premier Khrushchev:[84]

> He told me of their various long-range plans such as building canals and diverting the Pechora River to flow south to Kazakhstan instead of north to the Arctic. This gave me opportunity to express surprise that he had thrown out our article on peaceful uses... He said he opposed inclusion today since this reservation might be disappointing to world opinion and arouse suspicions. (This was undoubtedly a valid point.) He offered the view that we should have no difficulty in agreeing on such matters when the subject had been more carefully explored... when tensions were relieved by a test ban and reduced by other understandings. Peaceful uses... would then meet popular approval.

Following these exchanges, Harriman proposed to President Kennedy that the United States give up the PNE provision in the draft treaty in return for Soviet acceptance of the US draft withdrawal provision.[85] According to Freeman Dyson, Harriman cabled back to Kennedy: 'I think we shall have a treaty if I give way on this one.'[86] Kennedy reportedly picked up the phone and asked ACDA Director William Foster what he thought about it. Foster said he would like to talk it over with his experts. As Dyson recalls, it was then that he, Dyson, had his 'little moment of glory':[87]

> Foster called the ACDA Science and Technology Bureau and spoke with Al Wadman, one of the bureau staff. It was late in the afternoon and almost everyone had gone home. Wadman and I were the only people left in the office. Wadman came over to me and asked me if I thought we should stand firm on peaceful explosions. I was the only person in ACDA who had been at Livermore and knew something at first hand about Plowshare... I said to Wadman, 'Of course we should give way'. Wadman called back Foster and Foster called back Kennedy and the cable went back to Harriman and the treaty was signed.

Washington's cabled response to Harriman, while stating that PNEs had a 'constituency within the US Government', concluded that a withdrawal clause was far more important, since without it Senate ratification of the treaty might be jeopardised.[88]

Dyson admits that his account gives a misleading impression even if it happens to be true. Kennedy undoubtedly consulted others, including his Science and Technology Adviser, Jerome Weisner, who had earlier advised him that the prospects for Plowshare projects eventuating were remote (contrary to Seaborg's optimism) and that Plowshare should not be allowed to stand in the way of a test ban treaty.[89] As Dyson says, even if Wadman had given a different answer the treaty would still have been signed:[90]

> The treaty had lain for a long time in the womb of history and had come to the day of its birth. We were only the midwives.

One organisation Kennedy had not consulted was the AEC. Not only that, but Glenn Seaborg later learned that information about the progress of the negotiations had been deliberately withheld from the Commission so that, 'abetted by the JCAE', it would not 'intercede vigorously in favour of PNEs and possibly imperil the treaty President Kennedy so ardently wanted'.[91] To help placate the AEC, it was arranged that Harriman visit Seaborg on his return from Moscow to convey his confidence that in due time it would be possible to amend the treaty to permit Plowshare projects to proceed.[92]

Meanwhile back in Moscow the draft treaty was amended simply by including in Article 1 the phrase 'any other nuclear explosion' in the ban on nuclear weapon tests.[93] Parties to the treaty would thus be prohibited from conducting PNEs in outer space, under water and in the atmosphere, and from assisting others to conduct such explosions.[94] Underground PNEs were also banned, along with other underground tests, if they caused 'radioactive debris to be present outside the territorial limits of the State under whose jurisdiction or control' they were conducted.[95] The only concession to PNEs was a change in the amendment procedures to facilitate adoption of an amendment to permit future large-scale PNE projects. Instead of requiring a two-thirds majority of treaty parties for amendments, the treaty would now only require a simple majority.[96]

The treaty, officially styled the Treaty Banning Nuclear Weapon Tests in the Atmosphere, in Outer Space and Under Water, but known more commonly as the Partial or Limited Test Ban Treaty (PTBT or LTBT), was signed in Moscow by the United Kingdom, the United States and the Soviet Union on 5 August 1963. It was the first significant international legal constraint on the use of nuclear dynamite.

There was one further episode in the struggle between ACDA and the AEC over the treaty's implication for PNEs. In order to facilitate Senate approval of the treaty, Dyson and Wadman went to AEC headquarters to extract from Plowshare's director a written statement saying whether or not the program could continue within the terms of the treaty. As Dyson recalls:[97]

> The director was faced with a dismal choice. If he said yes, he was helping to ratify the treaty. If he said no, and the treaty was ratified in spite of him, his program would probably be closed down. Bureaucratic politics is a dirty game, even when the good guys are winning. We had him neatly skewered, and he knew it. He said yes, and we took his statement back in triumph to ACDA.

During its passage through the US Senate, the treaty was sharply criticised because of its likely effect on Plowshare. A particular target of criticism was the ban on underground explosions likely to release radioactivity into the atmosphere. Edward Teller opposed ratification, claiming that since most PNEs result in some measurable release of radioactivity, the treaty would severely wound Plowshare.[98] He proposed two reservations should US ratification proceed. One would state that the United States would only require the consent of states involved in a particular PNE project, such as the proposed second Panama Canal, for that project to be exempt from the provisions of the treaty regarding release of radioactivity.[99] The second would define 'in quantitative and reasonable' terms the amount of radioactivity that could be released by underground explosions under the treaty. Teller claimed that:[100]

> ... 2 years from now it will be possible to make an explosion that will have made a crater and land in this crater as soon as the dust has settled, in 15 minutes, without exposing ourselves to more radiation than we have taken year in and year out in our laboratories.

Teller also expressed doubt that the Soviets would agree to amend the treaty in the future to permit Plowshare projects to proceed and thereby end 'the hold that this treaty gives them over our negotiations with friends and other countries'.[101] On the contrary, Teller warned, the Russians were likely to attempt to steal a march on the United States in the PNE field.[102] Teller also predicted (with remarkable prescience as it turned out) that the PTBT 'will really, may really, make the difference between a most effective Plowshare and practically no Plowshare at all'.[103] Oddly, he declared that the most important Plowshare applications lay outside the United States and that these would be damaged most by the treaty.[104] For good measure, Teller claimed that Plowshare would be useful in colonising the moon![105]

Livermore director John Foster expressed substantially the same views as Teller. Former AEC Chairman Lewis Strauss urged the Senate to require a reservation to the treaty which would make special provision for Plowshare projects.[106] Strauss later endorsed the view of the *Washington Star* that the absence of PNE provisions was an 'absurdity' that would prevent 'the relatively cheap and swift digging of a new sea-level canal between the Atlantic and the Pacific Oceans'.[107] Norris Bradbury, director of Los Alamos, urged ratification on behalf of 'a majority of Los Alamos staff'.[108]

Glenn Seaborg, undoubtedly with his Plowshare director's dilemma firmly in mind, had little choice but to testify in favour of the treaty. He attempted to downplay the effect it would have on Plowshare, saying that 'when we in the AEC examined the technical facts... the situation seemed not so dark'.[109] Seaborg testified that:[110]

> Our present considerations lead us to believe that excavation experiments *or projects* [emphasis added] which have a downwind distance of several hundred miles from the project site to a territorial limit probably can be conducted and that these experiments will be sufficient to develop the excavation technology.

According to Seaborg:[111]

> To us in the AEC it seemed a matter of common sense that the treaty could

> not, as a practical matter, be enforced if escape of a minute amount of radioactivity — approaching for example, a single atom, which could cause harm to no one — were considered a transgression.

Although Seaborg later claimed that the future of nuclear excavation under the PTBT was contingent on the AEC perfecting 'clean' nuclear explosives and learning how to trap most of the remaining radioactivity underground, his testimony gave the clear impression that existing technology was sufficient to prevent a treaty violation during excavation experiments and during at least *some* actual projects. Seaborg stated that major projects like a new Panama Canal would, however, require a treaty amendment.

According to Seaborg, his testimony, which he notes 'was of course routinely cleared within the executive branch', strongly influenced at least one JCAE member, former Chairman Senator Clinton P. Anderson, and 'perhaps others' to support ratification.[112] President Kennedy also attempted to bolster Senate support for the treaty by giving an undertaking, albeit a carefully-worded one, that:[113]

> The United States will vigorously pursue its ... [Plowshare] programs *within the terms of the treaty* [emphasis added] and, when such developments make possible constructive uses of explosions... will seek international agreement under the treaty to permit such explosions.

The JCAE reported to Congress its understanding that 'the Plowshare program... will not be seriously inhibited by the Treaty'.[114]

The PTBT was approved by the US Senate by a resounding vote of 80–19 on 24 September 1963 and entered into force on 10 October. As of 1 January 1988 there were 116 states party to the treaty.[115]

Since 1963 there has been no international agreement as to what degree of venting of radioactivity from underground nuclear explosions constitutes a violation of the PTBT, whether as a result of weapon tests or PNEs. Both US and Soviet PNEs have vented radioactivity and both countries have registered protests about such venting. Neither party has sought amendments to the treaty to permit excavation projects as Kennedy and Khrushchev envisaged they might.

There has in fact been much debate over whether or not the Soviets would ever have been willing to amend the treaty. Congressman Hosmer claimed in 1969, on the basis of what he admitted was 'somewhat tenuous' evidence, that the Soviets were anxious to remove the PTBT 'handcuffs' from nuclear excavation. He arrogantly asserted that:[116]

> Since any treaty means precisely what the two most powerful signatory nations say it means, I am of the opinion that the [treaty] can be rapidly brought into line with the facts-of-peaceful-nuclear-explosions-life, if certain people in our government will stop throwing up artificial hurdles.

Contrary to his testimony in 1963, Edward Teller now also contends that the Soviets would have permitted the United States to undertake nuclear excavation projects, since the Soviets were interested in such projects themselves.[117] By contrast, Senator Anderson claimed in 1965 that the Soviets had threatened to withdraw from the treaty if an attempt was made to amend it.[118] Another commentator saw the PTBT as a Soviet plot designed to stifle Plowshare:[119]

> It would seem that the Soviet Union has successfully blocked American ambitions to put nuclear explosives to work for the betterment of mankind, and unless the treaty is amended the United States will be unable to go ahead with the new canal it wants in Central America or the new harbor once planned for Alaska. The dreams of vast water reservoirs in the desert, new methods of mining and releasing oil, and a source of water for future astronauts visiting the moon would also be still born [sic] because of the Test Ban Treaty.

In fact, because the amendment procedures for the PTBT require the agreement of a majority of the parties, as well as the original three nuclear weapon states, the non-nuclear weapon states also have a veto over whether nuclear excavation projects will ever be allowed. At least one observer has suggested that the developing countries, in order to secure the benefits of PNE technology, could use this leverage against the superpowers should they ever seek to amend the treaty.[120] As former AEC Chairman Strauss noted ruefully in 1965:[121]

> When we ratified the treaty, we allowed ourselves to be placed in a position where henceforth we have to go hat in hand to the other signatories to ask their consent before we can undertake an engineering project essential not only to our commerce but to our defense. If a Sukarno, a Nasser, a Nkrumah, or any one of a hundred signatories, not to mention the Soviets, should refuse, we are bound.

## *Notes*

1 US Congressional Record, 14 April 1969, US Government Printing Office, Washington DC, 1969, p. 8945.

2 R. L. Beckman, *Nuclear Non-Proliferation: Congress and the Control of Peaceful Nuclear Activities*, Westview Press, Boulder, Colorado, 1985, p. 116.

3 H. F. York (ed.), *Arms Control*, W. H. Freeman and Co., San Francisco, 1973, pp. 347–48.

4 *Status of Multilateral Arms Regulation and Disarmament Agreements*, United Nations, Second Edition, UN Department for Disarmament Affairs, New York, 1983, p. 14.

5 G. T. Seaborg and B. S. Loeb, *Stemming the Tide: Arms Control in the Johnson Years*, Lexington Books, Lexington, Mass., 1987, p. 314.

6 US Disarmament Administration, *Geneva Conference on the Discontinuance of Nuclear Weapon Tests: History and Analysis of Negotiations*, US Department of State, Washington DC, 1961, p. 266.
7 Seaborg and Loeb, *Stemming the Tide*, p. 313.
8 H. K. Jacobsen and E. Stein, *Diplomats, Scientists and Politicians: The US and the Nuclear Test Ban Negotiations*, University of Michigan Press, Ann Arbor, Michigan, 1976, p. 156.
9 R. Sanders, *Project Plowshare*, Public Affairs Press, Washington DC, 1962, p. 183. This was the so-called 'black box' proposal, not to be confused with the seismic black box proposal of the 1977–80 trilateral negotiations on a Comprehensive Nuclear Test Ban Treaty.
10 Ibid.
11 U. Ericsson, *The Question of Nuclear Explosions for Peaceful Purposes by Non-Nuclear Weapon States and the Possibility of Misuse of Such Technology for the Production of Nuclear Weapons*, Document A/CONF. 35/DOC. 3, Conference of Non-Nuclear Weapon States, Geneva, 1968, p. 24.
12 US Disarmament Administration, p. 47.
13 O. Dahlman and H. Israelson, *Monitoring Underground Nuclear Explosions*, Elsevier Scientific Publishing Co., Amsterdam, 1977, p. 304.
14 Ericsson, p. 16.
15 US Disarmament Administration, p. 47.
16 B. G. Bechhoefer, 'Plowshare Control' in B. Boskey and M. Willrich (eds), *Nuclear Proliferation: Prospects for Control*, Dunellen Publishing Co., New York, 1970, p. 108.
17 T. S. Lough, 'Peaceful Nuclear Explosions and Disarmament', *Peace Research Reviews*, Canadian Peace Research Institute, Ottawa, 1968, p. 46.
18 Sanders, p. 185.
19 Ibid. Such assistance would not have involved transferring PNE devices or design information to other states. Rather, the United States would conduct PNEs on their behalf.
20 US Congress, JCAE, Hearings on Frontiers in Atomic Energy Research, 22–26 March 1960, US Government Printing Office, Washington DC, 1960, p. 10.
21 Ibid.
22 Underground nuclear tests, like earthquakes, are detected by means of the seismic waves which are propagated through the earth's crust. There is continuing scientific dispute as to whether small explosions, those below 10 kilotons, can be adequately detected by seismic means.
23 US Disarmament Administration, p. 101.
24 Ibid.
25 L. A. Sobel (ed.), *Disarmament and Nuclear Tests 1960–1963*, Facts on File Inc., New York, 1964, p. 24.
26 Ibid.
27 US Disarmament Administration, p. 101; Sobel, p. 24.
28 US Disarmament Administration, p. 102.
29 Ibid., p. 103.
30. G. T. Seaborg and B. S. Loeb, *Kennedy, Khrushchev and the Test Ban*, University of California Press, Berkeley, 1981, p. 39.
31 Ibid., p. 40.
32 Sobel, p. 24.
33 Ibid.
34 Sobel, p. 25; US Disarmament Administration, p. 258.
35 US Disarmament Administration, p. 258.
36 Seaborg and Loeb, *Kennedy, Khrushchev and the Test Ban*, p. 37.
37 Ibid., p. 39.
38 Seaborg and Loeb, *Stemming the Tide*, p. 314.
39 Ibid.
40 Seaborg and Loeb, *Kennedy, Khrushchev and the Test Ban*, p. 40.
41 Seaborg and Loeb, *Stemming the Tide*, p. 314.
42 US Disarmament Administration, p. 259.
43 Sanders, p. 187; Seaborg and Loeb, *Stemming the Tide*, p. 314.
44 US Disarmament Administration, p. 137.
45 Ibid.

46 Ibid., p. 140.
47 Sanders, p. 186.
48 US Disarmament Administration, p. 131.
49 Ibid., p. 270.
50 Ibid.
51 Ibid.
52 Seaborg and Loeb, *Kennedy, Khrushchev and the Test Ban*, p. 86.
53 Ibid.
54 Ibid, p. 121.
55 *The United Nations and Disarmament 1945–1970*, Department of Political and Security Affairs, United Nations, New York, 1970, p. 222.
56 Seaborg and Loeb, *Kennedy, Khrushchev and the Test Ban*, p. 168.
57 Ibid., p. 169; Seaborg and Loeb, *Stemming the Tide*, p. 315.
58 *The United Nations and Disarmament 1945–1970*, p. 225.
59 D. L. Clarke, *Politics of Arms Control: The Role and Effectiveness of the US Arms Control and Disarmament Agency*, The Free Press, New York, 1979, p. 102.
60 Ibid., pp. 32–33.
61 Ibid., p. 102.
62 Seaborg and Loeb, *Kennedy, Khrushchev and the Test Ban*, pp. 195–96.
63 Seaborg and Loeb, *Stemming the Tide*, p. 315.
64 A chapter sub-title in Seaborg and Loeb's book, *Kennedy, Khrushchev and the Test Ban*, p. 195.
65 Seaborg and Loeb, *Kennedy, Khrushchev and the Test Ban*, p. 197.
66 F. Dyson, *Disturbing the Universe*, Pan, London, 1981, p. 138.
67 Seaborg and Loeb, *Kennedy, Khrushchev and the Test Ban*, p. 196.
68 Ibid.
69 Ibid., p. 197.
70 Ibid., p. 195.
71 Ibid., pp. 197–98.
72 Seaborg and Loeb, *Stemming the Tide*, p. 316.
73 Ibid.
74 Seaborg and Loeb, *Kennedy, Khrushchev and the Test Ban*, p. 198.
75 Ibid.
76 Seaborg and Loeb, *Stemming the Tide*, p. 316.
77 Seaborg and Loeb, *Kennedy, Khrushchev and the Test Ban*, p. 244.
78 Dyson, p. 138.
79 A. Carnesale and R. Haass (eds), *Learning From Experience With Arms Control*, Office of Sponsored Research, Presidents and Fellows of Harvard College, Cambridge, Mass., 1986, pp. 2–1 to 2–70.
80 Seaborg and Loeb, *Stemming the Tide*, p. 316.
81 Seaborg and Loeb, *Kennedy, Khrushchev and the Test Ban*, p. 244.
82 T. Ehrlich, 'Legal Problems (of PNEs)' in L. E. Weaver (ed.), *Education for Peaceful Uses of Nuclear Explosives*, University of Arizona Press, Tucson, Arizona, 1970, p. 224.
83 Seaborg and Loeb, *Kennedy, Khrushchev and the Test Ban*, p. 244.
84 Ibid., pp. 244–45.
85 The withdrawal provision permitted withdrawal from the treaty on only three months' notice, an unprecedented safeguard designed, like the American PNE position, to ensure Senate support for the treaty. It appeared that the Senate, preoccupied with growing Chinese nuclear capabilities, would insist on such a provision. In the event, as Seaborg points out, when the first Chinese nuclear test was conducted in 1964 not a single senatorial voice was raised in favour of withdrawal from the treaty (Seaborg and Loeb, *Stemming the Tide*, pp. 316–17).
86 Dyson, p. 138.
87 Ibid., pp. 138–39.
88 Seaborg and Loeb, *Kennedy, Khrushchev and the Test Ban*, p. 245.
89 Ibid.
90 Dyson, p. 139.
91 Seaborg and Loeb, *Stemming the Tide*, p. 317. The AEC was not the only agency to suffer as a result of President Kennedy's desire to limit bureaucratic 'involvement' in the negotiating process. While Harriman was in Moscow, cable traffic was restricted

to only eight copies (Carnesale and Haass, p. 2–15).
92 Ibid.
93 Jacobsen and Stein, pp. 456–57.
94 Ibid., p. 457.
95 *Status of Multilateral Arms Regulation and Disarmament Agreements*, p. 21.
96 G. Fischer, *The Non-Proliferation of Nuclear Weapons*, Europa Publications, London, 1971, p. 115.
97 Dyson, p. 139.
98 Seaborg and Loeb, *Kennedy, Khrushchev and the Test Ban*, p. 272.
99 US Congress, Senate Foreign Relations Committee, Hearings on the Nuclear Test Ban Treaty, 12–15 August, 19–23 August, 26–27 August 1963, US Government Printing Office, Washington DC, 1963, pp. 486–87.
100 Ibid., pp. 426–27. This was a rather odd way of making his point. Numerous nuclear researchers have died of cancer contracted as a result of contact in the laboratory with radioactivity, the most famous being Madame Curie.
101 Ibid., p. 247.
102 Ibid., p. 441.
103 Ibid., p. 433.
104 Ibid., pp. 433 and 441.
105 Ibid., p. 268.
106 Seaborg and Loeb, *Kennedy, Khrushchev and the Test Ban*, p. 274.
107 J. H. McBride, *The Test Ban Treaty: Military, Technological and Political Implications*, Henry Regency Co., Chicago, 1967, p. 92.
108 Seaborg and Loeb, *Kennedy, Khrushchev and the Test Ban*, p. 273.
109 Ibid., p. 267.
110 Seaborg and Loeb, *Stemming the Tide*, p. 317.
111 Seaborg and Loeb, *Kennedy, Khrushchev and the Test Ban*, p. 268.
112 Ibid., p. 267.
113 Seaborg and Loeb, *Kennedy, Khrushchev and the Test Ban*, p. 279.
114 Seaborg and Loeb, *Stemming the Tide*, p. 317.
115 *SIPRI Yearbook 1988*, Oxford University Press, Oxford, 1988, p. 552.
116 US Congressional Record, 14 April 1969, p. 8945.
117 T. C. Findlay, 'Twenty Minutes with Edward Teller', *Newsletter*, Peace Research Centre, Australian National University, Canberra, 2, 1, 1987, p. 2.
118 US Congress, JCAE, Hearings on AEC Authorizing Legislation for Fiscal Year 1965, 20 February 1964, US Government Printing Office, Washington DC, 1964, p. 19.
119 McBride, p. 92.
120 M. Willrich, *The Non-Proliferation Treaty: Framework for Nuclear Arms Control*, The Mitchie Company, Charlottesville, Virginia, 1969, pp. 138–39.
121 McBride, p. 92.

# 4

# 'Planetary Engineering': The Panama Canal Project

*Plowshare is a program which* ***might****, if it is allowed to test nuclear devices in ways which are* ***not*** *permissible under the test ban,* ***sometime*** *result in economical and safe nuclear excavation techniques. But it is not as if the test ban treaty would have to be modified to permit the building of a transisthmian canal; it would have to be modified to permit a long development program, including many tests, which would result in significant amounts of radioactive fallout.*[1]

*Thomas Lough, 1968*

THE day after the Partial Test Ban Treaty entered into force, 11 October 1963, the first Plowshare detonation under the new regime took place at the Nevada Test Site. Codenamed Tornillo, this device development experiment was a contained underground explosion. It caused the desert floor to collapse by just two feet. As far as is known, no radioactivity escaped into the atmosphere.[2] Several more device development shots would be conducted during the next twelve months. Plowshare was apparently plowing on, as President Kennedy had pledged during the PTBT ratification hearings.

In reality, no sooner had the Senate agreed to ratification than controversy erupted within the US Administration over the treaty's implications for Plowshare excavation experiments. In late September 1963 the AEC sought presidential approval for Project Schooner, a 100 kiloton cratering experiment in hard rock to be detonated in south-western Idaho.[3] Within a few days McGeorge Bundy, Special Assistant for National Security Affairs, informed the AEC that a memorandum was 'afloat' in the White House arguing that the entire excavation program be terminated because it was inconsistent with the PTBT.[4] In view of this agitation and the fact that the treaty had not yet entered into force, Bundy felt it was 'inopportune to ask the President to rule on so large an explosion as Schooner'.[5] Accordingly the AEC withdrew its request and re-oriented its program toward small-yield experiments. As Seaborg regretfully informed the Senate Committee on Commerce:[6]

> At the present time, there are both technical and legal uncertainties regarding the type of experiments and projects which might be undertaken under the [PTBT]. Therefore, in order to make the maximum possible progress to resolve the technical uncertainties, we have carefully reassessed our program of large-scale experiments and have decided for the time being to concentrate on the development of nuclear explosives which produce even less radioactive debris, on small-scale excavation experiments, and on fully contained experiments for scientific and other engineering purposes.

He expressed optimism that this would be only a temporary situation:[7]

> As the allowable experiments proceed, as present technical uncertainties are resolved, and as even cleaner explosives become available, we hope it will be possible to conduct, under the treaty, cratering experiments which will advance excavation technology further.

Despite this optimism, there ensued what Thomas Ehrlich, Professor of Law at Stanford University, has called 'a substantial struggle' within the Administration over the PTBT, a struggle which the AEC would lose.[8] The point of contention was the treaty's ban on any nuclear test that 'causes radioactive debris to be present outside the territorial limits of the State under whose jurisdiction or control such explosion is conducted'. Unfortunately, as Edward Teller had warned, the treaty provided no definition of 'debris'. ACDA and the State Department argued that *no* excavation experiments should be conducted because, as a matter of course, they would vent radioactivity which sooner or later would find its way across an international border, either north to Canada or south to Mexico. If detected, this would, at least technically, violate the new treaty.[9] ACDA argued that the Soviets had not agreed to anything but a strict interpretation of the treaty. Any attempt by the United States to unilaterally adopt a liberal interpretation could be seen by the Soviets and the rest of the world as bad faith. ACDA argued, as does Marvin Kalkstein, that 'it is hard to conceive that the Treaty was in any sense intended to be a license for cratering experiments where venting is expected'.[10]

Representative Hosmer accused ACDA and the State Department of having a 'paranoic distrust and abhorrence of Plowshare, which they cannot divorce in their minds from the weapons program'.[11] He charged them with promoting 'over-legalistic interpretations' of the PTBT 'calculated to eliminate the possibility of Plowshare excavations'. In Glenn Seaborg's view the 'very strict and confining' ACDA/State interpretation was in danger of stifling the *entire* Plowshare program.[12] The AEC was apparently not just concerned for its excavation shots, which would invariably vent radioactivity, but also for its supposedly fully contained explosions, which could accidently vent and, under a strict interpretation, violate the treaty. Seaborg confirms that one of the reasons the AEC opposed a strict interpretation was because underground testing was a relatively new and inexact discipline and the consequences of such explosions difficult to 'predict with precision'.[13] For Seaborg, interpreting the treaty as prohibiting the export of even the most minute detectable amount of radioactivity caused the treaty to be 'self-amending', its meaning changing with each advance in detection technology. Such advances were occurring at a pace which Seaborg claimed he had not been able to foresee when the treaty was being negotiated and which would lead to what he described as the 'scientific stunt' of atom-chasing at international borders.[14] In addition, the cessation of atmospheric testing had reduced the levels of man-made background radiation, thus making more noticeable any small increase in radioactivity.

Seaborg argued that in any event the word 'debris' in the treaty suggested a 'larger amount' of radioactivity than ACDA and the State Department were prepared to countenance.[15] He had difficulty however defining what this amount was. According to one version it was an amount 'delivered outside of the country's territorial limits ... sufficient to establish that such contamination resulted from a recent test within that country'.[16] In another version it was 'very minute quantities, bearing no foreseeable relationship to human health', an interpretation apparently first improvised by George Ball in response to questioning at a news briefing.[17] Plowshare officials were obviously confused by the varying interpretations, as revealed in testimony to the JCAE by Plowshare director John Kelly:[18]

> I *think* [emphasis added] we can do these experiments at places distant from national boundaries *possibly or even probably* [emphasis added] without an amendment to the Test Ban Treaty.

In his 1987 book *Stemming the Tide*, Glenn Seaborg reveals the real reason for the AEC's problem with the PTBT. Despite the fact that he had claimed in 1981 that the AEC's interpretation of the PTBT had been arrived at 'after careful analysis',[19] he now admits that: 'In all honesty we had been asleep at the switch, and the problem did not manifest itself until after the treaty was signed'.[20] The AEC, he concedes, had had 'ample opportunity' to foresee the problem, since the reference to 'radioactive debris' had been in the draft treaty tabled by the United States as early as August 1962. The position adopted by the AEC after ratification, Seaborg notes, was logically strong but legally weak, especially since he himself had testified to the Senate Foreign Relations Committee that the treaty would not permit tests resulting in '*detectable* [Seaborg's emphasis] amounts of radioactive debris ... outside of our continental limits'.[21] He also concedes that the AEC's view 'in truth ... had very little support inside the government'.[22]

On 31 October 1963 President Kennedy moved to resolve the controversy by establishing an inter-agency Review Committee on Underground Nuclear Tests to review any underground explosion, whether a weapon test or a PNE:[23]

> ... which by its nature could give rise to domestic or foreign charges of a violation of the Nuclear Test Ban Treaty [sic] ... even though the sponsoring agency believes that the charges would prove to be unfounded.

The AEC, for its part, was instructed to 'make every effort' to avoid such explosions.[24] The committee, under the chairmanship of the President's Special Assistant for National Security Affairs,[25] was to include representatives of the AEC, the Department of Defense, the State Department, ACDA, the Joint Chiefs of Staff, the CIA, the White House and the Office of Science and Technology.[26] It was required to examine each proposed test, in the light of time, place and circumstance and recommend to the President whether it should proceed. Decisions of the committee, which became known as the 269 Committee after the number of the National Security Action Memorandum that established it,[27] were to reflect all scientific, political and legal considerations.[28] Both the Kennedy and Johnson Administrations tried to keep the Committee's existence secret, especially from the JCAE, despite the AEC's obligation under the Atomic Energy Act to keep the JCAE 'promptly and fully informed'.[29] Bundy speculates that this was because both presidents wanted to stay in control of the review and approval process for nuclear tests. Certainly, Seaborg notes, 'they did not want to share control with a group as aggressive as the JCAE'.[30]

A natural bias apparently operated in the 269 Committee against Plowshare and in favour of the weapons testing program. After all, the testing of a nuclear weapon had national importance far beyond that of testing, as Ehrlich puts it, 'the feasibility of making a molehill out of a mountain'.[31] This may explain why the AEC, rather than run the risk of the committee recommending against Plowshare experiments, tried to consult committee members individually about what it called 'straightforward' Plowshare detonations. As AEC Commissioner Tape admitted in testimony to the JCAE in 1965, the inter-agency committee was only convened 'when there is a need, and I am not being facetious'.[32] Despite these ploys the 269 Committee proved to be the nemesis of nuclear excavation.

Plowshare's situation was not made any easier by the accidental venting of a nuclear weapon test at the Nevada Test Site, the Pike shot of 13 March 1964, which occurred when a crack developed as a result of a local weakness in the geological structure.[33] By 20 March a small amount of radioactivity had apparently crossed the Mexican border in the vicinity of Yuma, Arizona. According to Seaborg, the increase in radioactivity in Las Vegas, Boulder City, Yuma and elsewhere in Arizona, 'while measurable, was slight and considered not to be hazardous'.[34] Seaborg reports that the Soviet Union did not make 'a big issue' of the Pike incident, probably because it understood that its own nuclear tests could vent in similar fashion. However, according to Seaborg, 'the Pike episode would further weaken our ability to defend the Plowshare program against the onslaught of its critics'.[35]

The first Plowshare nuclear excavation shot for which the AEC sought approval from the 269 Committee was Sulky, a small detonation of only 0.1 kilotons. In his letter to Bundy of 15 January 1964, Seaborg described it as 'about the smallest-scale experiment from which useful cratering information [could] be obtained'.[36] He felt the risk of venting beyond US borders was 'minimal'. At the committee's meeting on 7 February the proposed test was supported by the Department of Defense and opposed by the State Department and ACDA. Seaborg recalls that four days later Bundy phoned him:[37]

> He said that unless I had an objection he proposed to issue a memorandum saying that in view of the delicacy of the US — USSR relationship, President Johnson had decided to defer consideration of Sulky, without prejudice, until next winter . . . I said I had no objection so long as this was only a postponement and did not preclude the shot.

Seaborg did in fact have an objection, confiding in his diary that 'this decision was a mistake'.[38]

In February 1964 the Joint Committee on Atomic Energy conducted the first of its yearly reviews of Plowshare as part of the annual AEC budgetary process.[39] The JCAE had by this stage, according to Peter Metzger, unilaterally extended its powers 'into the fiscal phase of government, taking over control of all spending in atomic matters, and at times even pressing unasked-for funds on an unwilling AEC or a reluctant President'.[40] Increasingly impatient for a major breakthrough in Plowshare, the JCAE was not concerned whether the program represented value for money or was environmentally and socially acceptable, but whether it should be pursued more aggressively. Its annual reviews of Plowshare would become a painful trial for Plowshare and AEC officials.

Chairman Pastore (later Holifield) and Committee members, including Representatives Craig Hosmer and Wayne Aspinall, and Senator Clinton P. Anderson, grilled such officials mercilessly. They berated them for timidity in planning and conducting experiments, for failure to promote Plowshare commercially, for conservatism in requesting funds, for caving in to budgetary cutbacks imposed by the Bureau of the Budget and for not being prepared to give categorical answers to the committee's technical enquiries. Seaborg himself was rudely treated. During one session, when he interrupted to ask whether he could answer a question about neutron fluxes, Chairman Pastore replied sarcastically: 'Yes, I was waiting for a Nobel Prize winner'.[41] Critics of Plowshare were also berated, *in absentia*, being rarely called to testify, other than inadvertently. Normally only the AEC, Plowshare officials and scientists, and the AEC's industrial partners were called to give evidence.[42]

At the February 1964 hearings, Seaborg was placed in the uncomfortable position of having to defend the Administration's interpretation of the PTBT, with which he disagreed, and the hiatus in excavation experiments, before what he described as 'a group of staunch Plowshare adherents'.[43] He stated in his diary: 'I had some difficulty'.[44] Some of the committee's hostility toward Seaborg was probably due to a feeling that he had duped them over the likely effect of the PTBT on Plowshare. As Seaborg himself now says, this:[45]

> ... caused me considerable personal discomfort in ensuing months and years as concern about violating the treaty began to impede the program. Certainly I had not intended to deceive any senators — yet there might have been an appearance that I had done so.

To make matters worse the committee, at its 1964 session, uncovered the existence of the 269 Committee and hence the major impediment to Plowshare's excavation program.[46] The committee also learned that while the PNE Division of the AEC had sought $27 million for Plowshare for 1965 and the Commission had reduced this to $21.5 million, the Bureau of the Budget had slashed this amount to just $11 million. In the Bureau of the Budget's view, Plowshare 'had no fixed national objective so far as a time date was concerned and in a year of financial stringency it should be cut back'.[47] In reaching this conclusion, it had undoubtedly noted the impact of the PTBT on Plowshare's excavation program.[48] Grilling by the JCAE drew from Plowshare officials an acknowledgement that the proposed budget cuts would retard the nuclear excavation program further.

Other attacks were being mounted against Plowshare on arms control grounds. Former Livermore director Herbert York and former Presidential science advisor Jerome Wiesner said of PNEs in an influential article in *Scientific American* (October 1964) that 'promising as these may be, the world could forgo them for a time in exchange for cessation of the arms race'.[49] In November 1964 President Johnson established the Gilpatric Committee, formally called the President's Task Force on Preventing the Spread of Nuclear Weapons.[50] The committee recommended that 'Plowshare must not be allowed to stand in the way of a comprehensive test ban'.[51] This recommendation, however, had little immediate impact outside a very small circle of Administration officials, since the report was kept secret. According to Spurgeon Keeny, currently head of the US Arms Control Association, the report was nevertheless 'very significant in sowing the intellectual soil for future actions' in the mind of President Johnson.[52]

## Sulky

The first Plowshare cratering shot since ratification of the PTBT, Sulky, was finally approved and conducted at the Nevada Test Site on 18 December 1964, after more than a year's delay. The experiment was intended to test the theory that by emplacing the cratering device at a lower depth than that for optimal crater size, radioactive venting would be considerably reduced. In fact the explosion caused a mound rather than a crater. According to Gerald Johnson this was because the yield was too low.[53] Under Craig Hosmer's persistent questioning during the JCAE's 1965 hearings,

Johnson denied that the explosive had been buried 'a little lower' than it would have been 'ordinarily'.[54] Glenn Seaborg, more than twenty years later, claims that, 'In order to be sure of not violating the test ban treaty, the AEC buried the device at an overly conservative depth'.[55]

Radioactivity from Sulky, according to Johnson, vented into the atmosphere 'as expected'.[56] Seaborg records that some radioactivity was detected off-site, although the amount was 'small' and dissipated long before it could reach a national border.[57] The usual precautions had been taken. Plowshare had a large 'gang' of meteorologists who closely monitored the weather prior to Plowshare shots being conducted.[58] In addition, such detonations were not conducted at all during the cattle feeding season, the preferred season being November-March, when the cattle were eating dry, stored food.[59]

## *The Panama Canal Project*

In December 1964, a beleaguered Plowshare, battered by controversy over the PTBT and arms control generally, straightjacketed by the 269 Committee, and threatened by looming financial stringencies, received a major political fillip. In that month President Johnson, in a directive to AEC Chairman Seaborg, established nuclear excavation as a 'national policy' of the United States.[60] This had less to do with the vaunted excavation capabilities of nuclear dynamite than with the political impact on Panama of the proposed construction of a new, sea-level, Panama Canal.

The President was faced with a serious crisis in United States/Panama relations. Violent anti-American riots had erupted in Panama in January 1964. The Panamanian government broke diplomatic relations with Washington. Gunfire was exchanged across the US-controlled Canal Zone border. In the United Nations Security Council Panama charged the United States with aggression and demanded that the 'inequitable' Canal Treaty be renegotiated.[61] In a surprise move on 18 December 1964, President Johnson agreed. He also announced that the United States was launching a search for a route for a new canal.

To conduct the search, Johnson authorised establishment of an Atlantic-Pacific Interoceanic Canal Study Commission (CSC). Congress quickly allocated funds, having anticipated the President's decision by authorising establishment of the Commission on 22 September 1964 with a mandate to:[62]

> ... make a full and complete investigation and study ... considering national defense, foreign relations, intercoastal shipping, interoceanic shipping ... for the purpose of determining the feasibility of, and the most suitable site for, the construction of a sea-level canal connecting the Atlantic and Pacific Oceans; the best means of constructing such a canal, *whether by conventional or nuclear excavation* [emphasis added], and the estimated cost thereof.

The Bill had been co-sponsored by JCAE Chairman Pastore and other JCAE members.[63] The CSC was requested to conclude its work by mid-1968. The Administration apparently considered that a serious study of a new canal, including routes outside Panama, would strengthen the President's hand in negotiating a new canal treaty with the Panamanians.[64] A Presidential directive to the AEC to develop nuclear cratering techniques for constructing such a canal made the threat even more credible.

The President designated for intensive consideration only four of the thirty alternative routes identified in 1947: Route 8, along the Nicaragua–Costa Rica border; Route 14, involving conversion of the existing lock canal to a sea-level canal; Route 17 in Panama, about 110 miles (180 kilometres) east of the existing canal; and Route 25 in Colombia.[65] Only the latter two were judged to require nuclear excavation. To these two routes the CSC added Route 23, which straddled Panama and Colombia, and Route 8, as being also possibly suited to the use of nuclear dynamite.[66] To facilitate its study of the PNE option, the Commission asked the AEC to conduct six to eight nuclear excavation experiments by June 1968.

By examining the routes most appropriate for nuclear excavation first, those necessarily in remote areas or outside Panama, the CSC appeared to confirm the suspicions of a number of Panamanians and US Congressmen opposed to the canal, that the Commission's real purpose was to 'bring the Panamanian Government in line to negotiate a more lenient treaty with the United States'.[67] As even Pastore admitted, the project had 'all kinds of political overtones to it...'[68] In the meantime, US negotiators offered Panama a draft treaty which specifically mentioned the possibility of building a new canal using PNEs. The treaty would give Panama a veto over the use of nuclear excavation and the right to approve the safety measures employed, but *not* if the canal was built outside Panama.[69]

For Plowshare enthusiasts the Panama Canal project promised not just a vindication of the hopes held for Plowshare, but also (if speedy Congressional funding of the CSC was a guide) a guaranteed and rising level of funding. The development of nuclear cratering and excavation techniques was indeed to be Plowshare's lifeblood for the next several years, representing two-thirds of all expenditure on the program between 1964 and 1971.[70] Gerald Johnson told the JCAE in 1965 that, 'From a personal standpoint, I feel the possibility of digging a sea-level canal across the American isthmus is extremely important to the Plowshare program'.[71] For JCAE Chairman Pastore, it was 'the one thing that has given this thing [Plowshare] life and the one thing that has more or less enthused this committee to provide the money for Plowshare'.[72]

At least one staff member at Livermore opposed the project from the outset. Medical physicist John Gofman, head of the bio-medical division, had been appointed to evaluate the health effects of radiation from AEC activities.[73] In an address to the 1964 Third Plowshare Symposium, he boldly declared that 'technological advances in the physical and engineering sciences have made a variety of projects feasible, while the biological and medical knowledge requisite to assess "biological cost" is far from adequate'.[74] The radiation standards for Plowshare were, Gofman said, 'painfully uncertain'.[75] Plowshare faced the possibility, he said, 'of delivery, rapidly or slowly, of *low* total doses of ionizing radiation to a small segment of the human population'.[76] Contrary to Edward Teller's blase attitude, Gofman stated frankly that:[77]

> ... we have no direct, valid information on the subject of injury to be anticipated in humans either for the generation receiving such radiation or for their descendants.

By 1965 Gofman had concluded that the Panama Canal project, using PNEs, was 'biological insanity'.[78] He recalls that this did not 'earn me a lot of favour' with the AEC. His bio-medical division 'became known in the lab as "the enemy within" because we opposed things like Plowshare'. This was, initially, according to Gofman, 'really fairly good-natured' and 'in no way did it interfere with my status in the lab'.[79] This good-naturedness soon changed. Gofman relates that after he and another AEC scientist, Arthur Tamplin, produced embarrassingly high predictions of mortality rates from nuclear testing and the nuclear power program, Livermore director Michael May, at AEC instigation, visited him. Gofman recalls what happened:[80]

> 'Jack,' he said, 'I defend absolutely your right, in fact your duty, to calculate that a certain amount of radiation will cause 32,000 extra deaths per year from cancer.' But to my disappointment, he then asked, 'What makes you think that 32,000 would be too many?' 'Mike,' I said, 'the reason is very simple. If I find myself thinking that 32,000 cancer deaths per year is not too many, I'll dust off my medical diploma, take it back to the Dean of the Medical School where I graduated, hand the diploma to the dean and say, "I don't deserve this" '.

Under intense pressure Gofman later resigned from the Livermore Laboratory.[81]

Meanwhile, the February 1965 hearings of the JCAE revealed growing impatience with the lack of progress in the excavation program, especially as no excavation experiment had been approved and conducted since Sulky in January 1964. Glenn Seaborg did much to trigger the Committee's ire by testifying that it would take five more years to develop the excavation technology necessary for constructing a new Panama Canal.[82] He added that the economics of nuclear excavation would then have to be weighed against those of conventional construction. After Seaborg outlined some of the problems facing Plowshare, Senator Anderson remarked exasperatedly, 'I am just wondering where Plowshare is going or is it going backward?'[83] Seaborg also confirmed that 'as far as the transisthmian canal is concerned, that would definitely require a modification of the limited test ban treaty'.[84] Chairman Pastore responded[85] with a veiled threat to cut off Plowshare funding:[86]

> Once you have ruled [the canal] out, I am afraid interest is going to drop off. I am perfectly willing to double, treble or quadruple the appropriation if we are going to use nuclear devices for the building of a canal. If we are not, I would like to take a second look at the size of the budget.

Later in the day, Seaborg records, he met with Gerald Johnson, who was 'somewhat discouraged' by the JCAE's attitude. Seaborg told Johnson that 'we were depending on him to carry on with his usual enthusiasm'. By the time Seaborg left, Johnson had indicated he would 'try to continue with the same interest'.[87]

The JCAE had also been keenly interested in the alleged arms control implications of Plowshare. Seaborg, by being extremely cagey about the similarities between nuclear weapons and PNEs, forfeited a perfect opportunity to educate the JCAE on the subject. When Representative Hosmer asked him whether PNEs had military applications the following exchange took place:[88]

> *Dr Seaborg*: You can use nuclear explosives for cratering, for demonstrating operations and so forth, but largely speaking, it is pointed toward peaceful applications.
>
> *Representative Hosmer*: You would not put one of these devices for cratering on the tip of an ICBM, would you? That is an entirely different breed of cat.
>
> *Dr Seaborg*: I don't know what the purpose of that question is.
>
> *Representative Hosmer*: What I am trying to establish, Dr Seaborg, is that if in fact this program is a totally peaceful one, then it has no bearing upon the general emotional approach to the nuclear weapons problem.
>
> *Dr Seaborg*: Yes.
>
> *Representative Hosmer*: It should be in an entirely different category.
>
> *Dr Seaborg*: But it is tied in this way, Congressman Hosmer, that the development of these advanced types of nuclear explosives might be considered to have relevance to the development of improved weapons, because the same principles might be used. There is a connection.

Seaborg failed to explain the implications of this last remark. Indeed he argued that a continuation of Plowshare was entirely compatible with a comprehensive test ban treaty:[89]

> In my opinion it would be possible to conduct Plowshare experiments for the improvement of devices for [their] use in Plowshare projects in such a way under a comprehensive test ban treaty that it would not be a danger to the other powers.

Seaborg did not say how this would be done, but said he nevertheless recommended that a CTB include an exemption for Plowshare. And with regard to nonproliferation:[90]

> I think it [Plowshare] would not have any appreciable effect on the non-proliferation problem.

Seaborg made no mention of the possibility that non-nuclear weapon states could use PNEs as a subterfuge for nuclear proliferation, with Plowshare as a model.

## Palanquin

In April 1965 Plowshare's second excavation shot since ratification of the PTBT was detonated at the Nevada Test Site. Codenamed Palanquin, the small 4 kiloton shot was expected to release very little radioactivity and, like Sulky, to create a mound rather than a crater. In fact it created a crater and resulted, according to the AEC, in the escape of 'a small amount of radioactivity which would normally have been filtered by the broken material in the dome'.[91] According to other reports, Palanquin released 'most of its radioactivity'.[92] Glenn Seaborg revealed in 1987 that the dust cloud rose to a height of 8000 feet (2.5 kilometres) and contained higher than expected levels of fallout.[93] The air mass moved slowly northwards, dispersing laterally. While well below any possible health hazard, even close to the test site, the radioactivity was sufficient to be readily detectable in Canada had the cloud crossed the border. When it appeared that this had not occurred, the AEC issued a 'low-key' press release, hoping, according to Seaborg, 'to soften the impact of any possible test ban violation'.[94]

On 29 April however the Soviets requested more information on Palanquin, accusing the United States of violating the PTBT. Despite the fact that the US Air Force had by now detected radioactivity over Bermuda, the State Department refused to acknowledge any possible violation of the PTBT, instead accusing the Soviet Union of violating the treaty with a nuclear weapon test it had conducted on 15 January 1965.[95] Seaborg says this was 'mischievous and damaging' and 'certainly not helpful to the future the AEC had in mind for international co-operation in PNEs, with major Soviet participation'.[96] The Palanquin experiment, like Sulky before it, was also unfortunate news for those like Edward Teller who expected the mysterious 'filtering' process, otherwise known as 'fallback', to be a major key to 'clean' nuclear excavation.

Palanquin was to be the last nuclear excavation experiment for almost three years, from April 1965 to January 1968. Plowshare director during the period, John Kelly, informed the JCAE that one of the major reasons was 'concern relating to the Test Ban Treaty and some uncertainties as to what is prohibited and what is permitted under that'.[97]

## Cabriolet

The major problem was Cabriolet, the first experiment requested by the Panama Canal Study Commission. Cabriolet would involve the detonation of a 2.7 kiloton explosive in hard rock at a depth of 170 feet (50 m) at the Nevada Test Site.[98] Authority was requested from the 269 Committee for this explosion to take place in 1966. The AEC frankly acknowledged that there would be a 50 per cent chance, after two days, of a 'miniscule' amount of radioactivity crossing the Canadian border. It argued however that if the experiment was not approved, 'the ill effects on the canal undertaking and on all future excavation projects would outweigh the diplomatic embarrassment of a *technical* [emphasis added] treaty violation'.[99]

ACDA, Plowshare's 'persistent antagonist', disagreed, arguing that it was 'an inopportune time to conduct an event which could very well create accusations of our not living up to our treaty commitments'.[100] ACDA director William Foster said he believed there were 'diplomatic tides stirring' that could lead to clarification of the PTBT's meaning 'within a few months'. Deputy Defense Secretary Cyrus Vance, according to Seaborg, 'weighed in with an even more severe opinion', recommending that Cabriolet be postponed and the entire Plowshare program reviewed. Vance thought it unwise, in view of the Vietnam War, to open another 'propaganda front' where the United States could be attacked. State Department opinion was divided. Defense Secretary McNamara, Secretary of State Dean Rusk and the new National Security Advisor, Walt Rostow, were also all opposed to Cabriolet. On 28 March the AEC received a letter from JCAE Chairman Holifield warning that if cratering experiments were not permitted, 'the entire Plowshare program would have to be reevaluated'.[101] Seaborg lamented that Plowshare now appeared to be threatened by friend and foe alike.

It was obvious that until actual excavation experiments could be conducted, the AEC and the Canal Commission would be operating from a limited data base in reaching any conclusions about nuclear excavation. According to Thomas Lough, only four nuclear surface bursts 'of importance to Plowshare' had been conducted prior to the 1958–61 moratorium[102] and only three more (Sedan, Sulky and Palanquin) by 1966.[103] While two non-Plowshare shots after the moratorium, Hardhat in February 1962, and Danny Boy in March 1962, were also of relevance, this still only amounted to nine shots. Plowshare scientists were of course able to use computer models and data from conventional explosive experiments.[104] In addition, fully contained nuclear detonations were continuing to be conducted, in order to perfect a low fallout device and refine entrapment techniques. These included the Templar, Saxon and Simms shots in 1966 and the Switch, Stoddard and Marvel shots in 1967.[105] However, as John Kelly, in a masterly display of bureaucratic obfuscation, told the JCAE in 1966, 'We were beginning to recognize that the simple empirical formula may not be adequate, and considering how to get something that is more adequate has entered into it, also'.[106] As he and other Plowshare officials conceded under questioning, nuclear excavation was more 'complex' than had been imagined and had not advanced as rapidly as anticipated at Plowshare's inception.

When the 269 Committee met in March 1966 to consider Cabriolet, Seaborg suggested that an attempt be made to reach an informal *modus vivendi* with the Soviet Union over the PTBT to permit large excavation experiments to proceed, since intelligence information indicated that the Soviets were beginning their own ambitious PNE program. Seaborg later contradicted this by noting that the Soviets would have little incentive to reach such an understanding since they were proceeding under a 'liberal' interpretation of the treaty, while the United States was 'hobbling' itself with a strict one.[107] Seaborg argued that Cabriolet should proceed since no canal route outside the present Panama Canal Zone was economically credible without nuclear excavation and the United States had a 'moral obligation' to develop excavation technology to honour its commitment to provide PNE 'services' to non-nuclear weapon states.[108] Such a commitment, which had not been solicited by such states, was of the AEC's own concoction. As Seaborg concedes, it was intended not as an altruistic gesture towards the world's have-nots, but to secure Plowshare's future back home.

ACDA Director Foster argued against Cabriolet by quoting Seaborg's 1963 testimony that the criteria for treaty violation would be detection and identification of radioactive debris, not the maintenance of health standards.[109] He also quoted President Johnson's assurance of 5 July 1966 that the United States intended 'to live up to the test ban treaty religiously and scrupulously follow it'. Finally, he pointed out that the potential for violating the treaty would increase with the more powerful excavation experiments scheduled to follow Cabriolet.

Despite Foster's 'powerful intervention', President Johnson decided that Cabriolet, on balance, should proceed. His Science Advisory Committee then recommended unanimously against this decision. At this point, the President, according to Seaborg, 'undoubtedly confused and probably exasperated by the conflicting advice he was getting', decided on the unprecedented step of submitting the decision to binding arbitration.[110] Under Secretary of State Nicholas Katzenbach, the arbitrator, after calling a meeting of the interested parties, recommended to Johnson that Cabriolet proceed.

Subsequently Cabriolet was postponed for technical reasons to 10 February 1967. On 9 February National Security Advisor Rostow advised Seaborg that the President had changed his mind again and decided to cancel Cabriolet 'for the present' because of negotiations, in which PNEs were an important issue, on the Nonproliferation Treaty and the Latin American nuclear-free zone.[111] Another factor, according to Rostow, was 'diplomatic pressures' over the bombing of North Vietnam and the possibility of peace negotiations.

This turn of events greatly concerned the JCAE and led to 'deep discouragement within the AEC family, particularly at Livermore'.[112] The director of the laboratory, Dr Michael May, even suggested abandoning nuclear excavation and concentrating on Plowshare's fully contained applications. Seaborg and other commissioners, however, felt that they had no choice but to continue, since nuclear excavation 'seemed the best hope of keeping Plowshare alive and we still deeply believed in the ultimate promise of the program'.[113]

Cabriolet continued to be postponed throughout 1967 because of the ongoing arms control negotiations and because both the Nonproliferation Treaty and the Latin American nuclear-free zone treaty were likely to be considered by the UN General Assembly in November 1967. As Rostow put it:[114]

> If Cabriolet is carried out at that time and, as seems probable, the US is charged with violating the Limited Test Ban Treaty, the US will be in the position of demonstrating its unwillingness to accept arms control restrictions imposed on it by existing international obligations at the very time we are asking other nations to undertake new treaty commitments of their own.

In April 1967 the Canal Commission, realising it could not conclude its report by its deadline of 30 June 1968, especially in regard to nuclear excavation, successfully sought legislation to extend its mandate to 1 December 1970.[115] Some observers suspected the delay was sought to prolong the negotiating pressure on the Panamanians.[116]

A further meeting on Cabriolet took place on 28 November 1967.[117] Dean Rusk noted that there was a 50:50 chance of radioactive debris from Cabriolet being detected at a national border. Rusk added that in his experience scientists were not always correct. He recalled that in 1962 scientists had substantially underestimated the efforts of a US atmospheric test on the upper atmosphere.[118] Seaborg again argued in favour of Cabriolet on the grounds of his 1963 testimony that Plowshare could survive the PTBT. Having given the Senate misleading advice to save Plowshare in 1963, Seaborg was again using it to save Plowshare. He now claimed his testimony had been necessary to convince Senators Anderson, Fulbright and Jackson to vote for the PTBT, and by implication, save the treaty. This was spurious. The Senate had voted 80–19 in favour of the treaty. While Senator Anderson was an ex-Chairman and a highly influential member of the JCAE, he was not adamantly opposed to the treaty, but only 'doubtful'. Moreover, in his 1981 book, *Kennedy, Khrushchev and the Test Ban*, Seaborg says his testimony only 'perhaps' influenced other senators.[119] Seaborg's real motivation for testifying as he did, had been the knowledge that if he had not so testified, Plowshare would have been finished. He would also surely have been reluctant to alienate President Kennedy.

Seaborg's strongest argument in favour of Cabriolet was that Dean Rusk had told the JCAE prior to ratification that the United States would interpret the treaty as broadly as the Soviets. Seaborg pointed out that the Soviets had conducted fourteen tests that had produced craters, that four had resulted in radioactivity being detected beyond Soviet borders and that two of these had produced levels of radioactivity that exceeded anything that could possibly emanate from Cabriolet.[120] At least one of these explosions had been a cratering experiment.[121]

The Chairman of the CSC, in his appearance before the 269 Committee, seemed to confirm the essentially political nature of the whole Panama Canal exercise by arguing that postponing Cabriolet would mean the end of the nuclear excavation program, since Panama would then know there was no alternative to an improved, conventionally-built canal in Panama and would therefore be more difficult than ever to deal with.[122] Colombia, for its part, might feel the United States had been dealing with it in bad faith, since a canal through Colombia was only likely if nuclear excavation was used.

On 19 December 1967 Seaborg was told that President Johnson had decided to proceed with Cabriolet, on the understanding that this did not mean approval for any further experiments until Cabriolet had been evaluated. After Dean Rusk concurred with the timing of the explosion, Cabriolet was finally detonated on 26 January 1968. As Seaborg recalls:[123]

> It created a crater about 400 feet [120 m] across and 125 feet [38 m] deep. The wind was right, blowing away from Mexico, and a snowstorm in northern Nevada apparently brought down much of the debris. The snowstorm was a stroke of good luck! No radioactivity attributable to Cabriolet was detected by the Canadians.

According to Mason Willrich, Cabriolet's mobilisation and remobilisation added $2.3 million to the cost of the experiment by the time it was finally conducted.[124] Glenn Seaborg puts the eventual total cost of Cabriolet at $5 million, nearly three times the original estimate.[125] He notes that the Administration's uncertainty and internal conflicts about nuclear excavation might have been resolved more readily by the exercise of presidential leadership, but that President Johnson was 'preoccupied elsewhere'.[126]

## Notes

1 T.S. Lough, 'Peaceful Nuclear Explosions and Disarmament', *Peace Research Reviews*, Canadian Peace Research Institute, Ottawa, 1968, p. 44.

2 B.A. Bolt, *Nuclear Explosions and Earthquakes: The Parted Veil*, W.H. Freeman and Co., San Francisco, 1976, p. 257.

3 G.T. Seaborg and B.S. Loeb, *Stemming the Tide: Arms Control in the Johnson Years*, Lexington Books, Lexington, Mass., 1987, p. 317.

4 Ibid., pp. 317–18.
5 Ibid., p. 318.
6 Quoted in Lough, p. 43.
7 Ibid.
8 T. Ehrlich, 'Legal Problems (of PNEs)' in L.E. Weaver (ed.), *Education for Peaceful Uses of Nuclear Explosives*, University of Arizona Press, Tucson, Arizona, 1970, p. 227.
9 At least one Los Alamos staff member, a Mr Walterscheid, agreed with the ACDA/State interpretation, as revealed during JCAE hearings on the 1967 AEC Authorizing Legislation (see US Congress, JCAE, Hearings on AEC Authorizing Legislation for Fiscal Year 1967, US Government Printing Office, Washington DC, 1966, p. 1367).
10 M. Kalkstein, *International Arrangements for the Peaceful Applications of Nuclear Explosives*, Stockholm International Peace Research Institute, Stockholm, 1970, p. 21.
11 US Congressional Record, 14 April 1969, US Government Printing Office, Washington DC, 1969, p. 8943.
12 G.T. Seaborg and B.S. Loeb, *Kennedy, Khrushchev and the Test Ban*, University of California Press, Berkeley, 1981, p. 268.
13 Seaborg and Loeb, *Stemming the Tide*, p. 208.
14 Ibid., pp. 208 and 329.
15 Seaborg and Loeb, *Kennedy, Khrushchev and the Test Ban*, p. 268.
16 Ibid., p. 267.
17 Seaborg and Loeb, *Stemming the Tide*, p. 208.
18 US Congress, JCAE, Hearings on AEC Authorizing Legislation for Fiscal Year 1967, p. 1348.
19 Seaborg and Loeb, *Kennedy, Khrushchev and the Test Ban*, p. 267.
20 Seaborg and Loeb, *Stemming the Tide*, p. 208.
21 Ibid.
22 Ibid., p. 328.
23 Ibid., p. 209.
24 Ibid.
25 The committee's role was later taken over by the National Security Council (US Congress, Senate Foreign Relations Committee, Hearings on the TTBT/PNET, 28 July, 3 August, 8 and 15 September 1977, US Government Printing Office, Washington DC, 1977, p. 92). During the Nixon Administration the President's Senior Science Advisor, Jerome Wiesner, advised the President on whether particular tests should proceed (from an interview with Ed Fleming, Livermore Laboratory, California, 3 December 1986).
26 US Congress, JCAE, Hearings on AEC Authorizing Legislation for Fiscal Year 1965, US Government Printing Office, Washington DC, 1964, pp. 1230–31.
27 Seaborg and Loeb, *Stemming the Tide*, p. 209.
28 Ehrlich, p. 227.
29 Seaborg and Loeb, *Stemming the Tide*, p. 319.
30 Ibid.
31 Ehrlich, p. 227.
32 US Congress, JCAE, Hearings on AEC Authorizing Legislation for Fiscal Year 1966, US Government Printing Office, Washington DC, 1965, p. 15.
33 Seaborg and Loeb, *Stemming the Tide*, p. 211.
34 Ibid., p. 210.
35 Ibid., p. 211.
36 Ibid., p. 318.
37 Ibid.
38 Ibid.
39 US Congress, JCAE, Hearings on AEC Authorizing Legislation for Fiscal Year 1965, p. 1196.
40 H.P. Metzger, *The Atomic Establishment*, Simon and Schuster, New York, 1972, p. 20. Unlike other Federal bodies, the AEC had to submit its budget to the JCAE's review process before it was sent to the usual Congressional Appropriations Committees.
41 US Congress, JCAE, Hearings on AEC Authorizing Legislation for Fiscal Year 1965, p. 1211.
42 A study of witnesses called before the JCAE in its 23 years of existence revealed over 3600 representatives of industry or government, but only 98 so-called public interest witnesses; see R.L. Beckman, *Nuclear Non-Proliferation: Congress and the Control of*

*Peaceful Nuclear Activities*, Westview Press, Boulder, Colorado, 1985, p. 161.
43 Seaborg and Loeb, *Stemming the Tide*, p. 319.
44 Ibid., pp. 318–19.
45 Ibid., p. 317.
46 Ibid.
47 US Congress, JCAE, Hearings on AEC Authorizing Legislation for Fiscal Year 1965, p. 1218.
48 Ibid.
49 J.B. Wiesner and H.F. York, 'National Security and The Nuclear Test Ban' in H.F. York (ed.), *Arms Control*, W.H. Freeman and Co., San Francisco, 1973, p. 133.
50 Seaborg and Loeb, *Stemming the Tide*, pp. 136–37. The committee included former disarmament negotiator Arthur Dean, Herbert York, and former National Security Advisor John McCloy.
51 Ibid., p. 325.
52 Ibid., p. 148.
53 From correspondence with Gerald Johnson, La Jolla, California, 28 September 1987.
54 US Congress, JCAE, Hearings on AEC Authorizing Legislation for Fiscal Year 1966, p. 13.
55 Seaborg and Loeb, *Stemming the Tide*, p. 319.
56 US Congress, JCAE, Hearings on AEC Authorizing Legislation for Fiscal Year 1966, p. 14.
57 Seaborg and Loeb, *Stemming the Tide*, p. 319.
58 From an interview with Fred Holzer, Livermore Laboratory, California, 3 December 1986.
59 Ibid. See also Kalkstein, p. 21.
60 Metzger, p. 223.
61 Ibid.
62 Quoted in Seaborg and Loeb, *Stemming the Tide*, p. 313.
63 US Congress, JCAE, Hearings on AEC Authorizing Legislation for Fiscal Year 1965, p. 1195.
64 Metzger, p. 223.
65 D.N. Farnsworth and J.W. McKenney, *US/Panama Relations 1903-1978: A Study in Linkage Politics*, Westview Press, Boulder, Colorado, 1983, p. 108.
66 Ibid., pp. 109–10.
67 Ibid., p. 110.
68 US Congress, JCAE, Hearings on AEC Authorizing Legislation for Fiscal Year 1966, p. 24.
69 Farnsworth and McKenney, p. 48.
70 *PNE Activity Projections for Arms Control Planning* (the *GURC Report*), Gulf Universities Research Consortium, Galveston, Texas, US Arms Control and Disarmament Agency Document, ACDA/PAB-253, Washington DC, 1975, p. 2.
71 US Congress, JCAE, Hearings on AEC Authorizing Legislation for Fiscal Year 1966, p. 27.
72 Ibid., p. 25.
73 L.J. Freeman, *Nuclear Witnesses: Insiders Speak Out*, W.W. Norton and Co., New York, 1981, p. 89.
74 J.W. Gofman, 'Hazards To Man From Radioactivity', paper presented to the Third Plowshare Symposium, reproduced in US Congress, JCAE, Hearings on AEC Authorizing Legislation for Fiscal Year 1966, p. 212.
75 Ibid.
76 Ibid., p. 213.
77 Ibid.
78 Freeman, p. 92.
79 Ibid.
80 P. Pringle and J. Spigelman, *The Nuclear Barons*, Avon Books, New York, 1981, pp. 317–18.
81 Ibid., p. 318.
82 Seaborg and Loeb, *Stemming the Tide*, p. 319.
83 US Congress, JCAE, Hearings on AEC Authorizing Legislation for Fiscal Year 1966, p. 22.

84 Ibid., p. 12.
85 Seaborg and Loeb, *Stemming the Tide*, p. 320.
86 US Congress, JCAE, Hearings on AEC Authorizing Legislation for Fiscal Year 1966, p. 25.
87 Seaborg and Loeb, *Stemming the Tide*, p. 321.
88 US Congress, JCAE, Hearings on AEC Authorizing Legislation for Fiscal Year 1966, p. 30.
89 Ibid., p. 31.
90 Ibid.
91 Lough, p. 20.
92 Freeman, p. 66.
93 Seaborg and Loeb, *Stemming the Tide*, p. 323.
94 Ibid.
95 Ibid., pp. 323–24
96 Ibid., p. 324.
97 Quoted in Lough, p. 20.
98 Seaborg and Loeb, *Stemming the Tide*, p. 327. The following account of the manoeuvrings over Cabriolet is based, except where otherwise noted, on Seaborg and Loeb's Chapter 25, 'Cabriolet, or How Not to Run a Government', pp. 327–39.
99 Ibid., p. 328.
100 Ibid.
101 Ibid., pp. 328–29.
102 Presumably including the Teapot ESS and Neptune shots conducted in 1955 and 1958 respectively (see L.J. Vortman in L.E. Weaver (ed.), p. 71).
103 Lough, p. 20. Gerald Johnson disputes Lough's claim.
104 Ibid.
105 J.R. Vogt, 'Radionuclide Production for the Nuclear Excavation of an Isthmian Canal', *Bioscience*, February 1969, p. 138; I.Y. Borg, *Nuclear Explosions for Peaceful Purposes*, Lawrence Livermore National Laboratory, Paper No. UCRL–95026, Rev. II, November 1986, p. 28.
106 US Congress, JCAE, Hearings on AEC Authorizing Legislation for Fiscal Year 1967, p. 1368.
107 Seaborg and Loeb, *Stemming the Tide*, p. 329. Seaborg says that there was evidence that three or four Soviet explosions had 'caused radioactive debris to be present' across the Chinese border. These may, however, have been accidents rather than excavation experiments.
108 Ibid., pp. 330–31; see Chapters 6 and 7.
109 Ibid., p. 331.
110 Ibid., p. 332.
111 Ibid., p. 333; see also US Congressional Record, 14 April 1969, p. 8945 and Chapters 6 and 7 for details of the negotiations.
112 Ibid., p. 334.
113 Ibid.
114 Ibid., p. 335.
115 US Congress, JCAE, Hearings on AEC Authorizing Legislation for Fiscal Year 1967, p. 15.
116 Farnsworth and McKenney, p. 114.
117 Seaborg and Loeb, *Stemming the Tide*, p. 336.
118 The 1962 'Starfish' test blacked out communications in a wide area of the Pacific Ocean, leading to the discovery of electromagnetic pulse (EMP).
119 Seaborg and Loeb, *Kennedy, Khrushchev and the Test Ban*, p. 267.
120 Seaborg and Loeb, *Stemming the Tide*, p. 336.
121 See Chapter 5.
122 Seaborg and Loeb, *Stemming the Tide*, p. 336.
123 Ibid., p 339.
124 M. Willrich, *Global Politics of Nuclear Energy*, Praeger, New York, 1971, p. 153.
125 Seaborg and Loeb, *Stemming the Tide*, p. 339.
126 Ibid., p. 325.

# 5

# Plowshareski: The Soviets Join In

*The Communists might develop Plowshare before we do. The time may be near when the Russians will announce that they stand ready to help their friends with gigantic nuclear projects. The consequences of such aid would be an economic penetration a hundred times more extensive than those following the Soviet offer to help Egypt construct the Aswan Dam... Cheap, safe, and 'clean' nuclear explosives in Communist hands would carry a most important implication: if the Soviet Union has surpassed America in the peaceful uses of the greatest force on the earth, Russia certainly must be ahead of the United States in military applications. As a propaganda weapon, Plowshare could finish the work begun with the launching of Sputnik.*[1]

Edward Teller, 1982

ON 12 August 1953 the Soviet Union detonated its first thermonuclear device. At least one Soviet scientist, like his American counterparts, foresaw the possibilities of nuclear dynamite. Professor G. Pokrovskiy, writing in a Soviet scientific journal in 1954, waxed lyrical about the potential uses of PNEs:[2]

> Progressive science claims that it is possible to utilize the noble force of the explosion builder for peaceful purposes... With the help of directional explosions one can straighten out the beds of large rivers to construct gigantic dams, to cut through canals literally in a few minutes whose construction by ordinary machines would be prolonged for years... Indeed, unlimited are the perspectives disclosed due to the new atomic energy. The cold current which washes our transpolar region can be made warmer than the Gulf Stream. The tundra will disappear, the subtopics will reach up to Moscow and in Kolyma there will be the climate of France...

In 1956, the year that Plowshare was being hatched, Pokrovskiy published another paper entitled 'On the Use of Nuclear Explosives for Industrial

Purposes', in which he analysed the potential of PNEs for excavation, especially overburden removal in open-cut mines.[3] Pokrovskiy concluded, like his American counterparts, that:[4]

> With data now available, we can say that radioactive contamination in a nuclear explosion should not be considered an insurmountable obstacle to the use of such explosives in mining and construction. On the basis of the many advantages of nuclear explosions, we conclude that the time is ripe to begin actual experiments in this field.

In 1956 the Soviets began an extensive program of conventional high explosive experiments, apparently in accordance with a directive of the 20th Party Congress held that year. The largest of these explosions were conducted in China in 1956 for the purpose of stripping overburden from iron-ore deposits. They were followed by another large test near Tashkent in December 1957, which was described as being part of the Soviet program for 'industrial' use of explosives. In a curious echo of Vishinsky's claims in 1949, a Moscow press release of 13 June 1958 claimed that 'persons in charge of blastings will soon be able to lift an entire mountain into the sky, converting mountains into plains'.[5] Another blast, described by *Pravda* as 'the largest rock clearing explosion in the history of the Soviet Union', was detonated on 25 March 1958 to help create a canal in the Urals.[6]

According to Gerald Johnson, writing in 1960, these explosions were 'precisely the kind of experiments that would be done in support of an aggressive Plowshare program'.[7] Describing them as being 'of fundamental interest to Plowshare', Johnson said the Soviet Union could make a 'major contribution to peaceful applications and science if it would provide the United States and the world with all the technical data from these important explosions'.[8] There is in fact some evidence that, along with their conventional explosive program, the Soviets had in 1956 also begun theoretical research into using nuclear explosives for 'industrial' applications, especially oil and gas stimulation.[9]

It appeared that the Soviets and Americans, as in so many other scientific fields, were about to embark on very similar research programs in the use of nuclear dynamite. This possibility abruptly ended in 1958, however, when the Soviets denounced Plowshare at the Second International Conference on the Peaceful Uses of Atomic Energy. The Soviet Chairman of the Conference, Professor Emilianov, claimed that whatever their alleged purpose, PNEs '... do not reach practical ends, but only political ends'.[10] Emilianov repeated his charge at a press conference, denouncing Plowshare as a trick designed to justify continued work on nuclear weapons.

When asked whether the Soviet Union had a similar program, Emilianov declared: '... I would not spend a kopek for such tests... We have no project or plans to use nuclear explosives this way'.[11] A reporter from the *Manchester Guardian* asked whether it was not a member of the Politburo who in 1949 had announced that the first Soviet nuclear explosion was 'devoted to peace, to the irrigation of deserts, the moving of mountains, the improvement of men's lives?' According to Gerald Johnson's account, Emilianov replied with the stunning *non sequitur*: 'I've never been in Novosibirsk!'[12] Edward Teller, who was not present, has recounted Emilianov as replying, 'Yes, but that was a politician speaking, and we Russian scientists are never influenced by politicians'.[13] Some years later, when the Soviet Union finally did begin its own PNE program, Edward Teller pleaded, tongue in cheek:[14]

> Don't expect me to explain this development. Maybe politicians have an influence in Russia after all. Whatever the reason, the Russians surely make up in flexibility what they may lack in consistency.

To this day Johnson remains puzzled about the original Soviet turnabout. There is some evidence, however, that Soviet nuclear scientists were divided over the value and practicability of nuclear dynamite. In 1958 the 'father of the Soviet H-bomb', Andrei Sakharov, published a detailed refutation of Edward Teller's predictions about the possibility of producing 'clean' thermonuclear devices. Sakharov concluded that the so-called clean bomb was 'a radioactive hazard'.[15] Such a view may have inclined Soviet scientists to be less enthusiastic about PNEs. Another reason for the Soviet change of heart may have been their wish to avoid having to compete in yet another field where the United States was clearly ahead.

The Soviets may have also recognised the implications of nuclear dynamite for arms control. With negotiations on a test ban treaty about to commence, they could have been genuinely interested in avoiding complicating the test ban issue. They may also have genuinely believed that the Americans would try to use Plowshare to mask an intensified nuclear weapon test program, as Arthur Dean has alleged.[16] Their suspicions, however unjustified, would not have been allayed by the fact that until August 1961, Plowshare was run by the Division of Military Applications, where it had originated, and actual detonation of Plowshare devices was handled by the same body that conducted nuclear weapon tests, the Nevada Operations Office (NEVOO).[17] As Glenn Seaborg notes, while this was 'logical in the sense that PNEs and nuclear weapons employed basically the same technology', it was 'not very smart public relations'.[18] According to Seaborg, it was only after the AEC recognised this 'error' that it established a separate Division of Peaceful Nuclear Explosions.[19] Gerald Johnson regarded this move as itself a 'mistake', presumably because it gave credence to the idea that there was such an animal as a 'peaceful nuclear explosive'.[20]

Whatever the reasons, the Soviet Union did not, as far as is known, conduct any PNEs or PNE experiments before the mid-1960s. It is not exactly clear when the first Soviet PNE experiment was conducted. If detonated at the Soviet nuclear weapon test sites at Semipalatinsk or Novaya Zemlya, such an experiment would have been largely indistinguishable from a nuclear weapon test unless used for an identifiable purpose such as excavation. Small underground PNEs could in any case have gone completely undetected.

By 1964 it was apparent that the Soviet Union was again interested, at least in theory, in nuclear dynamite. In marked contrast to 1958, the American presentation on Plowshare at the 1964 Third International Conference on the Peaceful Uses of Atomic Energy elicited favourable comment from the Soviet Union. Its delegate reported that his country was considering the use of PNEs in irrigation, excavation and gas stimulation projects. He even suggested the possibility of joint US/Soviet studies.[21]

Just as it is not known why the Soviet Union abandoned the idea of PNEs in 1958, it is also not known why they again became interested in the mid-1960s. Perhaps it was the fear of being outshone by the United States. The Americans were, after all, about to begin studies on the Panama Canal proposal, the biggest planetary engineering project of all time. Maybe, having mastered the basic techniques of nuclear weapons production, the Soviets now had spare scientific capacity to investigate peaceful uses. Maybe Plowshare propaganda was working, as it was beginning to do with the developing countries. Or perhaps with the PTBT signed and ratified and a comprehensive ban on testing a distant possibility, the Soviets no longer saw PNEs as a major threat to testing limitations.

In any event, the first large PNE experiment conducted by the Soviet Union was explosion No. 1004, detonated on 15 January 1965 at or near the Semipalatinsk range.[22] A cratering explosion of approximately 125 kilotons, it was used to dam a stream to create a reservoir. Some of the rock later slumped back into the reservoir.[23] According to Herbert Scoville, while the formation of the dam was carried out successfully, 'the water was later allowed to escape for reasons which have not been explained'.[24] Gerald Johnson points out that this was done because the dam lacked a spillway.[25] As a result of the experiment the Soviets described nuclear excavation technology as 'proven'.[26] However, like the first Plowshare shot, explosion 1004 vented radioactive debris into the atmosphere. Some of this later fell over Japan and Canada, leading to accusations that the USSR had violated the PTBT.[27]

On 12 February 1965 Glenn Seaborg reported to Under Secretary of State Ball, on the basis of 'bits of intelligence that came our way', that the Soviet attitude towards nuclear excavation had 'materially improved since the period prior to the execution of the Limited Test Ban Treaty'.[28] Seaborg proposed that, as a first step in securing international co-operation in Plowshare, talks should be initiated with the Soviets, since their attitude would 'have an important bearing on the evolution of the program'.[29] The Soviets were thus to be enticed into the PNE business to boost international co-operation in PNEs and to make them more likely to support an amendment to the PTBT, both of which, in Seaborg's view, were vital to Plowshare's future.

The first nuclear detonation conducted well outside the Soviet Union's nuclear weapon test sites took place on 22 April 1966 just north of the Caspian Sea.[30] Its purpose is unknown, although it may have been intended to create a gas storage cavity or regulate water flow. On 30 September 1966, a nuclear explosion was used to extinguish a runaway gas fire by plugging the well, at the Urtabulak gas field near Bukhara in Uzbekistan, an application of PNEs described by Edward Teller as 'one of the most impressive'[31] but by Gerald Johnson as 'nutty'.[32]

According to Bruce Bolt, once the Soviets began their PNE program they pursued it aggressively, in contrast to the 'rather slow' development of Plowshare.[33] From information available in the public domain, it appears that there have been three major applications for Soviet PNEs: the creation of underground storage cavities; the stimulation of oil and natural gas; and as a source of strong signals for deep seismic sounding.[34] The Soviets have, in addition, studied nuclear excavation techniques for the diversion of rivers and the creation of dams and canals.[35] Like the Americans, they have also envisaged more fantastic uses for PNEs, such as melting glaciers.[36]

## *Deep Seismic Sounding (DSS)*

One of the most impressive uses of PNEs by the Soviet Union has been its deep seismic sounding (DSS) program for investigating the earth's geology. This involves using underground nuclear explosions to produce shock waves that bounce off underground rock strata. Seismologists can record and analyse these signals to build up a multi-dimensional view of the structure, thickness and type of rock in a particular area.[37] In 1971 Soviet geologists began using nuclear explosives along traverses thousands of kilometres long for a seismic survey extending from Vorkuta in the northern Urals to Kineshma, north-east of Moscow. In some instances a combination of nuclear and chemical explosions appear to have been used.[38] By 1984 there was evidence that nine long DSS lines had been generated in the Soviet Union in the previous twelve years. These involved the use of at least 29 nuclear explosions, most of them in the oil- and gas-rich provinces of Siberia and the relatively unexplored areas of the Soviet Far East.[39] Initially the Soviets failed to reveal much information about their DSS technique. This, in the view of Scheimer and Borg, may have been related to 'initial difficulties in reducing and synthesizing the vast amount of data acquired'.[40] Today, in Borg's view, 'there can be little doubt as to the uniqueness and value of the DSS nuclear program...'[41]

The reason why the United States has never been interested in DSS techniques with PNEs is that it is unlikely that the detonation of a string of nuclear explosions across the country would have been politically acceptable. American seismologists have also regarded such a method as seismic overkill. American efforts have been directed towards better resolution of seismic signals from much lower energy sources, such as small conventional explosions.[42] However, as a 1975 report compiled by the Gulf Universities Research Consortium (the *GURC Report*) noted, there is no reason why all nuclear shots, for whatever purpose, cannot be used for seismic studies. While this was at one stage discussed with US energy exploration companies, 'practically no interest was expressed' by them.[43] However, the US seismic community has used seismic signals from nuclear tests at Nevada since the 1950s for investigating the earth's structure.

## Underground Storage Cavities

The Soviet PNE program has also focused on the creation of underground storage cavities, particularly for the storage of natural gas. A project at the giant Orenburg gas-condensate deposit north of the Caspian Sea, the principle source of gas for the Soyuz (Friendship) pipeline to Eastern Europe, involved two nuclear explosions, in 1971 and 1973.[44] Between 1980 and 1982 seven underground explosions attended the development of the Astrakhan gas-condensate field, near the mouth of the Volga river. In 1983 there were three nuclear explosions at the new Karachaganak gas-condensate field near the Orenburg field, and six more north of Astrakhan.[45] Borg estimates all of these explosions to have been around 15 kilotons.[46]

In all three fields the nuclear explosions preceded the arrival of teams of Western experts contracted to design and build collection facilities and processing plants. On the basis of exchanges with the Soviets in the 1970s and information from US engineers who worked on the construction of one of the gas-processing plants at Orenburg, Iris Borg believes that the PNE-created cavities are used primarily to store the highly corrosive sour condensate before it is piped to the processing plants — either to provide 'surge' gas capacity during periods of peak demand or to 'de-gas' or stabilise the condensate before it is processed. According to Borg, the relatively large number of cavities associated with the Astrakhan field suggests that some of them may also be used for chemical waste disposal.[47]

Borg reports that the Soviets have broadly described the advantages of PNE storage cavities for gas-condensate fields, including preservation of agricultural land and safety issues. According to Borg, however, their principal rationale is the lower capital cost of such cavities, which are estimated to be three to five times cheaper than surface storage, reducing construction costs and eliminating the need for expensive corrosion-resistant steel tanks and related equipment.[48] At a 1986 meeting of the Nuclear Technology, Science and Health Seminar at the University of California, Berkeley, Jack Evernden of the US Geological Survey was asked if the Soviets worried about radioactive contamination of the material stored in their underground cavities. Evernden replied:[49]

> No ma am! They do not worry a bit, and the radioactivity must come out into the gas stoves of the Soviet homes.

The Soviet program for producing underground cavities using PNEs has direct arms control implications. Some American scientists have suggested that by conducting clandestine nuclear tests inside such cavities the Soviets could circumvent a comprehensive nuclear test ban treaty (CTB).[50] The theory is that seismic signals, by which underground tests are normally detected at long distances, could be 'decoupled' or muffled by the use of such a technique. The Soviets deny this possibility, pointing to the limited decoupling effect achieved in the 1966 Salmon experiment in the Tatum Salt Dome in Mississippi, when a 0.38 kiloton explosion was detonated in a 35-metre diameter cavity created by a previous 5.4 kiloton explosion.[51] Whether or not effective decoupling is possible, the controversy on the subject constitutes yet another barrier to the negotiation of a comprehensive test ban treaty. Dr Jeremy Leggett, formerly of the London-based Verification Technology Information Centre (VERTIC), has proposed that a register of PNE cavities be compiled, to provide assurance that they would not be used for 'decoupling' clandestine nuclear weapon tests under a CTB.[52]

## *Oil Stimulation*

Unlike the Americans, the Soviets have also carried out at least two field experiments in stimulating oil deposits.[53] The AEC told the JCAE in 1970 that the Soviets had successfully carried out an oil 'recovery' experiment in an oilfield with a declining rate of production.[54] This was of particular interest because the nuclear detonations were conducted only 1.5 kilometres from a village of about 2000 people, and 13–14 kilometres from a town of about 24 000 people, with buildings in it as high as five stories. Representative Hosmer asked tongue in cheek: 'Do they have a Sierra Club over there?'[55]

According to the AEC, the yields of the blasts were quite low. In the first detonation, two 2.6 kiloton devices were used, with a delay of 100 milliseconds between them. While the delay had been intended to reduce possible ground motion, the Soviets reported that this probably had little or no effect. In the second shot a single 7.6 kiloton explosive was used. The Soviets reported a general increase in oil production of 30–50 per cent from the ten to twelve wells in the region. In the words of Dr T.J. Thompson of the AEC, 'this indeed was a very, very useful type of experiment'.[56] According to the *GURC Report*, however, the data on these tests are difficult to interpret.

## *Extinguishing Oil and Gas Fires*

The Soviet Union has also claimed success in the use of PNEs for extinguishing fires at oil and gas wells. The United States has eschewed such a technique, largely because its conventional technology in this field is highly developed.[57] A further consideration is that a runaway fire is an emergency situation requiring immediate action. Such a use of PNEs would require setting aside all the normal regulatory and judicial procedures, a politically impossible feat in the United States.[58]

## Nuclear Excavation

An aspect of the Soviet PNE program which mirrored that of the United States was nuclear excavation. The Soviets for many years contemplated using PNEs to divert the northward-flowing Pechora River by means of canals to join the southward-flowing Kama River, a tributary of the Volga. This would raise the level of the Caspian Sea, drain the Siberian marshes and provide water for irrigation. Up to 250 explosions were envisaged.[59] The nuclear excavation would stretch over 65 kilometres and involve explosions with yields between 40 and 60 kilotons. The Soviets estimated in 1974 that the use of PNEs would reduce construction costs by a factor of 3 to 3.5 in comparison with conventional construction methods.[60]

Experiments conducted for the Pechora-Kama Canal Project included explosion No. 1004 in 1965 and explosion No. 1003, described in a Soviet paper in 1974. The latter, a very small 1.1 kiloton explosion,[61] produced a crater 'in terms of its initial surface' with a diameter of 124 metres and a depth of 20 metres. Conventional explosives were later used to construct an approach canal for water to enter the crater. The paper reported that 'years of investigation of this model of a man-made reservoir demonstrated that it is feasible to build such reservoirs with nuclear explosions'.[62] Wolfgang Panofsky commented to the US Senate Foreign Relations Committee that such reports 'are very similar in flavor to those prevalent in the United States concerning the construction of a second Trans-Isthmian (Panama) Canal by nuclear means . . .'[63]

At a January 1975 technical meeting of the International Atomic Energy Agency (IAEA) on the subject of PNEs, the Soviets reported an experiment conducted in March 1971 involving a 'row-charge' of three 15 kiloton explosions along the actual route of the proposed Pechora-Kama Canal.[64] The tests were conducted in a low-lying water-saturated section of the canal route and produced only a very shallow cut because of slumping of the soft surface material. As a report for ACDA later commented:[65]

> Since the important utilization of PNEs will be for cutting through the hard rock section of the highlands which separate the two drainage basins, not in excavating the shallow portions, the choice of the test area for this experiment seems rather strange. Even so, the Soviet engineers reported satisfaction with their experiment.

Radioactive particles from the explosion were detected in Sweden.[66]

From 14–16 April 1969 the United States and the Soviet Union held their first bilateral, technical meeting in Vienna on the use of nuclear explosives for peaceful purposes. In a joint communique issued after the meeting they expressed the view that underground explosions could be used 'in the not so far off future' for peaceful purposes.[67] Dr Tape of the AEC later testified to the JCAE that the Soviet and American PNE programs were essentially the same: 'I see no essential differences between the two of us in terms of what the potential applications might be around the world . . . it was gratifying to me to know that they have a peaceful nuclear explosions program, and they, too, are exploring the same sorts of questions'.[68] So useful had the Soviets found these talks that they pressed for another round.

Further technical talks were held in Moscow in February 1970 to exchange views on the 'possibility of the effective application' of PNEs.[69] At a subsequent multilateral meeting, organised under the auspices of the IAEA in Vienna in March 1970, the Soviets volunteered a considerable amount of detail about their PNE program.[70] The delegation's spokesman on nuclear engineering, O.L. Kedrovskiy, claimed that:[71]

> Thanks to a large volume of construction, the possession of vast, sparsely populated territories with bleak climatic conditions, and the distribution over these territories of large deposits of valuable minerals, the Soviet Union enjoys unlimited possibilities for the successful application of underground nuclear explosions for industrial purposes.

The Soviets also showed dramatic film footage of explosion No. 1004 in 1965 which had created a reservoir in a river bed. The film showed people subsequently boating and swimming in the lake.[72] Commenting on the meeting, the *New York Times* noted that in contrast with the Soviet program, Plowshare officials 'have found their program coming under attack as a radiation health hazard'.[73] Such officials, the *Times* reported, concluded that the Soviet Union was 'drawing ahead' in PNE experimentation.

According to Herbert Scoville, the fact that fallout from PNEs would pass over large uninhabited areas of Siberia before reaching other countries was one advantage the Soviets had over the Americans in using nuclear dynamite for excavation projects.[74] This may explain Soviet persistence with the idea long after the United States abandoned it. Nonetheless, the Soviet excavation program was also eventually terminated. Iris Borg speculates that this was in fact due to Soviet concern over their inability to contain radioactivity from large excavation experiments within their borders.[75] In addition, experiments for the Pechora-Kama Canal Project had aroused 'an outcry of public opinion, because many feared that without the warm water [the Pechora] the Arctic Sea would have frozen, and the project was eventually abandoned'.[76]

After 1974 the Soviets released little additional information on their PNE program, having, according to Iris Borg, 'little to gain' by doing so.[77] Most of what is known of the Soviet program since then has been 'inferred from often fragmentary disclosures by Soviet authorities and pieced together with information obtained in other contexts' — much of it by Iris Borg.[78] The *GURC Report* cautioned that in any case the information that the Soviets had released 'must be regarded with considerable caution... since only very meager data is available... and some of the data is regarded as obscure or contradictory'.[79] Despite this, Iris Borg believes that the Soviet Union has pursued many of the PNE applications it claimed were 'proven technology' in earlier international exchanges.[80] According to Borg, the Soviet PNE program has been 'large, multifaceted, apparently successful and, judging from its size, of great value to the Soviet economy'.[81] In August 1973 the scientific attache at the Soviet Embassy in Washington told Glenn Seaborg that the Soviet Union considered PNEs to have considerable future importance.[82] In Seaborg's view, the Soviet Union 'deserves credit for being the first to apply nuclear explosives to some of the world's practical problems in planetary engineering'.[83] In contrast with Plowshare, which in total managed only six experiments outside the Nevada Test Site, the vast majority of Soviet PNEs have been conducted in the field.[84]

The size of the Soviet program is indicated by the fact that by the early 1980s probably more than half the annual numbers of Soviet nuclear tests in some years were PNEs. In 1982, for example, 16 out of 31 nuclear tests were conducted outside the known Soviet weapon test sites, indicating that they were probably PNEs.[85] In 1984, 10 out of 17 Soviet nuclear tests were assumed to have been PNEs.[86] According to the US National Resources Defense Council, the Soviets exploded 110 PNEs between 1966 and 1985, when they began their unilateral testing moratorium. This amounted to 19 per cent of their total number of nuclear tests.[87] Iris Borg puts the figure at more like 100.[88] Borg warns, however, that her estimates must be considered a minimum, given that some of the smaller Soviet PNEs may not have been correctly identified as nuclear in origin. In addition, it is likely that other explosions have not been detected for technical reasons such as the overlapping of their seismic signals with those of earthquakes,[89] although Gerald Johnson says he knows of no evidence to corroborate this.[90]

As to whether or not the Soviet PNE program has been controversial among Soviet scientists, a 1975 ACDA report noted disagreement about the technical utility of PNEs for various applications.[91] Certainly in 1974 two widely differing Soviet views were published. The first, expounded by Y.A. Izrael in an article entitled *Peaceful Nuclear Explosions and the Environment*, claimed that PNEs would be 'used extensively in the near future for economic purposes'.[92] As for the environmental drawbacks of PNEs:[93]

> The problem involving the possible fallout of small amounts of radioactive contaminants in the environment during explosions will not differ too much from other problems associated with the curtailment of the discharge of ordinary industrial pollutants.

This view contrasted sharply with that of Academician V.S. Emelyanov in the Stockholm International Peace Research Institute publication *Nuclear Proliferation Problems*. Emelyanov claimed that:[94]

> ... none of the [PNE] projects under investigation has yet reached the stage of wide and practical application. The advantages of using nuclear explosions for such projects lie chiefly in saving labour and therefore money. However, the danger of subsequent radioactive contamination of the environment is very real; the problem of designing a 'clean' explosive has still not been solved. It is concluded that, at present, peaceful nuclear explosions are advisable only for exceptionally urgent problems which cannot otherwise be solved.

This may, however, have been a case of the Soviet Union speaking with one voice for domestic consumption and one for international consumption. While anxious to investigate and use PNEs themselves, the Soviets have never been keen to share this technology with other states. One exception was their announcement at the IAEA in the early 1970s that their method of using PNEs to extinguish gas well fires was available commercially.[95] Glenn Seaborg says he is not aware of anyone taking advantage of this offer. The Soviets have also been acutely aware of the arms control implications of non-nuclear weapon states acquiring the means to manufacture their own PNEs.

The Director of the Stanford Linear Accelerator Centre, Wolfgang Panofsky, testified in 1977 to the US Senate Foreign Relations Committee, that he believed there was a 'fair amount' of dissension within the USSR on the question of the economic merits of PNEs. He noted that the PNE question was one of the few issues where dissension on a vital policy issue had surfaced in public in the Soviet Union. According to Panofsky the Soviets had not resolved the matter and were keeping their options open.[96] Professor Frank Long of Cornell University has noted that 'clearly the Soviet approach to economic analyses is different to ours and I am not sure we will ever get an analysis that we would recognise as useful'.[97] Another American observer has commented that 'from available data, it is impossible to itemise cost factors for nuclear devices or site development within the Soviet Union'.[98]

Among the generalisations about the future of the Soviet PNE program made by the *GURC Report* in 1975 were the following:[99]

- The USSR has about the same capability for developing PNE technology as the United States
- The USSR would presumably not be delayed by regulatory problems to the degree that Western industrialised states would
- The USSR possesses enormous undeveloped natural resources, partly located in very thinly populated areas, which might alter the economic analysis in favour of PNE technology.

A later ACDA report corroborated this assessment thus:[100]

> The apparently more favourable assessment of the utility of PNEs in the USSR may be a result of different geography, topography, or resource opportunities that make PNEs more useful than they seem to be in the United States. Alternatively, or in addition, different standards for public health and safety, different levels of public access to the decision-making process, different methods for computing costs and benefits, and different criteria for commercial use by private industries as compared to government agencies may account for the difference.

## *Other Nuclear Weapon States*

Despite the interest displayed by the Soviet Union and the United States, there has been little or no interest in peaceful nuclear explosions on the part of the other nuclear weapon states.

### The United Kingdom

In the early 1970s the United Kingdom Atomic Energy Authority (UKAEA) commissioned several theoretical studies on the use of nuclear dynamite from the Atomic Weapons Research Establishment (AWRE) at Aldermaston. These studies included such titles as 'Radioactivity and PNE', 'Ground Motion and Seismic Damage' and 'Containment of Nuclear Experiments'.[101] Britain has not however, as far as is known, conducted any PNE experiments.[102] Indeed the British have indicated that they do not consider themselves as either users of nuclear dynamite or as suppliers of PNEs to other countries.[103]

### France

The first book on the subject of PNEs was written, not by an American or Soviet scientist, but by the Frenchman Camille Rougeron. Published in Paris in December 1956 it was entitled *Les Applications de L'Explosion Thermonucleaire*.[104] Despite this trail blazing effort, official French interest in nuclear dynamite only developed during the series of underground nuclear weapon tests conducted by France in Algeria from 1961 to 1966. Although the French Foreign Ministry announced in 1961 that France would conduct a nuclear explosion to test the use of PNEs in excavation, there is no evidence that this was actually carried out.[105] It was therefore on the basis of thirteen fully-contained underground nuclear *weapon* tests in Algeria that France presented a report on nuclear dynamite to the first IAEA Panel Meeting on PNEs in 1970. The report provided the panel with what was described as 'a unique body of information on contained explosions in a very dry, homogeneous, strong medium'.[106]

At the time France expressed considerable interest in becoming a PNE supplier.[107] French scientists hoped to undertake PNE projects with financial support from French industry.[108] In 1969 an organisation called APEX (*applications des explosions*) was created to carry out French PNE projects. There is however no evidence that any industrial support for PNEs eventuated. After a 4–5 year official program of research and study, the French abandoned the idea of PNEs in the mid-1970s.[109] France has apparently never conducted an actual PNE experiment.

In view of the relatively small land area and dense population of both the United Kingdom and France, it is unlikely that either country would have as many opportunities for using nuclear dynamite as the two superpowers, even if all other economic and technical factors were favourable. It will be recalled that when Australian Foreign Minister Hayden suggested that the French conduct their nuclear weapon tests in France rather than in the South Pacific, the French pointed to the damage that ground motion from underground nuclear blasts could cause to historic French buildings, particularly churches.[110]

## China

No information is available on whether China has studied PNEs or conducted a PNE test.[111] In view of the relatively small number of nuclear tests conducted by the Chinese it is unlikely to have been able to afford the luxury of a PNE program.[112]

# *Notes*

1 E. Teller, *The Legacy of Hiroshima*, Doubleday and Co., New York, 1982, p. 87.
2 US Disarmament Administration, *Geneva Conference on the Discontinuance of Nuclear Weapon Tests: History and Analysis of Negotiations*, US Department of State, Washington DC, 1961, p. 267.
3 G.W. Johnson, 'The Soviet Program for Industrial Applications of Explosions', *Bulletin of the Atomic Scientists*, November 1960, p. 366.
4 Ibid.
5 Ibid., p. 368.
6 Ibid.
7 Ibid., p. 367.
8 Ibid., p. 369.
9 This is according to a statement by E. Bakirov, a Soviet expert in geological and mineralogical sciences, on 19 April 1987, reported in the *Arms Control Reporter*, 1987 edition, Institute for Defense and Disarmament Studies, Brookline, Mass., p. 605. B.61.
10 R. Sanders, *Project Plowshare*, Public Affairs Press, Washington DC, 1962, p. 183; J. Koop, 'Plowshare and the Nonproliferation Treaty', *Orbis*, XII, 3, Fall 1968, p. 806.
11 Johnson, 'The Soviet Program for Industrial Applications of Explosions', p. 369.

12 From an interview with Gerald Johnson, Washington DC, 10 December 1986.
13 E. Teller, *Energy From Heaven and Earth*, W.H. Freeman and Co., San Franciso, 1979, p. 153.
14 Ibid.
15 P. Pringle and J. Spigelman, *The Nuclear Barons*, Avon Books, New York, 1981, p. 252.
16 T. Ehrlich, 'Legal Problems (of PNEs)' in L.E. Weaver (ed.), *Education for Peaceful Uses of Nuclear Explosives*, University of Arizona Press, Tucson, Arizona, 1970, p. 224.
17 A. Kramish, *The Peaceful Atom in Foreign Policy*, Harper and Row, New York, 1963, p. 117.
18 G.T. Seaborg and B.S. Loeb, *Stemming the Tide: Arms Control in the Johnson Years*, Lexington Books, Lexington, Mass., 1987, p. 310.
19 Ibid.
20 From correspondence with Gerald Johnson, La Jolla, California, 28 September 1987.
21 Koop, p. 806.
22 B.A. Bolt, *Nuclear Explosions and Earthquakes: The Parted Veil*, W.H. Freeman and Co., San Francisco, 1976, p. 276; O. Dahlman and H. Israelson, *Monitoring Underground Nuclear Explosions*, Elsevier Scientific Publishing Co., Amsterdam, 1977, p. 298.
23 I.Y. Borg, 'Nuclear Explosives — The Peaceful Side', *New Scientist*, 8 March 1984, p. 13.
24 H. Scoville Jr., 'Peaceful Nuclear Explosions — An Invitation to Proliferation' in A.W. Marks (ed.), *NPT: Paradoxes and Problems*, Arms Control Association, Washington DC, 1975, p. 48.
25 From correspondence with Gerald Johnson, La Jolla, California, 28 September 1987.
26 I.Y. Borg, *Nuclear Explosions for Peaceful Purposes*, Lawrence Livermore National Laboratory, Paper No. UCRL-95026, Rev. 11, November 1986, p. 13.
27 Ibid.
28 Seaborg and Loeb, *Stemming the Tide*, p. 321.
29 Ibid.
30 Bolt, p. 276; J.I. Sands, R.S. Norris and T.B. Cochran, *Nuclear Weapons Data Book* Working Paper NWD86-3, 'Known Soviet Nuclear Explosions, 1949–1985 Preliminary List', February 1986, p. 18.
31 Bolt, p. 276. Confirmed in an interview with Edward Teller, Livermore Laboratory, California, 9 December 1986.
32 From correspondence with Gerald Johnson, La Jolla, California, 28 September 1987.
33 Bolt, p. 187.
34 Borg, 'Nuclear Explosives — the Peaceful Side', p. 11.
35 Bolt, pp. 190-92.
36 E.H. Voss, *Nuclear Ambush: The Test Ban Trap*, Henry Regency Co., Chicago, 1963, p. 460.
37 Borg, 'Nuclear Explosives — the Peaceful Side', pp. 11–12.
38 Ibid., p. 12.
39 Ibid.
40 J.F. Scheimer and I.Y. Borg, 'Deep Seismic Sounding with Nuclear Explosives in the Soviet Union', *Science*, Vol. 226, No. 4676, 16 November 1984, p. 787.
41 Borg, *Nuclear Explosions for Peaceful Purposes*, p. 16.
42 *PNE Activity Projections for Arms Control Planning (*the *GURC Report)*, Gulf Universities Research Consortium, Galveston, Texas, US Arms Control and Disarmament Agency Document, ACDA/PAB-253, Washington DC, 1975, p. 133.
43 Ibid.
44 Borg, 'Nuclear Explosives — the Peaceful Side', p. 13.
45 Ibid.
46 I.Y. Borg, 'Peaceful Nuclear Explosives in Soviet Gas Condensate Fields', *Energy and Technology Review*, Lawrence Livermore National Laboratory, May 1983, p. 37.
47 Ibid.
48 Ibid., p. 35.
49 *Interdisciplinary Seminar on Nuclear Technology, Society and Health*, University of California, Berkeley, Transcript of seventh meeting (unpublished), 30 October 1986, p. 10.
50 'The Livermore Lab's Nuclear Test Detectives', *The Washington Post*, Washington DC, 10 June 1986, p. 17.
51 O. Kedrov, *Capabilities for Verificiation of a Nuclear Test Ban*, paper presented at the

International Symposium on Seismic Monitoring, Australian National University, Canberra, August 1986, p. 5.

52 J. Leggett, R. Clark and R. Westaway, *The Possible Role of Confidence-Building Measures in Arriving at a Ban on Nuclear Weapons Testing*, paper presented at the Soviet Academy of Sciences' Forum for Peace, Moscow, 15 February 1987, p. 3.
53 Thc *GURC Rcport*, p. 131.
54 US Congress, JCAE, Hearings on AEC Authorizing Legislation for Fiscal Year 1971, 3 and 5 March 1970, US Government Printing Office, Washington DC, 1970, p. 638.
55 Ibid.
56 Ibid., p. 639.
57 The *GURC Report*, p. 134.
58 Ibid.
59 Bolt, p. 192.
60 Y.A. Izrael, *Peaceful Nuclear Explosions and The Environment*, US Joint Publications Research Service (JPRS 63502), Washington DC, 22 November 1974, p. 7.
61 Dahlman and Israelson, p. 298.
62 Izrael, p. 8.
63 Testimony of Wolfgang Panofsky to the US Senate Foreign Relations Committee, 12 September 1974, reprinted in US Congress, Senate Foreign Relations Committee, Hearings on the TTBT/PNET, 13 and 15 January 1987, US Goverment Printing Office, Washington DC, 1987, p. 268.
64 *An Analysis of the Economic Feasibility, Technical Significance and Time Scale for Application of Peaceful Nuclear Explosions in the US, with Special Reference to the GURC Report Thereon (The Long Report)*, Program on Science, Technology and Society, Cornell University, US Arms Control and Disarmament Agency, Washington DC, 1975, p. 28; Dahlman and Israelson, p. 298.
65 *The Long Report*, p. 28.
66 Dahlman and Israelson, p. 298; Borg, *Nuclear Explosions for Peaceful Purposes*, p. 13.
67 *The United Nations and Disarmament 1945–1970*, Department of Political and Security Affairs, United Nations, New York, 1970, p. 249.
68 US Congress, JCAE, Hearings on AEC Authorizing Legislation for Fiscal Year 1970, US Government Printing Office, Washington DC, 1969, p. 317.
69 *The United Nations and Disarmament 1945–1970*, p. 375.
70 Bolt, p. 187.
71 Ibid., p. 181.
72 US Congress, JCAE, Hearings on AEC Authorizing Legislation for Fiscal Year 1971, p. 640.
73 Ibid., p. 641.
74 Scoville, p. 49.
75 Borg, *Nuclear Explosions for Peaceful Purposes*, p. 13.
76 *Interdisciplinary Seminar on Nuclear Technology, Society and Health*, p. 11.
77 Borg, 'Nuclear Explosives — The Peaceful Side', p. 10.
78 Borg, 'Peaceful Nuclear Explosions in Soviet Gas Condensate Fields', p. 31.
79 The *GURC Report*, p. 188.
80 Borg, *Nuclear Explosions for Peaceful Purposes*, p. 20.
81 Ibid., p. 21.
82 Seaborg and Loeb, *Stemming the Tide*, p. 351.
83 G.T. Seaborg and W.R. Corliss, *Man and Atom: Building a New World Through Nuclear Technology*, E.P. Dutton and Co., New York, 1971, p. 191.
84 Borg, *Nuclear Explosions for Peaceful Purposes*, p. 18.
85 *SIPRI Yearbook 1983*, Taylor and Francis, London, 1983, p. 97.
86 *SIPRI Yearbook 1985*, Taylor and Francis, London, 1985, p. 81. The figures for Soviet PNEs are politically significant in that if the number of presumed Soviet PNEs is subtracted from the total number of Soviet tests, it can be readily seen that in recent years the United States and the Soviet Union have conducted roughly the same number of nuclear *weapon* tests.
87 *Arms Control Reporter*, Institute for Defense and Disarmament Studies, Brookline, Mass., 1987, p. 605. A.4.
88 Borg, *Nuclear Explosives for Peaceful Purposes*, p. 1. This is the number detected by the Swedish National Defence Research Institute, the US Department of Energy and

the National Earthquake Information Service of the US Geological Survey.

89 Ibid., p. 14.

90 From correspondence with Gerald Johnson, La Jolla, California, 28 September 1987.

91 *The Long Report*, p. 33.

92 Izrael, p. 2.

93 Ibid.

94 V.A. Emelyanov, 'On the Peaceful Use of Nuclear Explosions' in *Nuclear Proliferation Problems*, SIPRI, Almqvist and Wiksell, Stockholm, 1974, p. 215.

95 Seaborg and Loeb, *Stemming the Tide*, p. 351.

96 US Congress, Senate Foreign Relations Committee, Hearings on the TTBT/PNET, 28 July, 3 August, 8 and 15 September 1977, US Government Printing Office, Washington DC, 1977, p. 148.

97 Ibid., p. 129.

98 The *GURC Report*, p. 131.

99 Ibid., p. 183.

100 *The Long Report*, p. 33.

101 *Review of the Peaceful Uses of Nuclear Explosives (PNE)*, Aldermaston Weapons Research Establishment (AWRE) Document 10/74, June 1974, HM Stationary Office, London, p. 2.

102 Dahlman and Israelson, p. 294.

103 The *GURC Report*, p.179.

104 Seaborg and Corliss, p. 100.

105 Sanders, p. 191.

106 A.R.W. Wilson, 'Current Status of Civil Engineering and Mineral Resources Development Applications of Peaceful Nuclear Explosions' in Proceedings of the Fourth International Conference on the Peaceful Uses of Nuclear Energy, 6–16 September 1971, IAEA, New York and Vienna, 1972, p. 214.

107 M. Willrich, *Global Politics of Nuclear Energy*, Praeger, New York, 1971, p. 152.

108 M.I. Shaker, *The Nuclear Non-Proliferation Treaty: Origin and Implementation 1959–1979*, Vol. I, Oceana Publications, London, 1980, p. 394.

109 'Peaceful Nuclear Explosions and Arms Control', *Backgrounder*, US Information Service, Canberra, 16 August 1988, p. 2.

110 B. Danielsson and M. Danielsson, *Poisoned Reign: French Nuclear Colonialism in the Pacific*, Penguin, Harmondsworth, 1986, p. 301.

111 The *GURC Report*, p. 179.

112 Willrich, p. 152.

# 6

# Australia, Argentina and Brazil Discover Nuclear Dynamite

*I feel that if we can go to our friends abroad and tell them, 'Here it is, what we are doing. You see the tools. You see the operations. You know how it works. We can tell you why we have confidence that it will work.' I think that in this way it is much more likely we will have their wholehearted acceptance and support.*[1]

Edward Teller, 1960

IN apparent disregard of the difficulties being encountered by Plowshare, the US Atomic Energy Commission continued to publicise the program, both domestically and internationally, throughout the 1960s. In April 1964 a Third Plowshare Symposium was held at the University of California, Davis, on 'Engineering with Nuclear Explosives'.[2] It was attended by representatives of the United Kingdom, France, Australia, Canada, Mexico, Switzerland, South Africa, Israel and the IAEA. A fourth symposium was held in 1970, appropriately in Las Vegas, Nevada.[3] Sponsored by the American Nuclear Society, it was attended by sixteen countries, with France for the first time presenting data on PNEs.

The AEC also assisted colleges and universities in developing special Plowshare curricula and in integrating Plowshare technology into existing engineering programs. By 1968 at least eight American universities were offering full or partial courses in nuclear explosives engineering.[4] At the populist level the AEC mounted what it described as 'a very attractive exhibit' at the New York World's Fair in 1965, which described the 'Gnome and Sedan stories'.[5] A 'Plowshare Today' exhibit was also shown in several venues, including Gimbel's department store in New York.[6] In 1966 a booklet issued by the AEC described Plowshare experiments as:[7]

> ... little different from that of an apprentice carpenter learning to use a saw. Properly applied, the saw can facilitate the building of a home, a boat or a beautiful piece of furniture. Improperly used, it can produce inferior work or harm its user.

The AEC issued frequent news releases announcing progress in Plowshare. In addition, an advertising campaign was conducted on the benefits of 'underground engineering', including a full-page advertisement in the *New York Times* by the El Paso Oil Company.[8]

The Atomic Industrial Forum Inc., a grouping of companies involved in the US nuclear industry, also helped promote Plowshare. In July 1973, in an effort to assure the public that PNEs were safe, it issued a pamphlet which pointed out that only one person had ever been injured as a result of a nuclear blast conducted outside the Nevada Test Site. This was during the Salmon detonation (not a Plowshare experiment) in Hattiesburg, Mississippi in 1966, when a woman standing outside her house was grazed on the shoulder by a brick falling from the chimney. She was taken to hospital, given minor treatment, and released.[9]

Internationally, the AEC's exhibit and demonstration programs for Latin America, Western Europe and Asia included Plowshare components.[10] In addition the US delegation gave detailed presentations on Plowshare, including an exhibit and a film, at the Third International Conference on the Peaceful Uses of Atomic Energy in Geneva in 1964.[11] In an address to the Conference, Glenn Seaborg claimed that, through international support and co-operation, nuclear weapons technology could be converted into a valuable research and engineering tool for the benefit of mankind.[12] With this kind of glowing recommendation, William Epstein notes, 'it is not surprising... that both the rich and the poor countries were dazzled' by the promise of nuclear dynamite.[13] Among such countries were Australia, India, Romania, Egypt[14] and Israel. Egyptian and Israeli interest centred on the use of PNEs to excavate canals to enable sea water to enter large, natural depressions below sea level, generating electricity in the process. In Israel, for instance, it was envisaged that the Dead Sea could be connected to the Gulf of Aqaba.[15]

After the warm reception received by Plowshare at the 1964 conference, Glenn Seaborg began to see co-operative ventures with other countries as being Plowshare's 'international road to salvation'.[16] Seaborg reported to the JCAE in early 1965 that:[17]

> Our opinion that there is international interest in Plowshare was strengthened by the discussions on Plowshare and international co-operation in Plowshare which occurred in Geneva during the Third International Conference on the Peaceful Uses of Atomic Energy. We probably should begin to give serious consideration to some form of international co-operation in Plowshare.

Seaborg hoped that such co-operation, based on international interest in PNEs stimulated by the AEC, would stem opposition to Plowshare within the United States, facilitate amendment of the PTBT and secure Plowshare's future:[18]

> At this point, I, among others, felt that [the] AEC's best chance of engendering enough enthusiasm for PNEs to make revision of the test ban treaty possible was to develop cooperative projects with other countries. [The] AEC did its best to stimulate international interest and enthusiasm.

Seaborg seemed oblivious to the arms control implications of this plan. He also seems not to have considered whether the real interests of other countries lay in the use of PNE techniques. His sole aim seems to have been the survival of a program which was apparently unable to stand on its merits in the country that had given it birth.

Not everyone in the US Administration agreed with him. When Seaborg suggested in a letter to Under Secretary of State Ball on 12 February 1965 that a program of international co-operation in PNEs be initiated as soon as possible, the State Department's reply, signed by Secretary Rusk, was cautious.[19] Rusk suggested that 'a few initial experiments with international participation... be conducted [in this country] under substantially the present constraints as to permissible radioactive debris'. He held out the hope that eventually 'this course of action could lead to such amendment of the treaty [the PTBT] or other appropriate international understanding as might be necessary to permit major excavation projects'.[20] Clearly the State Department was alive to Seaborg's plans. This manifested itself most clearly in regard to Egyptian interest in PNEs.

The Egyptians, who had become 'quite excited' by American enthusiasm for nuclear dynamite,[21] seized on the idea of an 80-kilometre canal to bring Mediterranean seawater to the Qattara Depression. The project called for the use of at least 181 nuclear explosions in the 150–600 kiloton range, or 439 of 150 kilotons. The cost was estimated to be $510–820 million, depending on the number of explosives used.[22] The Egyptians expected to increase the rainfall of the area, generate hydroelectricity, establish tourist resorts, extract minerals from the water, and establish agriculture sustained by irrigation. They broached the idea with US delegates at the 1964 peaceful uses conference. AEC Commissioner Palfrey, on a visit to Egypt later in 1964, mentioned the possibility of a visit by Plowshare experts to discuss the project in greater detail.[23] Palfrey was acting, according to Glenn Seaborg, in accordance with State Department advice that it was appropriate in discussions with other governments to mention US interest in co-operative PNE endeavours. Subsequently, while Seaborg himself was in Egypt, Egyptian officials requested that Palfrey's suggestion be implemented. In a letter to National Security Advisor Bundy, Seaborg supported the idea and asked for comment. Bundy replied:[24]

> I think it is clear that such a visit at this time would inevitably focus considerable attention on the project and encourage the belief that the US was prepared to support it. I do not believe we have yet really thought through all the

> political, economic, and arms control implications that would be involved in either encouraging or undertaking this project.

Seaborg notes, with apparent disapproval, that, 'evident in Bundy's response... was a persistent skittishness in the administration about risking the future of arms control endeavors on the as yet unproven prospects of PNEs'.[25]

Despite this setback, and obvious disagreement within the Administration over the wisdom of promoting Plowshare internationally, the AEC continued to do so at every opportunity. Teller, Seaborg and the AEC had long identified at least two of the largest countries in the southern hemisphere, Australia and Brazil, as well as Latin America in general, as being particularly suited to Plowshare techniques.

## *Australia*

The Australians had been among those who had responded most favourably to the idea of nuclear dynamite. Gerald Johnson told the JCAE in 1964 that Australia was 'a notable example' of a foreign country interested in Plowshare.[26] With its vast, unpopulated but mineral-rich spaces, Australia seemed ideally placed to take advantage of 'geographical engineering'. In Edward Teller's view, if PNEs were 'natural' for the United States, they were 'even more natural for Australia'.[27]

While the Australian Atomic Energy Commission (AAEC) and the Department of National Development[28] had followed the progress of Plowshare from its inception, Australia's interest blossomed after July 1963 when Glenn Seaborg invited Australia to send a group of technical officers to the United States to study the program. The team, which visited the United States from September to October 1963, comprised officials of the AAEC, the Snowy Mountains Hydroelectric Authority and the Bureau of Mineral Resources. Its leader was Dr Alan Wilson, a physicist with the AAEC and at that time a Plowshare enthusiast, although not an uncritical one. Despite the practical problems with PNEs that Wilson identified, he believed that:[29]

> A rising world population and the rapid depletion of the easily won reserves of some resources provide compelling reasons for pressing ahead with the development of a viable PNE technology.

The Australian government was apparently less sanguine, specifying that Wilson's team limit itself to 'seeking a first hand appreciation of the scientific, engineering, and safety aspects of the use of nuclear explosives without any suggestion that Australia's interest would be extended further'.[30] The government was apparently conscious of the implications of PNE technology for the PTBT, which Australia had just signed. While the mission was expected to report on the status and usefulness of PNE techniques, 'having regard to general Australian conditions', no attempt was to be made to study or evaluate particular projects where nuclear explosives might have some application.

The report of the mission, issued in March 1964, concluded that Australia was 'well placed to exploit applications of nuclear explosions':[31]

> In many areas where major works might be contemplated the population density is extremely low and safety considerations would not rule out the use of nuclear explosives or render projects uneconomic because of the cost of damage. Further, apart from the manufacture of the necessary nuclear devices, Australia has the resources of technologists, the engineering capacity, and the financial and management skills needed to carry out major projects involving the use of nuclear explosives.

Not only was Australia ready for PNEs, but the mission was also confident that certain applications of PNEs were ready for Australia:[32]

> ... in the range of less sophisticated engineering applications where the final work is basically 'rough finished', such as the removal of navigation hazards and construction of breakwaters, nuclear methods could probably be successfully employed on some projects even on the basis of current technology.

Despite the fact that the United States itself had not yet employed PNEs in such an application and that its only planned excavation experiment in the field, Project Chariot, had been cancelled the year before, the mission concluded that 'harbour construction appears to be one of the most promising applications for nuclear excavation'.[33]

While optimistic about the potential uses of PNEs in Australia, the mission found that all was not well with Plowshare. It discovered, for instance, that the cost of nuclear excavation projects was 'subject to major uncertainties' due to the unpredictability of excavation explosions, which was in turn due to lack of experience.[34] The mission was sceptical about Plowshare's ability to resolve such difficulties without using nuclear dynamite in the field, a learn-as-you-go approach reminiscent of Project Chariot.[35]

Finally, the Australians could not help remarking on the inter-agency wrangle that had erupted in the US Administration over the PTBT. They noted that, while official statements suggested acceptance of the 'practical' interpretation of the treaty, the postponement of Plowshare's excavation program suggested that the US government was 'unwilling to risk committing itself on this interpretation at present'.[36] The mission refrained from saying how this would affect Australia's use of nuclear dynamite:[37]

> Any special limitations which may result from interpretation placed on the Nuclear Test Ban Treaty [sic] are not considered... in this report.

The Americans took great interest in the report, it being the first relatively objective assessment made of Plowshare and the first by a foreign country. The US Atomic Energy Commission even reprinted it, with a foreword describing it as a 'milestone' in the history of Plowshare. In retrospect Alan Wilson sees the report as representing the best and worst of Australian science and engineering — careful, honest and technically sound but lacking in vision.[38] Indeed it was to be another five years before Australia would seriously consider a PNE project.

Only one specific proposal appears to have been made during the early 1960s. Two oil companies, Ex–Oil and Magellan Oil, expressed interest in using PNEs to increase gas flow in the Amadeus Basin in central Australia.[39] Certainly there was no serious consideration given to Australia developing its own PNEs. Only one person connected with an Australian government seems to have advocated such a move. A renegade member of the governing Liberal Party, Mr St John, told the Australian Parliament in October 1968 that:[40]

> With the development of nuclear power stations we shall be in a position in future to develop the use of nuclear explosives in Australia for peaceful purposes without having to rely on the United States or any other country to provide us with the necessary supply of nuclear explosives.

Mr St John's statement, while totally ignored in Australia, was a reminder that the international promotion of Plowshare, preoccupied as it was with securing the program's future, had failed to convey any hint of the arms control implications of nuclear dynamite. Indeed it seemed almost deliberately designed to obfuscate the message that the development and use of PNEs by a non-nuclear weapon state would amount to the proliferation of nuclear weapons. The promotion of Plowshare, by exciting keen interest in an unproven technology, sowed the seeds of a long-running international dispute between the nuclear 'haves' and the 'have-nots'. This controversy, over the arms control implications of PNEs and the right of non-nuclear weapon states to acquire the full range of nuclear technology, including PNEs, first erupted during negotiations on a nuclear weapon-free zone in Latin America.

## *Argentina, Brazil and the Treaty of Tlatelolco*

The states of Latin America were a particular target of Plowshare publicity, especially as one of the major motivations behind the program was the construction of a new transisthmian canal through either Panama, Nicaragua or Colombia. Plowshare enthusiasts had long since identified South America, with its vast, unexplored and unpopulated interior, as being, like Australia, particularly suited to the use of nuclear dynamite. Ralph Sanders, for one, proposed using the Inter-American Atomic Energy Commission to develop a co-ordinated plan for training and research into Plowshare operations throughout the region.[41]

Enthusiasm for Plowshare in Latin America was stimulated by the AEC through international conferences, touring exhibitions and its links with atomic energy commissions in several countries. Also influential was the Puerto Rico Nuclear Centre, founded in 1957 as part of the US Atoms for Peace Program, primarily to train Latin American students in nuclear technology. According to its director, Henry J. Gomberg, the Centre saw Plowshare as 'a possibly inexpensive source of very large amounts of energy to be used for engineering works of a fairly familiar nature, such as digging holes and trenches'.[42] It was the Centre's hope, he said, 'to participate in the training and the first applications suitable to conditions in developing world areas in general and Latin America in particular'.[43]

Of all the Latin American states it was Brazil which seized upon the idea of PNEs with the greatest enthusiasm. Influential Brazilians became convinced that nuclear dynamite would be important for the country's economic development. One of Brazil's most distinguished lawyers, Clovis Ramalhete, argued, for instance, that PNEs could be used for extracting oil from shale, for linking the Plata, Orinoco and Amazon Rivers, and for the construction of ports and canals.[44] Plowshare was clearly responsible for this situation. Glenn Seaborg notes that the Brazilians 'apparently interpreted an over-enthusiastic Plowshare booklet on our own Plowshare program' as indicating that PNEs had been developed to the point where they were ready for use in actual applications.[45] Seaborg should not have been surprised. As head of the AEC he helped instigate, participated in, and was ultimately responsible for the Commission's over-zealous international promotion of Plowshare.

Unlike the Australians however, the Brazilians concluded that they should develop and use their own 'peaceful' nuclear explosives. One reason for this was undoubtedly their long-standing belief that Brazil was destined to become a world power. Hence, refraining from the acquisition and use of nuclear dynamite made no more sense than not producing and using conventional dynamite. Both were seen as essential for Brazilian economic development.[46] Even if Brazil could not yet produce the necessary nuclear devices, an announced intention to keep its PNE option open was perceived to be a demonstration of Brazilian independence.[47] As ACDA put it:[48]

> ... Brazil feels that with her size... she should be able to develop a peaceful nuclear capability. That means explosive devices... We have no indication that they are really able to do so but they want to maintain that option...

A further motive was fear of Argentina, its old rival for influence in Latin America. Argentina's nuclear industry was older and more advanced than Brazil's. Behind Brazil's fears was not so much Argentine interest in PNEs, as its interest in nuclear weapons. Evidence of such an interest surfaced in 1951 when Argentine dictator Juan Peron erroneously announced that his scientists had succeeded in producing the world's first controlled thermonuclear reaction. Although Peron said that Argentina needed nuclear power for economic development, he ominously added that 'foreign technicians should know that in the course of the experiments the problems of the hydrogen bomb were studied'.[49] By keeping its PNE option open, Brazil could provide itself with some insurance against the nightmare of an Argentine bomb.

Prior to 1964, Brazilian governments were content to stress arms control as a solution to nuclear rivalries in the region. Brazil initiated the idea of a Latin American nuclear weapon-free zone in 1962, only days after the Cuban missile crisis. Brazil's president joined those of Bolivia, Chile, Ecuador and Mexico in declaring, on 29 April 1963, that their governments were willing to sign a multilateral Latin American agreement which would commit them not to 'manufacture, store or test nuclear weapons or devices for launching nuclear weapons'.[50] There was notably no reference to PNEs. In December 1964 Brazil ratified the PTBT.

Brazil's posture changed dramatically after the 1964 military takeover. Its foreign policy became 'that of a nationalistic barracks officer, quite ready to defy the United States' lead on any question affecting Brazilian interests'.[51] To Brazil's existing motivations for pursuing the PNE option, the military would add intense nationalism, anti-Americanism, and a barely concealed desire for the military power and prestige represented by nuclear weapons and all forms of nuclear technology. For the Brazilian military, attempts by the United States and other nuclear weapon states to prevent Brazil from acquiring nuclear explosives were simply attempts to preserve the international *status quo*.

Preliminary talks on a Latin American nuclear weapon-free zone treaty, to be known as the Treaty of Tlatelolco after the suburb of Mexico City in which the talks were held, began in early 1965. Initially there was no mention of PNEs by any participant, including Brazil. By March 1965 however, determined to obstruct rather than facilitate the establishment of a zone, Brazil was insisting that the treaty permit the use of PNEs in Latin America.[52] There was still, however, no proposal that the treaty allow Latin American countries to develop and use their own PNEs. According to Glenn Seaborg this only occurred in May 1966 when tentative agreement was reached on a definition of a nuclear weapon that included the idea of intent: a nuclear explosive was to be considered a weapon only if it was 'intended to be used for a military purpose'.[53]

The United Kingdom, which was observing the negotiations, promptly protested that a PNE was in effect a weapon and that at least in terms of existing technology there could be no possible distinction. The treaty was subsequently modified to incorporate this idea and to remove the notion of 'intent'. The United States agreed with the British. It wrote to the Preparatory Commission urging that the treaty parties not be permitted access to any kind of nuclear device. To do so would, in the US view, be 'inconsistent with the whole scheme of keeping knowledge of nuclear weapons design, or nuclear explosive design, from them'.[54] To avoid exacerbating differences over the issue, the US postponed the detonation of Plowshare's Cabriolet shot for the duration of the Tlatelolco negotiations. US policy had in this instance clearly come down on the side of arms control.

The AEC, however, influenced US policy towards the new treaty in at least one respect. As Seaborg says, 'with an eye to Plowshare's future' the AEC convinced the Administration that the treaty, while banning all nuclear explosions in Latin America, should make an exception for PNE 'services' obtained from a nuclear weapon state.[55] Here was another case of the AEC having as its first concern the survival of Plowshare, rather than nuclear proliferation in Latin America, or with the bad example that Plowshare might set for ambitious Third World nations. With the possibility of enshrining the idea in an international treaty, Seaborg's vision of co-operative Plowshare ventures now became one in which the United States would provide a 'service' for other countries on a bilateral basis. This would have many benefits for Plowshare. It would help offset costs, assuage the JCAE, which had been clamouring for a 'commercial' PNE service, and boost Plowshare's domestic profile and support.

While it is not clear whether the US offer of PNE services had any causal effect, Brazil thereafter demanded that the treaty permit Latin American states to develop and use their own PNEs. While Brazil's principal motivation was undoubtedly a desire to keep its nuclear weapon option open *vis-à-vis* Argentina, it was also resentful that the United States, after vigorously promoting Plowshare technology for a number of years through its AEC, was now warning of the proliferation dangers of PNEs. Indeed, while the AEC had promoted and would continue to promote Plowshare with little concern for the arms control implications, the US State Department and ACDA were increasingly determined to pursue nonproliferation objectives at the expense of the promotion of PNEs. Brazil, like Egypt, was to some extent a victim of this schizophrenic US policy, born of divided Administrations and disinterested Presidents.

Brazil was also probably motivated partly by a fear, to some extent justified, that the United States would set 'humiliating' conditions and charge excessive payments for any PNE services it provided.[56] As Glenn Seaborg discovered during his visit in 1967, the Brazilians did not seem to understand that the US did not intend to include in the prices charged for its PNE services any part of the research and development costs connected with its nuclear weapon program.[57] Hence the service could be provided at a fraction of what it would cost Brazil to manufacture PNEs itself.

Brazil's demand that it be allowed to produce its own PNEs vastly complicated the negotiation of the Treaty of Tlatelolco. According to David Fischer, former IAEA Director of External Affairs, such a demand was difficult to counter at a time when 'the benefits of PNEs were being trumpeted from the rooftops'.[58] According to Glenn Seaborg, 'the right of nations to carry out their own peaceful nuclear explosions, and to develop their own explosives for that purpose', became the 'principal bone of contention in the negotiations...'[59] Apart from Brazil, Panama and Nicaragua also sought assurances that the treaty would not interfere with the possible construction of a new inter-oceanic canal using nuclear dynamite.[60] Argentina, suspicious of Brazil's motives,[61] soon echoed Brazil's demands, presumably on the grounds that if it could not beat them, then for reasons of continental prestige and power, it would join them.[62]

In July 1967 Glenn Seaborg visited Brazil, adding fuel to the PNE controversy. Seaborg was reported as stating that US uranium enrichment services to Brazil would continue in the event that it pursued its plan to develop PNEs.[63] Apparently this report arose from Seaborg's attempt to rectify a previous Brazilian press report that misquoted him as advocating the opposite — that the United States should *halt* all nuclear assistance to Brazil if it attempted to develop nuclear devices. Seaborg attributes the earlier story to the unwillingness of certain Brazilian officials to let the Brazilian public hear the US position on nonproliferation.[64]

Later Seaborg met a large group of Brazilian military leaders for a 'frank and full discussion of the potential of nuclear explosives for both military and civilian use on the South American continent'.[65] The AEC had reviewed possible PNE applications in Brazil and focused quickly on oil shale, since Brazil had large oil shale resources but few other oil reserves. Seaborg offered to include one or two Brazilians in the AEC/Bureau of Mines joint study into this application of nuclear dynamite, which was then in its early stages. The reaction he got could best be described, he says, as 'a polite brush-off'. Seaborg claims this was 'clear evidence that the Brazilians' avowed interest in peaceful nuclear explosions was mainly a cover to keep alive a nuclear weapons option'.[66]

On 13 September 1967 Seaborg participated in a meeting in Washington with Brazilian Foreign Minister, Sergio Correa da Costa, at which the subject of PNEs was raised. Seaborg recalls:[67]

> I mentioned that the United States could help Brazil with peaceful nuclear explosions under proper circumstances. Da Costa mentioned a specific excavation project — the construction of a 5 to 7 kilometre canal linking the Amazon River basin with the River Platte basin. It was his hope that the US would provide assistance in performing the necessary feasibility studies on this project. He did not mention our providing the explosion service, however, which was what we had in mind. As I had done when I visited Brazil two months earlier, I explored in some detail Brazil's expressed interest in building its own nuclear explosives for peaceful purposes. I pointed out that it would be an enormous economic burden for Brazil to take this path, whereas the United States could furnish the service through an appropriate international organization at our direct cost, without any charge for research and development. I emphasized that we had been working on this technology for at least twenty years [sic] and still had at least five years to go before we would have the proper kind of explosives. This was by way of indicating what Brazil would have to face if it tried to go it alone. Da Costa nevertheless clung to the position that Brazil wanted to build its own nuclear explosives for peaceful purposes 'as a matter of principle'.

It was only through determined Mexican opposition, with American and British encouragement, that Brazil and Argentina failed to have the Treaty of Tlatelolco openly permit PNEs in Latin America.[68] While Mexico's position was at least partly attributable to its fear of losing Anglo-American support for the treaty, it must nonetheless be given credit for being remarkably strong in opposing Brazil and Argentina. Fortunately Mexico was supported by all the remaining Latin American states except Nicaragua. According to Alfonso Garcia Robles, former Mexican Foreign Minister and Chairman of the treaty's Preparatory Commission:[69]

> ... when drafting the provisions which would later be included in Article 18 dealing with nuclear explosions for peaceful purposes — which although they still are at an experimental stage, could have future importance — special care was exercised to avoid [legalising] any attempts to test or manufacture nuclear weapons under the pretext of carrying out such explosions for peaceful purposes, attempts which would completely negate the fundamental purpose involved, the very *raison d'être* of the treaty.

Unfortunately this attempt was not entirely successful. The treaty's PNE provisions are open to two incompatible interpretations. As Seaborg notes, 'the final treaty contained so many compromises as to seem self-contradictory'.[70] While Article 1 bans only the acquisition and testing of 'nuclear weapons' and not 'other nuclear devices' as in the PTBT, Article 5 defines a nuclear weapon in a way that appears to include PNEs in the ban:[71]

> ... a nuclear weapon is any device which is capable of releasing nuclear energy in an uncontrolled manner and which has a group of characteristics that are appropriate for use for warlike purposes.

Article 18 (see Annex III) then creates confusion by specifically permitting treaty parties to carry out 'explosions of nuclear devices for peaceful purposes — including explosions which involve devices similar to those used in nuclear weapons... provided they do so in accordance with.. articles 1 and 5'.[72] Somewhat naïvely Garcia Robles claims that an 'objective analysis' of Article 18, read in conjuction with Articles 1 and 5, 'would not seem to lend itself to conflicting interpretations'.[73] According to him the treaty permits a state party to carry out a PNE itself only if it is able to 'prove previously that a nuclear weapon will not be required for that explosion'. Since, according to Garcia Robles 'the consensus of the experts' is that this is impossible to prove, then 'it must obviously be concluded that the states parties to the treaty will not be able to manufacture or acquire nuclear explosive devices, even though they may be intended for peaceful purposes...'[74]

Argentina, Brazil and Nicaragua disagree. All have put on record their own interpretations, the first two when signing the treaty, Nicaragua when ratifying. The attitudes of Argentina and Brazil are identical. Brazil stated on signing the treaty on 14 February 1967 that:[75]

> It is the understanding of the Brazilian Government that said Article 18 permits the States signatories [sic[76]] to carry out, by themselves or in association with third parties, nuclear explosions for peaceful purposes, including those which presuppose devices similar to those used for military armaments.

Argentina did not sign the Treaty of Tlatelolco until 27 September 1967, five months after Brazil. Its statement at the time of its signature reads as follows:[77]

> The Government of the Argentine Republic, upon signing the Treaty for the Prohibition of Nuclear Weapons in Latin America... wishes to express its satisfaction at the inclusion in that instrument of provisions which protect the peaceful development of nuclear energy and, among them, the provisions of Article 18 which recognize the right of the Contracting Parties to carry out explosions of nuclear devices for peaceful purposes — including explosions which involve devices similar to those used in nuclear weapons — or to collaborate with third parties for the same purpose.

It is likely that Brazil's statement was influential in inducing Argentina to make such a statement. Although Argentina's position on PNEs is in substance the same as Brazil's, Brazil has traditionally been more vocal about its perceived rights. As a member of all the multilateral disarmament bodies since the ENDC was established in 1962, Brazil has also had more opportunity. As George Quester has noted:[78]

> Argentina's diplomatic style has indeed been much less eloquent and explicit than that of its neighbour; Argentina has been more cautious and reserved...

As to Nicaragua, it obviously had an eye to the possibility of a new Panama Canal through its territory when it declared on ratifying the Treaty that it reserved its:[79]

> ... sovereign right to employ, according to its own judgement, nuclear energy for peaceful purposes, such as for the removal of great quantities of earth for the construction of inter-oceanic or any other kind of canals, irrigation works, electric plants etc...

If the interpretation of the treaty advocated by Argentina, Brazil and Nicaragua is accepted, the Treaty of Tlatelolco nonetheless does contain certain 'safeguards' against the acquisition of military benefits from the use of PNEs. Article 18 requires detailed information to be provided to the verification agency for the treaty, OPANAL,[80] and the IAEA, namely:[81]

(a) the nature of the nuclear device and the source from which it was obtained
(b) the nature and purpose of the planned explosion
(c) the procedures which will be employed to comply with international observation
(d) the expected force of the device, and
(e) the fullest possible information on any possible radioactive fallout and measures taken to avoid danger to the population, flora, fauna and territories of any other treaty party.

Article 18 also specifies the type of observation that OPANAL and the IAEA can carry out at the PNE detonation site. This includes all preparations for the detonation and the detonation itself. The two agencies would also have 'unrestricted access to any area in the vicinity of the site of the explosion in order to ascertain whether the device and the procedures followed during the explosion are in conformity with the information supplied . . .'[82]

These so-called safeguards are clearly inadequate. In the first place they would fail to prevent a treaty party obtaining knowledge of nuclear weapons through the development of its own PNE device prior to any detonation being carried out. The only requirement is that OPANAL and the IAEA be notified of the source of the nuclear device involved in each PNE application. Second, even if the device is obtained from a source outside Latin America, the safeguards envisaged would clearly be unsuccessful in preventing military information being obtained from a PNE detonation. Article 18 says nothing about who is to carry out the explosion, or how closely the Latin American state concerned can be involved in preparations for the detonation, or what types of diagnostic equipment are permitted or prohibited. Moreover, the information required to be submitted to OPANAL and the IAEA would be practically useless for determining whether the underlying purpose of a PNE explosion was the development of a nuclear weapon or the carrying out of some genuine 'peaceful' project.

International uneasiness about the PNE provisions of the Treaty of Tlatelolco was demonstrated when Mexico tabled the text of the treaty at the 1967 United Nations General Assembly. Those expressing concern included Canada, Pakistan, Poland and the Soviet Union. It was partly because of the PNE provisions that the Soviets abstained in the vote on the treaty.[83] The United States, the United Kingdom and the Soviet Union all found the ambiguities regarding PNEs sufficiently worrying to oblige them to make clear their own interpretations when signing and/or ratifying the Additional Protocols to the Treaty.[84] All made it clear that they interpreted the treaty as prohibiting the acquisition and use of PNEs by the regional treaty parties.

The United Kingdom, while declaring that in its view the treaty prohibits PNEs, echoed Mexico in adding the qualification, 'unless and until advances in technology have made possible the development of devices for such explosions which are not capable of being used for weapons purposes'.[85] The United States said it wished to point out again:[86]

> . . . the fact that the technology of making nuclear explosive devices for peaceful purposes is indistinguishable from the technology of making nuclear weapons and the fact that nuclear weapons and nuclear explosive devices for peaceful purposes are both capable of releasing nuclear energy in an uncontrolled manner and have the common group of characteristics of large amounts of energy generated instantaneously from a compact source.

The United States cited the negotiating history of the treaty and 'the prevailing view of the parties' in support of its position.[87] It also reiterated its offer to make available nuclear explosive 'services' for peaceful purposes, on a non-discriminatory basis under 'appropriate international arrangements'. The PNE devices involved would of course remain under strict US control.[88] This reassurance would not have comforted Argentina and Brazil. It was the very sort of paternalism in nuclear matters they abhorred.

The Soviet Union was blunter. It declared, upon signing Protocol II on 18 May 1978, that the carrying out of PNEs by any treaty party 'would constitute a violation of its obligations as defined in Article 1 and would be incompatible with its non-nuclear status'.[89] The Soviets lamely pointed to Article V of the Nonproliferation Treaty (NPT) as providing a 'solution to the PNE question'. By this time the NPT had been in existence for ten years and no implementation of its PNE provisions was in prospect.[90]

In view of the controversy about its provisions, does the Treaty of Tlatelolco have any restraining effect on the possible development and use of PNEs by the states of Latin America? The first point to be noted is that the treaty is not yet in force for Latin America as a whole, but only for those states which, having ratified the treaty, have then deposited a declaration of waiver which brings it into force for them. Brazil and Chile have not done this, while Argentina has signed but not ratified the treaty. However almost all other Latin American states, including three that have been involved in seriously considering a PNE project — Nicaragua, Panama and Colombia, which were involved in considering the construction of a new Panama Canal — are full parties. Venezuela, which along with Colombia at one stage gave fleeting consideration to a nuclear-blasted canal for connecting the Orinoco and Rio Negro Rivers, is also a full party. In addition, most full parties have concluded safeguards agreements with the IAEA, as required by the treaty, which forbid the use of safeguarded nuclear material in any kind of nuclear explosion.[91] Hence, at least as far as their safeguarded nuclear material and facilities are concerned, these parties have had any ambiguities about PNEs in the treaty itself removed for them. In addition the Secretary-General of OPANAL declared in 1974 that in terms of the Treaty of Tlatelolco, India's so-called 'peaceful nuclear explosion' was a nuclear weapon test.[92]

Meanwhile the pro-PNE positions of Brazil and Argentina have continued into the 1980s. At the Conference on Disarmament in Geneva, the UN General Assembly and other international forums, both countries continue to uphold their right to develop and use PNEs. While claiming that Tlatelolco does not ban PNEs, they remain unwilling to test their interpretation by becoming parties to the treaty. Both countries have also attempted, not always successfully, to avoid other legal restrictions on their right to conduct PNEs. The 1972 US/Brazil nuclear co-operation agreement, under which the United States supplies enriched uranium to the Angra dos Reis power plants, in theory permits Brazil to use nuclear material for the fabrication of PNEs.[93] By contrast, Brazil's 1975 nuclear technology agreement with the Federal Republic of Germany contains a solemn commitment by Brazil not to use the nuclear equipment, installations, materials and technological information supplied by the FRG for the production of nuclear weapons or 'other nuclear explosives'.[94]

As late as 1983, after Argentina announced that it had succeeded in producing enriched uranium, Brazil's military leaders openly proclaimed Brazilian ambitions in the PNE field. In an interview on 10 December 1983 the chief of the Brazilian Armed Forces General Staff, Waldyr Vasconcellos, stated that Brazil 'is preparing to master nuclear technology completely' and 'will be able to produce armaments, including the atomic bomb', but that the goal of the effort 'is for peaceful purposes'.[95] In 1986, former Brazilian Navy Minister and head of the Brazilian state oil company, Admiral Fonseca Maximiano, said he favoured a nuclear test explosion for political reasons: 'My personal opinion is that Brazil should explode a device to show we have the capacity. Let's show we're not bluffing'.[96] Also in 1986 a local Brazilian newspaper published photographs of a remote military base in the Amazon state of Para, where it said an underground test site for nuclear weapons was finished in 1984 or earlier.[97] The armed forces issued a cryptic denial, saying the test site was for trials of 'equipment and materials for development of aerospace capability in the interests of national security'.[98]

As for Argentina, the PNE issue became a 'complete sticking point', according to David Fischer, in several years of desultory negotiations between the Argentinians and the IAEA on a so-called 'full-scope' nuclear safeguards agreement.[99] Argentina flatly rejected as models the IAEA agreements with Colombia and Panama under Tlatelolco, maintaining that they took little account of the 'singularity' of the treaty, a clear reference to PNEs.[100] It produced a draft text of its own, unacceptable to the IAEA, which would have permitted the withdrawal of safeguarded material for use in PNEs. The draft also provided for the unsatisfactory type of observation of PNEs described in Article 18 of the Tlatelolco Treaty.[101] Fischer says that in an effort to help the Argentinians save face, an agreement was offered involving a broad 'no military uses' pledge and a protocol which would spell out that this meant no PNEs. The draft agreement, which was intended to last for twenty-five years — the same duration as the 1968 Nuclear Nonproliferation Treaty, which Argentina also has refused to sign — was rejected by Buenos Aires. Argentina made clear it would neither ratify the Treaty of Tlatelolco nor bring it into force for Argentina until a safeguards agreement was settled.

In 1981 the then head of the Argentine nuclear program, Vice-Admiral Castro Madero, stated that Argentina had no intention of developing PNEs and that the development of nuclear weapons by any Latin American nation would serve no purpose and only trigger a wasteful nuclear arms race.[102] Shortly after retiring he confirmed, in December 1983, that there had never been any concrete applications envisaged for PNEs in Argentina and that there never would be.[103] Argentina was simply attempting to keep its PNE option open as insurance against a Brazilian nuclear explosive capability.

With the advent of the new democratic government of Raul Alfonsin, hopes were high that Argentina would change its position. However, in a letter to Senator Charles Percy, Chairman of the US Senate Foreign

Relations Committee, in early 1984, Foreign Minister Caputo echoed the view of previous Argentinian governments. He stated that before Argentina ratified the Treaty of Tlatelolco there was a 'need to clarify Argentina's safeguards agreements with the IAEA' to reflect the 'particularities' of the treaty, again a reference to PNEs.[104] Opposition from Peronist and other right-wing parties appears to be the principal factor, according to nuclear proliferation expert Leonard Spector, in restraining the Alfonsin government from ratifying Tlatelolco, even though Alfonsin himself is said to favour ratification.[105] In 1986 Argentina ratified the Partial Test Ban Treaty, twenty-two years after Brazilian ratification. The Argentine Deputy Foreign Minister declared that, 'no-one in Argentina is thinking to develop PNEs. [Their proposed uses are] not convincing enough to make them'.[106] In July 1986 Argentina signed a nuclear co-operation agreement with the new democratically-elected Sarney Government in Brazil. At the same time Brazil announced it was scaling down its plans for reprocessing spent fuel to obtain weapons-grade plutonium, due to 'high cost and lack of need in the medium term'.[107] None of these developments, however, provides any guarantee that either Argentina or Brazil have given up the PNE option.

The Brazilian journal *O Estado de Sao Paulo* on 22 November 1983 summed up the dilemma of Argentina and Brazil in regard to their mutual assertion of the right to develop and use PNEs, when it declared that:[108]

> Since the purpose of the use (peaceful or military) will not be known until the moment of the use of the explosive, it is valid to suppose that the two governments are conscientiously pursuing the objective of mastering the technology that will permit them access to the bomb.

The tragedy is that the pro-PNE positions of Brazil and Argentina have had effects far beyond their own region, disrupting international efforts to close off the PNE option by means of a nonproliferation treaty, safeguards agreements and a CTB, and raising the level of interest in so-called peaceful nuclear explosions well above what it would otherwise have been.

# Notes

1 US Congress, JCAE, Hearings on Frontiers in Atomic Energy Research, 22–26 March 1960, US Government Printing Office, Washington DC, 1960, p. 53.

2 *PNE Activity Projections for Arms Control Planning* (the *GURC Report*), Gulf Universities Research Consortium, Galveston, Texas, US Arms Control and Disarmament Agency Document, ACDA/PAB-253, Washington DC, 1975, p. 4.

3 Ibid.

4 P. Kruger, 'A Survey of University Courses in Nuclear Civil Engineering', in L.E. Weaver (ed.), *Education for Peaceful Uses of Nuclear Explosives*, University of Arizona Press, Tucson, Arizona, 1970, p. 239.
5 F.J. Clark Jr., 'Plowshare Program Development Activities' in Weaver (ed.), p. 235.
6 Ibid., p. 236.
7 G.T. Seaborg *et al.*, *Plowshare*, Division of Technical Information, US Atomic Energy Commission, Oak Ridge, Tennessee, 1966, p. 36.
8 M. Kalkstein, *International Arrangements for the Peaceful Application of Nuclear Explosives*, Stockholm International Peace Research Institute, Stockholm, 1970, p. 9.
9 J.T. Arnett, *Industry and Nuclear Explosives: Facts About An Emerging Technology and Its Immense Potential for Benefit*, Atomic Industrial Forum Inc., New York, July 1973, p. 2.
10 Clark, p. 236.
11 G.T. Seaborg and B.S. Loeb, *Stemming the Tide: Arms Control in the Johnson Years*, Lexington Books, Lexington, Mass., 1987, p. 321.
12 Clark, p. 232.
13 W. Epstein, *The Last Chance*, The Free Press, New York, 1976, p. 173.
14 Between 1958 and 1971 Egypt was in a confederation with Syria, known as the United Arab Republic. As Syria never evinced any interest in PNEs, the term 'Egypt' will be retained throughout for the sake of consistency.
15 J. Koop, 'Plowshare and the Nonproliferation Treaty', *Orbis*, XII, 3, Fall 1968, p. 806.
16 Seaborg and Loeb, *Stemming the Tide*, p. 321.
17 US Congress, JCAE, Hearings on AEC Authorizing Legislation for Fiscal Year 1966, US Government Printing Office, Washington DC, 1965, p. 13.
18 Seaborg and Loeb, *Stemming the Tide*, p. 321.
19 Ibid., pp. 321–22.
20 Ibid., p. 323.
21 From an interview with Dr Bob Mikulak, Arms Control and Disarmament Agency, Washington DC, 9 December 1986.
22 I.Y. Borg, *Nuclear Explosions for Peaceful Purposes*, Lawrence Livermore National Laboratory Paper No. UCRL-95026, Rev. II, November 1986, p. 10.
23 Seaborg and Loeb, *Stemming the Tide*, p. 325.
24 Ibid.
25 Ibid.
26 US Congress, JCAE, Hearings on AEC Authorizing Legislation for Fiscal Year 1965, US Government Printing Office, Washington DC, 1964, p. 1241.
27 T.C. Findlay, 'Twenty Minutes with Edward Teller', *Newsletter*, Peace Research Centre, Australian National University, Canberra, 2, 1, 1987, p. 2.
28 Dr Harold Raggart of that department was a particular enthusiast.
29 A.R.W. Wilson, 'Current Status of Civil Engineering and Mineral Resources Development Applications of Peaceful Nuclear Explosions' in Proceedings of the Fourth International Conference on the Peaceful Uses of Nuclear Energy, 6–16 September 1971, IAEA, New York and Vienna, 1972.
30 Australian Atomic Energy Commission, *Peaceful Uses of Nuclear Explosions — An Evaluation, for Australian Purposes, of Proposed Civil Engineering and Mining Applications*, AAEC, Sydney, 1964, p. 1.
31 Ibid., p. 169.
32 Ibid., p. 172.
33 Ibid., p. 153.
34 Ibid., p. 141.
35 Ibid., p. 219.
36 Ibid., p. 5.
37 Ibid., p. 6.
38 From an interview with Alan Wilson, Sydney, 7 August 1987.
39 B.A. Bolt, *Nuclear Explosions and Earthquakes: The Parted Veil*, W.H. Freeman and Co., San Francisco, 1976, p. 189.
40 Australian Parliamentary Hansard, 15 October 1968, p. 1974.
41 R. Sanders, *Project Plowshare*, Public Affairs Press, Washington DC, 1962, p. 191.
42 Weaver (ed.), p. 278.
43 Ibid.

44 H.J. Rosenbaum, 'Brazil's Nuclear Aspirations' in O. Marwah and A. Shulz (eds), *Nuclear Proliferation and the Near-Nuclear Countries*, Ballinger Publishing Co., Cambridge, Mass., 1975, p. 263.
45 Seaborg and Loeb, *Stemming the Tide*, p. 258.
46 Rosenbaum, p. 269.
47 Ibid.
48 US Congress, Senate Foreign Relations Committee, Hearings on the Treaty of Tlatelolco, 22 September 1970 and 23 February 1971, US Government Printing Office, Washington DC, 1971, p. 15.
49 P. Pringle and J. Spigelman, *The Nuclear Barons*, Avon Books, New York, 1981, p. 200.
50 W.H. Courtney, 'Brazil and Argentina: Nuclear Choices for Friendly Rivals' in J.A. Yager (ed.), *Nonproliferation and US Foreign Policy*, Brookings Institution Press, Washington DC, 1980, p. 255.
51 G.H. Quester, *The Politics of Nuclear Proliferation*, Johns Hopkins University Press, Baltimore, Maryland, 1973, p. 143.
52 L.S. Spector, *Nuclear Proliferation Today*, Vintage Books, New York, 1984, p. 238; Seaborg and Loeb, *Stemming the Tide*, p. 342.
53 Seaborg and Loeb, *Stemming the Tide*, p. 342.
54 US Congress, Senate Foreign Relations Committee, Hearings on the Treaty of Tlatelolco, p. 12.
55 Seaborg and Loeb, *Stemming the Tide*, p. 342.
56 Quester, p. 145.
57 Seaborg and Loeb, *Stemming the Tide*, p. 258.
58 D. Fischer, 'The Role of the IAEA' in J. Simpson and A.G. McGrew (eds), *The International Nuclear Non-Proliferation System*, Macmillan, London, 1984, p. 66.
59 Seaborg and Loeb, *Stemming the Tide*, p. 341.
60 Ibid., p. 342.
61 I. Smart, 'Nonproliferation Treaty: Status and Prospects' in A.W. Marks (ed.), *NPT: Paradoxes and Problems*, Arms Control Association, Washington DC, 1975, p. 24.
62 Spector, p. 202.
63 Rosenbaum, p. 265.
64 Seaborg and Loeb, *Stemming the Tide*, pp. 258–59.
65 Ibid., p. 259.
66 Ibid.
67 Ibid., p. 343.
68 Courtney, p. 225; Seaborg and Loeb, *Stemming the Tide*, p. 342.
69 A. Garcia Robles, 'The Tlatelolco Treaty', in *SIPRI Yearbook 1969–70*, Taylor and Francis, London, 1970, p. 227.
70 Seaborg and Loeb, *Stemming the Tide*, p. 334.
71 *Status of Multilateral Arms Regulation and Disarmament Agreements*, United Nations, Second Edition, UN Department for Disarmament Affairs, New York, 1983, p. 49.
72 Ibid., p. 55.
73 Garcia Robles, p. 227.
74 Ibid.
75 *Status of Multilateral Arms Regulation and Disarmament Agreements*, p. 63.
76 A treaty is not legally binding on signatories but only on states which have ratified the treaty.
77 *Status of Multilateral Arms Regulation and Disarmament Agreements*, p. 63.
78 Quester, p. 148.
79 *Status of Multilateral Arms Regulation and Disarmament Agreements*, p. 64.
80 OPANAL is the Spanish acronym for the Agency for the Prohibition of Nuclear Weapons in Latin America.
81 *Status of Multilateral Arms Regulation and Disarmament Agreements*, p. 55.
82 Ibid., p. 55–56.
83 *The United Nations and Disarmament 1945–1970*, Department of Political and Security Affairs, United Nations, New York, 1970, p. 344.
84 Additional Protocol I applies the treaty's provisions to territories in the zone for which external powers have responsibility. The Netherlands, which at that stage had two colonies within the zone, specifically declared upon signing Protocol I that the Treaty's PNE provisions would apply to Surinam and the Netherlands Antilles (see *Status of Multilateral*

*Arms Regulation and Disarmament Agreements*, p. 65). Protocol II commits the nuclear weapon states not to use nuclear weapons against states in the zone.

85 *Status of Multilateral Arms Regulation and Disarmament Agreements*, p. 65.

86 Ibid., p. 69.

87 US Congress, Senate Foreign Relations Committee, Hearings on the Treaty of Tlatelolco, p. 25.

88 By this time the United States had made the same offer to all non-nuclear weapon states in the negotiations in the ENDC on a nonproliferation treaty.

89 *Status of Multilateral Arms Regulation and Disarmament Agreements*, p. 68.

90 See Chapters 7 and 14.

91 D. Fischer and P. Szasz, *Safeguarding the Atom*, Stockholm International Peace Research Institute, Stockholm, 1985, p. 77.

92 See Chapter 12.

93 Rosenbaum, p. 265.

94 Ibid., p. 258.

95 Spector, p. 259.

96 *Washington Post*, 28 August 1986, p. E-1.

97 Ibid.

98 Ibid.

99 From an interview with David Fischer, Canberra, 4 September 1986, confirmed by Spector, p. 208. A full-scope safeguards agreement is one that applies IAEA safeguards not just to nuclear material but to all nuclear installations within a particular country, whether such material and installations are imported or of local origin. Argentina has however negotiated a more limited form of safeguards agreement with the IAEA, relating to nuclear cooperation with Canada, which provides that none of the items subject to the agreement 'shall be used for the manufacture of any nuclear weapon or to further any other military purpose or for the manufacture of any other nuclear device'. (See 'Text of the Agreement of 22 July 1977 Between Argentina and the Agency for the Application of Safeguards in Connection with a Cooperation Agreement Between Argentina and Canada', INFCIRC/251, 25 November 1977.) Argentina strongly protested the introduction of this language on the grounds that it was foreign to the IAEA Statutes and a surrogate way of enforcing on the entire IAEA membership principles agreed to by only some of its members. (L. Scheinman, *The IAEA & World Nuclear Order*, Resources for the Future, Washington DC, 1987, pp. 244–5 and 264.)

100 Fischer and Szasz, p. 91.

101 Ibid., p. 98.

102 Spector, p. 211.

103 J.A. Aja Espil, 'Argentina' in J. Goldblat (ed.), *Non-proliferation: The Why and Wherefore*, SIPRI, Taylor and Francis, London, 1985, p. 78.

104 Spector, p. 226.

105 Ibid., p. 277.

106 P. Lomas, 'Attitudes of the Nuclear Threshold Countries' in J. Goldblat and D. Cox (eds), *Nuclear Weapon Tests: Prohibition or Limitation?*, SIPRI, Oxford University Press, Oxford, 1988, p. 314.

107 *Washington Post*, 28 August 1986, p. E-1.

108 Spector, p. 257.

# 7

# PNEs Go International: The Nonproliferation Treaty

*Beat your plowshares into swords, and your pruning hooks into spears: let the weak say, I am strong.*

Joel 3:10

AT the same time as the Latin Americans were debating PNEs in Mexico City, the rest of the world was beginning to address the issue in the negotiations on a nuclear nonproliferation treaty. These negotiations, conducted in the ENDC and the UN General Assembly from 1965 to 1968, would result in the third major international treaty to deal with nuclear dynamite, the Nuclear Nonproliferation Treaty. Initiated by Ireland in 1958, the idea of such a treaty was to prevent the spread of nuclear weapons to states not already possessing them. This would be achieved by preventing the diversion to explosive uses of nuclear materials used for peaceful purposes, such as generating electricity in nuclear power plants.

The original United States draft, submitted to the ENDC on 17 August 1965, ignored PNEs altogether, referring in Articles I and II simply to 'nuclear weapons'.[1] Non-nuclear weapon states would not manufacture or otherwise acquire nuclear weapons, while nuclear weapon states would not transfer them. By not specifically mentioning PNEs the draft appeared to states such as India and Brazil to leave open the possibility of non-nuclear weapon states developing their own nuclear dynamite. The Soviet Union's competing draft treaty, tabled in the United Nations General Assembly on 24 September 1965, was also silent on the PNE issue.[2]

The US draft was supposed to have dealt with PNEs in an article defining the term 'nuclear weapons'. It was, however, left blank because of disagreement within the US Administration over how to handle the PNE issue.[3] According to Glenn Seaborg:[4]

> Predictably, the disputants were... ACDA and the AEC and the issue was the survival of Plowshare.

ACDA wanted to define nuclear weapons as encompassing all nuclear explosives, whatever their purpose, thereby including both weapons and PNEs in the prohibitions outlined in Articles I and II. While willing to include 'other nuclear explosives' in Articles I and II, the AEC wanted a separate article that would permit nuclear weapon states to provide PNE services for other states under some sort of international arrangement. This was an extension of Seaborg's idea of Plowshare services being provided through bilateral agreements between the United States and any state requiring such services. Now there was to be an *international* service under NPT auspices.

Seaborg claims that the AEC's proposal resulted from complaints by several ENDC members, particularly Egypt, about the injustice of a treaty that would deny them the benefits of PNEs. According to Seaborg, 'it was part of [the] AEC's effort to remove every possible excuse for nations not to join the NPT'.[5] This is almost certainly an *ex post facto* rationalisation. No ENDC member, not even Brazil, had yet mentioned PNEs in the NPT context. It was to be the United States which would introduce the subject. Certainly Egypt did not yet know how PNEs would be treated under the NPT. The real reason for the AEC's proposal, as Seaborg acknowledges elsewhere, was that, 'we felt that the future of the Plowshare program rested on our ability to carry out such an international service'.[6] Again the AEC was not concerned about the NPT or about other countries' ability to obtain PNE services, but about the future of its beloved Plowshare. In addition, since the Soviet PNE program had only just begun, the AEC proposal would have given the United States a monopoly in the provision of PNE services to the non-Soviet world.

In this instance the AEC's view prevailed within the Administration. On 30 June 1966 an informal memorandum was sent to the Soviet ENDC delegation to ascertain its attitude towards providing an international PNE service under the NPT. In Seaborg's words, 'little came of this probe, but at least it did not provoke an openly hostile reaction'.[7] The United States thus went ahead and introduced the PNE proposal, and with it the PNE issue itself, into the ENDC discussions, a paradoxical move in view of its later efforts to dampen enthusiasm for PNEs.

Its statement of 9 August 1966 to the ENDC brusquely noted the 'inescapable technological fact' that 'the technology of making nuclear explosive devices for peaceful purposes is essentially indistinguishable from the technology of making nuclear weapons'.[8] Hence 'any nonproliferation treaty would be ineffective indeed if it permitted the development by non-nuclear weapon states of the nuclear explosive devices which are necessary for nuclear explosions for peaceful purposes'. The statement, delivered by Ambassador Fisher, pointed to American experience with Plowshare, noting (probably to the chagrin of the AEC) that, 'we still have several difficult technical problems to solve before we shall be able to demonstrate applications which are both technically sound and economically feasible'.[9] According to Fisher, the difficulties facing non-nuclear weapon states would be even greater, because the projects of most interest to them — 'nuclear earthmoving projects, such as digging canals or building dams' — were feasible only with highly sophisticated fusion devices. The development of such PNEs by a non-nuclear weapon state could not be justified economically and could only be viewed as an attempt to acquire nuclear weapons. In the United States' view therefore, the constraints imposed by a nonproliferation treaty on the spread of nuclear weapons technology should also apply to PNEs.

Then came the sweetener to this bitter medicine. The United States proposed that:[10]

> ... if and when peaceful applications of nuclear explosives that are permissible under test ban treaty limitations prove technically and economically feasible, nuclear weapon states should make available to other states nuclear explosives for peaceful applications.

Such a service would consist of 'performing the desired nuclear detonation under appropriate international observation, with the nuclear device remaining in the custody and under the control of the state which performs the service'.

The effect of the American statement, which was devoted entirely to PNEs, was to introduce the issue into the NPT negotiations as a major bone of contention. It also aroused interest in PNEs on the part of countries as diverse as Bolivia, Burma, Chile, Ethiopia, Italy, Nigeria, Malta, Thailand, Tanzania and South Africa, none of which had previously expressed any interest whatsoever. It has been noted of the NPT negotiations as a whole that they may have been 'a learning process in the opposite direction for those states most crucial for its success'.[11] Iris Borg has noted that one effect of the NPT was 'to encourage non-nuclear states such as Thailand and Egypt to consider the nuclear option in solving long-standing development problems and to expect co-operation from the nuclear nations'.[12] While it was probably inevitable that the PNE issue would arise at some stage, given that Brazil and India had erroneously interpreted the US draft treaty as permitting PNEs, the US initiative gave PNEs a high profile from the outset. This distorted the treaty, from being one that should have dealt only with nuclear *explosives* by banning their acquisition in any form by non-nuclear weapon states, to one that also dealt with one type of nuclear *explosion* — PNEs. It was only Sweden that consistently argued that the PNE issue should have been left for resolution by a treaty dealing exclusively with all nuclear explosions: a comprehensive test ban treaty.

The Soviet Union's reaction, in keeping with its previous lack of response to US overtures, was to largely ignore the American proposal. The Soviets simply lumped it in with others which, they charged, were intended to prevent agreement on the NPT. Initially the rest of the ENDC also ignored the US statement. Seaborg describes the immediate reaction as 'sparse and mixed',[13] which was as good an indication as any of the lack of clamour from the developing countries for the proposed PNE service. Ambassador Fisher had to appeal two weeks later to the committee to face up to the PNE problem and the 'implications it has for our work'.[14] Among the few countries to comment in detail was Canada. While describing itself as 'a country with large natural resources requiring development', Canada disclaimed any desire to use nuclear dynamite. Nevertheless it welcomed the US proposal, calling it a 'starting point' for development of a PNE 'regime'. The Canadians proposed a system of prior notification of PNEs as well as observation by an international agency.[15]

Even at the 1966 session of the UN General Assembly which immediately followed the ENDC meeting, there was little debate on the US proposal. India was the first to declare itself opposed, claiming the proposal meant not only nonproliferation but non-dissemination of science and technology, which was 'unacceptable'.[16] Canada on the other hand elaborated its ideas for international control of PNEs, suggesting, for the first time, that the IAEA should be the organisation responsible for running an international PNE service. The Agency could, the Canadians suggested, establish machinery to examine the feasibility of proposed PNE projects, determine the appropriate price for the service, act as an intermediary between user and supplier, and supervise each project to ensure that it served only peaceful purposes.[17]

It was however the 1967 session of the ENDC that would see the first multilateral debate and detailed consideration of the PNE problem. The United States began with a message from President Johnson on 21 February, offering to 'make available nuclear explosive services for peaceful purposes on a non-discriminatory basis under appropriate international safeguards'. It proposed that the nonproliferation treaty clearly state the intention of the parties to 'make available the full benefits of PNE technology, including any benefits that are the by-product of weapons research'.[18]

Also on 21 February, Ambassador Garcia Robles of Mexico reported on the successful negotiations just concluded on the Treaty of Tlatelolco. He highlighted the PNE problem and proposed, as 'an immediate remedy':[19]

> ... the establishment, within an existing international organization such as the IAEA, or independently, of a special program similar to those for economic development, to help all states to carry out PNEs on their territories once it has been shown, of course, that absolutely no danger is involved.

He later suggested that this proposal be contained in an article in the nonproliferation treaty.[20]

The United Kingdom also joined the PNE debate, in particularly scathing terms. Its ambassador, Lord Chalfont, speaking of what he described as 'so-called peaceful nuclear explosions', asserted bluntly that 'a device which moves a million tons of earth to dig a canal or create an oil deposit can just as easily pulverize a city of a million people'.[21] He went on presciently to describe the effect that the development of PNEs could have on 'suspicion between pairs of states which are on bad terms with each other'. He also expressed the United Kingdom's scepticism about the effectiveness and economics of PNEs:[22]

> ... it does not seem at all likely that the uncontrolled use of nuclear energy... will ever constitute an everyday industrial technique... It is likely to be hazardous — at least above ground — expensive, and of strictly limited application.

Nevertheless Britain would support 'workable arrangements, possibly of an international kind', either as part of or along with a nonproliferation treaty, to make PNE benefits available to non-nuclear weapon states.

Sweden declared that the PNE issue should be tackled not in the NPT context but during negotiations on a comprehensive nuclear test ban treaty, because 'the control of PNEs made by nuclear weapon states cannot very well be discussed under the heading of nonproliferation'.[23] Sweden's ambassador, Mrs Alva Myrdal, suggested that negotiations on a CTB proceed in tandem with those on the nonproliferation treaty. The ENDC studiously ignored this proposal.

Brazil, having quietly reserved its position on a previous US statement which interpreted the Treaty of Tlatelolco as banning the acquisition and use of PNEs by Latin American states, now delivered an elaborate rationalisation of its pro-PNE stance. Brazil conceded that there was no difference between nuclear weapon technology and PNE technology. It nevertheless argued, somewhat illogically, that states should not be prevented from conducting their own nuclear *explosions*, because such a capability did not amount to possessing nuclear weapons. According to Brazil, the development of nuclear energy 'inevitably' involved, at a certain stage, the use of explosions. But after attaining this capability, 'non-nuclear weapon states would still have to take several additional steps to embark on the manufacture of nuclear weapons'.[24] To bar access to nuclear explosions, would, in Brazil's view, 'amount to hindering the development of the peaceful uses of nuclear energy'. In a tribute reminiscent of early American and Soviet enthusiasm for the wonders of PNE technology, Brazil asserted that PNEs 'may provide a solution to many of the serious problems which confront Latin American countries and developing countries generally in the economic field'.

These tendentious verbal gymnastics placed Brazil in the virtually unique position of questioning the arms control benefits of a nuclear test ban treaty. Clearly it ran counter to all international opinion (except India's) to suggest that the detonation of a nuclear device did not necessarily transform a country into a nuclear weapon state. Later, in a transparent attempt to bolster its by now shaky pro-disarmament credentials, Brazil claimed that the only alternative to each state acquiring its own PNEs was for all nations to renounce the acquisition of all nuclear explosive devices and for the authority to manufacture them to be concentrated in a single international organisation.[25] This proposal amounted to nuclear disarmament, a proposition which Brazil knew was not immediately achievable, and which would not therefore threaten its own nuclear aspirations.

Not surprisingly, India completely endorsed the Brazilian stand, even though it had for years been a staunch advocate of a nuclear test ban.[26] While also conceding that the technology of nuclear weapons and PNEs was the same, the Indians asserted that 'a weapon has many characteristics which are not present in a peaceful device', a claim later hotly contested by the United States. India then added the argument that nuclear explosives were not in themselves dangerous: like conventional dynamite, the danger depended on the intentions of the user. This was to become a recurring theme in India's defence of PNEs. Finally, in a pointed rejection of President Johnson's offer of PNE 'services', India said it would not accept an 'atomic super-monopoly'. India claimed, however, to recognise the need for 'adequate safeguards' on PNEs and said it was prepared to work with others in evolving such a system.

On 14 March 1967 the Soviet Union ended its long silence in the ENDC on the PNE question. Ignoring its own nascent PNE program, it described the use of PNEs as 'not a practical issue' at the 'present stage'.[27] Even more curiously, it accused the West Germans of whipping up the issue to hinder the nonproliferation negotiations. Adding to the confusion, the Soviets conceded that the non-nuclear weapon states might just have a legitimate interest in PNEs. The Soviet delegate proposed that the question of 'the procedure and conditions' governing PNEs be settled in an international agreement separate to the NPT.

The attack on the Federal Republic of Germany over the PNE issue was at first sight inexplicable. The FRG was not even a member of the ENDC, much less involved in the PNE controversy. The Soviet Union's real target however was not West German policy on PNEs, but its flirtation with US plans to arm with nuclear weapons a Multilateral Force (MLF) of NATO countries.

Soviet use of the PNE issue for harassing the West Germans was facilitated by an ambiguous statement by Foreign Minister Brandt on 10 February 1967, during a visit to Washington. While Brandt told reporters that West Germany was a small country and 'we don't need [peaceful nuclear] explosions on our soil', he added that his country *was* interested in research and industrial development.[28] The implication appeared to be that the FRG did not want to rule out experimenting with PNEs. On 7 April 1967 the West Germans circulated a memorandum which stated that PNEs should be made available to the non-nuclear weapon states as soon as technical developments allowed.[29] Once the MLF problem was resolved however, no more was heard from the Soviets on the FRG attitude towards PNEs.

Meanwhile, on 16 March 1961, Egypt proposed that the NPT specifically nominate the IAEA to manage an international PNE service, and require the nuclear weapon states to supply PNEs for such a service 'without political conditions'.[30] Speaking for the United States, ACDA director William Foster agreed that the IAEA *could* be the body responsible for PNEs. He also concurred with the Soviet view that the PNE issue should be settled by a separate international agreement.[31] Foster then outlined five principles for sharing the potential benefits of PNEs:

(1) If and when PNEs proved to be technically and economically feasible, nuclear weapon states should make PNE services available to other states. The nuclear weapon states would perform the required detonations under 'appropriate international observation', with nuclear devices remaining under the custody and control of the nuclear weapon state.
(2) Non-nuclear weapon states should be able to request such services through an international body in which they would be represented. This body might consider the priority and feasibility of proposed PNE projects and necessary safety precautions, thereby ensuring that the nuclear weapon states could not withhold services from others because of 'extraneous' considerations.
(3) Costs to non-nuclear weapon states should be as low as possible and not include research and development costs.
(4) There should be full consultation between nuclear weapon and non-nuclear weapon parties to the PTBT about any amendment needed to permit major PNE projects to proceed.
(5) Conditions and procedures for international collaboration in PNE projects should also be developed in full consultation with the non-nuclear weapon states.

Sweden, delicately suggesting that no non-nuclear weapon state would wish to manufacture its own PNEs, said that the NPT should of course ban the *manufacture* of peaceful nuclear explosives by non-nuclear weapon states. Sweden suggested however that the *use* of peaceful nuclear *explosions* for the benefit of such countries should be handled through a comprehensive test ban and a separate PNE agreement. Such an agreement would allow exemptions from the test ban, under international management and control, for peaceful purposes.[32] To ensure 'equity', Sweden also proposed that the nuclear weapon states contribute to a PNE 'pool', from which devices would be allocated to particular PNE projects by an international body.

On 24 August 1967 the United States and the Soviet Union submitted a joint draft NPT to the ENDC. The text specifically mentioned PNEs in its preamble:[33]

> ... Declaring their intention that potential benefits from any peaceful applications of nuclear explosions should be available through appropriate international procedures to non-nuclear weapon States Party to this Treaty on a non-discriminatory basis and that the charge to such Parties for the explosive devices used should be as low as possible and exclude any charge for research and development...

The draft also specifically banned the acquisition by non-nuclear weapon states of both nuclear weapons and *other nuclear explosive devices*, the wording of the PTBT. As the United States explained, this was done not because of a change in US policy, but because 'other states', presumably Brazil and India, interpreted the previous US text as permitting PNEs.[34] That the new draft achieved its aim is illustrated by India's complaint that the old draft had been preferable because it dealt only with nuclear weapons.[35]

On 12 September 1967 Canada presented to the ENDC an outline of a PNE treaty to be negotiated separately to the NPT. It proposed that:[36]

(1) non-nuclear weapon states wishing to obtain the benefits of PNEs would apply to an international body, such as the IAEA, which would manage the PNE service and any bilateral arrangements between the applicant state and one or more nuclear powers
(2) nuclear weapon states would agree to provide on request, through the international body, explosive devices, technical advice and assistance in carrying out PNE projects
(3) PNEs would be subject to international observation and appropriate international safeguards to ensure that they were directed to peaceful applications and were consistent with the PTBT, the IAEA statute and other international agreements
(4) the nuclear devices involved would remain under the control of the nuclear weapon state providing the service
(5) the nuclear weapon states would make available full information about the potential uses of PNEs.

Mexico agreed with Canada's proposal. However, undoubtedly in a vain attempt to make Brazil and India more amenable to eventually supporting the NPT, Mexico proposed that the treaty contain an *operative* paragraph on PNEs, which would have more legal force than the preambular paragraph proposed in the US/Soviet draft. The paragraph would legally commit the nuclear weapon states to make PNEs available to non-nuclear weapon states. According to Mexico:[37]

> If the non-nuclear countries have to undertake not to develop or explode nuclear devices — since such devices, despite their peaceful purpose, are nothing other than nuclear bombs — they must be assured in return, categorically and unequivocally, that they will not thereby be deprived of the possible benefits of such explosions.

The Mexican delegation subsequently drafted the text of a PNE article, borrowing heavily from the five principles enunciated by the United States on 21 March 1967 and the preambular paragraph contained in the joint US/Soviet draft treaty.[38]

Egypt, Burma, Canada, Nigeria, Sweden and Switzerland supported Mexico. Sweden still stressed the relevance of a CTB. It proposed that under such a treaty a licensing system should operate to enable both nuclear and non-nuclear weapon states to obtain the benefits of PNEs, thereby putting all countries on an equal footing.[39] The United Kingdom, while supporting Mexico's idea of an operative paragraph on PNEs in the body of the NPT, expressed concern that the Mexican proposal implied that Britain would be obliged to develop and share PNE technology, even if it had no need itself for such technology. As the British delegate pointed out, 'the geography and population density of Great Britain make it uncertain whether PNEs will be an industrially-useful technique in our own islands in the foreseeable future'.[40] Nigeria suggested that scientists from non-nuclear weapon states should have opportunities to develop fully their 'intellectual capabilities' in nuclear explosive technology.[41] It subsequently tabled draft treaty language to this effect, which was universally ignored, except by Canada, which pointed out that the proposal would amount to nuclear proliferation.[42]

Brazil continued passionately to advocate the right of states to manufacture and conduct their own PNEs, albeit under 'international control', but nevertheless using indigenous resources. Brazil said it would not sign the NPT unless it permitted non-nuclear weapon states to carry out nuclear explosions for 'great engineering works'.[43] Only India unequivocally supported this view. On 31 October 1967 Brazil tabled amendments to the US/Soviet draft NPT designed to rid the treaty of any restrictions on PNEs. Articles I and II would refer only to nuclear weapons, not 'other nuclear explosive devices', and a reworded Article IV on peaceful uses would specifically give states the unrestricted right to develop, produce and use PNEs without restriction.[44] Presumably the safeguards envisaged in Article IV of the draft NPT (to prevent the illicit diversion of nuclear materials to nuclear weapons) would remain, although Brazil never explained how such safeguards were meant to distinguish between nuclear material intended for nuclear weapons and material intended for PNEs. The Canadians commented sharply that the Brazilian proposals would 'make a treaty so amended complete nonsense'.[45]

By the end of the 1967 ENDC session there was a clear majority, even among the non-aligned developing countries, in favour of preventing the acquisition and use of PNEs by non-nuclear weapon states. Not only had the ENDC debates highlighted the dangers of PNEs, but an informal group of experts convened by the ENDC on the PNE issue in March 1967 had concluded that there was no fundamental difference between nuclear weapons and PNEs.[46] Only Brazil, India and a confused Nigeria held out. There was also, however, a swelling tide of opinion that the nuclear weapon states should give firmer guarantees that they would make PNE services available cheaply, and on a non-discriminatory basis, to non-nuclear weapon states. During consultations with the Americans throughout 1967, several US allies, including Japan, West Germany and Italy, had privately expressed concern that the treaty as drafted could circumscribe their ability to obtain the benefits of PNE technology. Even Soviet allies, such as Czechoslovakia and Poland, were concerned.[47] The British, on the other hand, were opposed to a PNE article in the body of the treaty on the grounds that it might undermine its 'stability'.[48]

By the time the ENDC assembled for its 1968 session, the United States and the Soviet Union had agreed to accede to non-aligned demands for an operative paragraph on PNEs. On 18 January 1968, at the opening session, the two superpowers surprised the non-aligned by tabling revised identical drafts. One of the most important revisions was a new Article V, which read as follows:[49]

> Each Party to this Treaty undertakes to co-operate to insure that potential benefits from any peaceful applications of nuclear explosions will be made available through appropriate international procedures to non-nuclear weapon States Party to this Treaty on a non-discriminatory basis and that the charge to such Parties for the explosive devices used will be as low as possible and exclude any charge for research and development. It is understood that non-nuclear weapon States Party to this Treaty so desiring may, pursuant to a special agreement or agreements, obtain any such benefits on a bilateral basis or through an appropriate international body with adequate representation of non-nuclear weapon States.

The United States and the Soviet Union claimed that their revised texts took account, to a great extent, of the positions supported by a majority of the ENDC membership. Indeed the PNE article was largely based on Mexico's draft.[50] However, while Mexico wanted the nuclear weapon states to undertake to 'adopt the pertinent measures' to ensure the availability of the potential benefits of PNEs, the US/Soviet draft merely required them to undertake to 'cooperate to insure' that such benefits were made available.[51] Brazil, whose proposed amendments had been totally ignored, charged that the proposed treaty would impose 'a new form of dependence which is certainly inconsistent with our aspirations for development', since it would 'freeze all technological development that might be connected, even remotely, with the specific technology of nuclear explosive devices intended for civil use'.[52] Burma expressed concern about the financial aspects of the US/Soviet proposal.[53]

Surprisingly, Italy, whose position on the NPT swung from strong advocacy to severe criticism and back again,[54] called for an amendment to guarantee the right of all parties, nuclear and non-nuclear, to obtain nuclear devices for peaceful purposes.[55] Quester attributes this proposal to the new Italian delegation leader and Italy's realisation that the Soviet Union and the United States were close to agreement on the NPT.[56]

Sweden expressed concern that the draft PNE language might be incompatible with a CTB. It proposed amendments to ensure that a PNE agreement could be negotiated as part of a future CTB without affecting the NPT. Mrs Myrdal suggested that bilateral arrangements for PNE projects between a nuclear weapon state and a non-nuclear weapon state be placed under international supervision to prevent any suspicion that they were weapon tests.[57] She criticised bilateral arrangements on the grounds that favoured treatment could be given to the allies of nuclear weapon states, mostly developed countries, leaving the developing countries disadvantaged.[58] Egypt, with an eye to its Qattara Depression proposal, supported Sweden.[59] Sweden also proposed that Article V omit any reference to non-nuclear weapon states, so that 'the benefits in question' could be made available to all states.[60] The British welcomed this idea as having potential significance for the United Kingdom.

The provision for bilateral arrangements had been included in the draft as a result of strong US representations, instigated by the AEC, to the Soviets. The AEC's argument was that, in order to preserve the viability of projects such as a new Panama Canal, the Commission should not be saddled with the need to seek multilateral approval.[61]

The Americans and Soviets refused to accept any of the Swedish amendments. The United States pointed out that PNEs under bilateral arrangements would take place under the 'appropriate international procedures' envisaged in Article V. Further, there would be no incompatibility between the NPT and a future CTB because, 'if under a CTB, international approval were needed for a PNE, such approval would constitute an "appropriate international procedure" for services conducted bilaterally or through an appropriate international body'.[62] Soviet Ambassador Roshchin denied there was a PNE 'loophole' in the draft treaty, noting that if any changes in PNE procedures were necessary as a result of a CTB this could be 'duly taken into account at the proper time'.[63]

Swedish behaviour during the NPT negotiations was difficult to fathom. Although a champion of disarmament, Sweden seemed unwittingly to assist the anti-NPT states by its often obscure criticisms of the US/Soviet drafts. A striking example was Mrs Myrdal's comment that:[64]

> The nuclear weapons states are the sole possessors of the know how and the means necessary to use nuclear devices for such potentially very lucrative activities as digging canals and extracting oil, gas and minerals. Thus, the mighty have a virtual control over the application or non-application of these techniques, a fact that cannot but cast ominous shadows over the future.

This was later used by India to support its view of the superpower PNE 'monopoly'. According to George Quester, seen in its best light, Sweden may have:[65]

> ... astutely sensed a minimum of concessions that other near-nuclear states would require before being reconciled to the NPT. By becoming the spokesman for the NPT-recalcitrants, perhaps Sweden could in the end lead them happily back into the fold.

If this was the Swedish plan it failed miserably.

By the time the ENDC concluded its discussions on the NPT on 14 March 1968, only five members, the United Kingdom, Bulgaria, Canada,

Czechoslovakia and Poland, had endorsed the US/Soviet draft. Brazil, India, Romania and Italy had reserved their positions but were known to have strong objections. Of the remaining six active members of the ENDC, Mexico, Sweden and Egypt appeared to be favourably disposed towards the treaty, while Burma, Ethiopia and Nigeria remained undecided. France, which had not participated in the session, was considered opposed to it.[66]

Unable to reach agreement by the deadline of 15 March imposed by the United Nations General Assembly, the ENDC submitted to it the co-Chairmen's draft, noting that it was not necessarily supported by all delegations. According to SIPRI, the nuclear weapon states apparently 'hoped that the respect of other nations for the ultimate aims of the Treaty and for the work already done on preparing the draft would ensure a majority vote for a resolution endorsing the draft as it stood'.[67] This proved to be more difficult than anticipated.

After extensive and often bitter debate, and with only 80 of the 124 members of the UN likely to vote in favour of the treaty, the Soviet Union and the United States reluctantly agreed to further revisions. Changes were made to Article V on the basis of suggestions worked out by Mexico in consultation with Colombia and Chile,[68] after both superpowers had tried unsuccessfully to persuade Mexico to drop its proposals.[69] Mexico's suggestions were as follows:

(1) The undertaking which the nuclear weapon states should assume regarding the provision of PNE 'services' should be strengthened.
(2) Work on the international agreement embodying the statute of the 'appropriate international organ' for PNE services should begin as soon as possible. The agreement should be approved by a body representing the world community.
(3) The provision of multilateral assistance through the international organ should be given pride of place, with the resort to bilateral arrangements being secondary.
(4) PNEs should be carried out under 'appropriate international observation'.[70]

(The first two of these, at least, would not have been displeasing to the AEC, which was keen to begin its commercial PNE service as soon as possible.)

In accordance with Mexico's first proposal, the beginning of Article V was changed from, 'Each party to this treaty undertakes to co-operate to insure...' to 'Each party to the treaty undertakes to take appropriate measures to ensure, in accordance with this treaty.' The reference to 'in accordance with this treaty' was added to ensure that the obligations of Article V were in conformity with Articles I and II banning the acquisition of nuclear weapons by non-nuclear weapon states. According to Mohamed Shaker, author of the seminal work on the NPT negotiations, while this may have seemed superfluous, it was a useful change in view of the objections raised by such countries as India and Brazil to Articles I and II.[71]

Incredibly, it was also only as a result of the Mexican proposal that the caveat 'under appropriate international observation' was added to Article V.[72] This was despite the attention given to the subject in the ENDC, especially by the United States. Presumably it was assumed by the Americans and Soviets that they would simply refuse to provide any PNE service unless such observation was provided for, and that this did not need spelling out in the NPT.

The new version of Article V, and the one that eventually appeared in the NPT, reads as follows:[73]

> Each Party to the Treaty undertakes to take appropriate measures to ensure that, in accordance with this Treaty, under appropriate international observation and through international procedures, potential benefits from any peaceful applications of nuclear explosions will be made available to non-nuclear weapon States Party to the Treaty on a non-discriminatory basis and that the charge to such Parties for the explosive devices used will be as low as possible and exclude any charge for research and development. Non-nuclear weapon States Party to the Treaty shall be able to obtain such benefits, pursuant to a special international agreement or agreements, through an appropriate international body with adequate representation of non-nuclear weapon States. Negotiations on this subject shall commence as soon as possible after the Treaty enters into force. Non-nuclear weapon States Party to the Treaty so desiring may also obtain such benefits pursuant to bilateral agreements.

Through its amendments Mexico achieved the strengthening of Article V that the superpowers had resisted in the ENDC. Mexico's role, not only in bringing Article V into being, but in developing the article to the satisfaction of the large majority of non-nuclear weapon states, was a crucial one.

In seeking to rally support for the new Article V, the United States stressed that it would continue PNE research and development and that all benefits from the program would be available under the treaty's provisions to non-nuclear weapon states without delay.[74] The Soviet Union concurred and suggested that preparatory work on a PNE agreement could start even before the NPT entered into force. However, while the United States had always identified itself as a supplier of PNE services, the Soviets, although defending the merits and benefits of Article V in principle, had not. They offered to do so for the first time in April 1969 during the first US/Soviet experts meeting on PNEs, held in Vienna.[75]

In commenting on the revised draft article, Sweden proposed that any international body administering PNEs should carry out feasibility studies of proposed projects, observe and control their execution, and help finance them in developing countries.[76] It argued that while the IAEA would be suitable for the first two tasks, the third should be entrusted to the UN

Development Program (UNDP) or the International Bank for Reconstruction and Development (IBRD).[77] Sweden asserted that the recent creation of a multinational company, Nobel Paso Geonuclear, to supply commercial PNE services, was inconsistent with providing such services at as low a charge as possible and implied that the profit motive was not excluded from PNE activities.[78]

Japan suggested that if and when the distinction between military and peaceful nuclear explosions became possible, the restrictions of the NPT should no longer apply to PNEs.[79] South Africa expressed scepticism about the supposed 'non-discriminatory basis' of future PNE services, citing its own bitter experience with international organisations.[80] Concern over the charges for PNE services was expressed by Tanzania, which found the expression 'as low as possible' too vague. Ethiopia wondered to what extent the price 'could be made to depart from conformity with the well-known principle of monopolistic price mechanism'.[81]

The United States and the Soviet Union tried to dissipate these concerns. The US representative denied that the cost of PNE services would be determined by any 'monopolistic price mechanism'. The Soviet representative was even more outspoken, asserting that Article V would make it 'impossible for any nuclear power in any way to manipulate prices for the purpose of gaining a profit from the carrying out of peaceful explosions in non-nuclear countries'.[82]

On 12 June 1968, with debate on the treaty concluded, the UN General Assembly voted 95 to 4 (Albania, Cuba, Tanzania and Zambia), with 21 abstentions, to 'commend' the NPT and support the 'widest possible adherence to the treaty'.[83] Prominent among the abstainers were Argentina, Brazil and India. The treaty entered into force on 5 March 1970. By January 1988, 134 non-nuclear weapon states were party to the accord.[84] In effectively barring such states from acquiring nuclear dynamite, the treaty represents a considerable additional constraint on the use of PNEs.

In another sense, however, the treaty was a positive encouragement to the development and use of PNEs. Indeed, the three nuclear weapon states party to the NPT, the United States, the United Kingdom and the Soviet Union, far from being constrained in their use of PNEs by the treaty, were actively encouraged by Article V to provide PNE services to the world. This was exactly what Edward Teller and Glenn Seaborg had in mind. In fact Herbert York attributes the inclusion of a Plowshare provision in the NPT largely to Edward Teller, presumably through his influence with Seaborg, the AEC, the US Administration generally, and through his world-wide promotion of the wonders of PNE technology. York states that:[85]

> In my opinion, he deserves very much of the credit (or blame), probably more than any other single individual... for the inclusion in the 1968 Non-Proliferation Treaty of the provision making a special place for the so-called peaceful uses of nuclear explosives, a provision which in the long run will

probably prove the undoing of that treaty if something else does not do so first.

While this undoubtedly exaggerates the influence of Edward Teller, by implication it correctly identifies Plowshare as being, in the final analysis, the progenitor of Article V of the NPT.

# Notes

1 *The United Nations and Disarmament 1945-1970*, Department of Political and Security Affairs, United Nations, New York, 1970, p. 270.
2 Ibid., p. 280.
3 G.T. Seaborg and B.S. Loeb, *Stemming the Tide: Arms Control in the Johnson Years*, Lexington Books, Lexington, Mass., 1987, p. 364.
4 Ibid.
5 Ibid., p. 331.
6 Ibid., p. 364.
7 Ibid.
8 Eighteen-Nation Disarmament Committee Document, ENDC/PV. 280.
9 Ibid.
10 Ibid.
11 O. Marwah and A. Shulz (eds), *Nuclear Proliferation and the Near-Nuclear Countries*, Ballinger Publishing Co., Cambridge, Mass., 1975, p. 5.
12 I.Y. Borg, *Nuclear Explosions for Peaceful Purposes*, Lawrence Livermore National Laboratory Paper No. UCRL-95026, Rev. 11, November 1986, p. 21.
13 Seaborg and Loeb, *Stemming the Tide*, p. 365.
14 ENDC/PV. 284.
15 ENDC/PV. 285.
16 *The United Nations and Disarmament 1945-1970*, p. 284.
17 Ibid.
18 ENDC/PV. 287.
19 Ibid.
20 ENDC/PV. 295.
21 ENDC/PV. 288.
22 Ibid.
23 Ibid.
24 ENDC/PV. 293.
25 ENDC/PV. 310.
26 ENDC/PV. 298.
27 ENDC/PV. 293.
28 V. Mastny (ed.), *Disarmament and Nuclear Tests 1964-69*, Facts on File Inc., New York, 1970, p. 131.
29 W. Epstein, *The Last Chance*, The Free Press, New York, 1976, pp. 111-12.
30 ENDC/PV. 294.
31 ENDC/PV. 295.
32 ENDC/PV. 302.

33 ENDC/PV. 192.
34 ENDC/PV. 330.
35 ENDC/PV. 334.
36 ENDC/PV. 329.
37 ENDC/PV. 331.
38 M.I. Shaker, *The Nuclear Non-Proliferation Treaty: Origin and Implementation 1959-1979*, Vol. I, Oceana Publications, London, 1980, p. 382.
39 Epstein, p. 75; ENDC/PV. 335.
40 ENDC/PV. 337.
41 ENDC/PV. 327.
42 ENDC/PV. 344.
43 Mastny, p. 130.
44 ENDC/PV. 343.
45 ENDC/PV. 345.
46 ENDC/PV. 293.
47 Epstein, p. 71.
48 Seaborg and Loeb, *Stemming the Tide*, p. 365.
49 ENDC/PV. 193/Rev. 1.
50 Shaker, p. 383.
51 M. Willrich, *The Non-Proliferation Treaty: Framework for Nuclear Arms Control*, The Mitchie Company, Charlottesville, Virginia, 1969, p. 138; Shaker, p. 418.
52 Mastny, p. 162; *The United Nations and Disarmament 1945–1970*, p. 291.
53 *The United Nations and Disarmament 1945–1970*, p. 292.
54 G.H. Quester, *The Politics of Nuclear Proliferation*, Johns Hopkins University Press, Baltimore, Maryland, 1973, p. 240.
55 Mastny, p. 163.
56 Quester, p. 190.
57 *Negotiations on A Comprehensive Test Ban 1963–71*, US Arms Control and Disarmament Agency, Unclassified Document Series, DDS Ref. #587, 22 December 1971, p. 62.
58 ENDC/PV. 373.
59 *The United Nations and Disarmament 1945–1970*, p. 292.
60 Shaker, p. 408.
61 Seaborg and Loeb, *Stemming the Tide*, p. 366.
62 *Negotiations on a Comprehensive Test Ban 1963–71*, p. 63.
63 Ibid.
64 K. Subrahmanyam, 'India's Nuclear Policy' in Marwah and Shulz (eds), p. 143.
65 Quester, p. 133.
66 Mastny, p. 164.
67 *SIPRI Yearbook 1968-69*, Almqvist and Wiksell, Stockholm, 1969, p. 156.
68 Shaker, p. 383.
69 Epstein, p. 119.
70 Conference of Non-Nuclear Weapon States, Geneva, 1968, Document A/CONF. 35/DOC. 15, p. 1.
71 Shaker, p. 422.
72 Ibid., p. 428.
73 *Status of Multilateral Arms Regulation and Disarmament Agreements*, United Nations, Second Edition, UN Department for Disarmament Affairs, New York, 1983, p. 74.
74 *The United Nations and Disarmament 1945–1970*, p. 296.
75 Shaker, p. 405.
76 *The United Nations and Disarmament 1945–1970*, pp. 296–97.
77 Ibid., p. 297.
78 Shaker, p. 425.
79 *The United Nations and Disarmament 1945–1970*, p. 294.
80 Shaker, p. 423.
81 Ibid., p. 424.
82 Ibid., p. 425.
83 *The United Nations and Disarmament 1945–1970*, p. 298.
84 *SIPRI Yearbook 1988*, Oxford University Press, Oxford, 1988, p. 552.
85 H.F. York, *The Advisors: Oppenheimer, Teller and the Superbomb*, W.H. Freeman and Co., San Francisco, 1976, p. 145.

# 8

# Dangling Morsels: The IAEA and the PNE 'Regime'

*It is essential that the policy of providing these [PNE] devices be developed concurrently with the technology of producing them, or earlier, lest we find ourselves in the uncomfortable position of dangling before the other nations an attractive morsel which they cannot possess.[1]*

Arnold Kramish, 1963

WITH the NPT agreed and open for signature, enthusiasm among the non-nuclear weapon states for obtaining the benefits of PNEs was running high. Pressure arose immediately, from both pro- and anti-NPT states, for establishment of the PNE service or 'regime' promised in Article V of the treaty.

The most obvious candidate for organising such a regime was the Vienna-based International Atomic Energy Agency. A role for the IAEA in regard to PNEs had been suggested during the test ban negotiations during the late 1950s. It was argued then that investing the IAEA with the task of safeguarding 'black boxes' containing PNE devices and observing PNE detonations might hasten the development of the fledgling agency.[2] The IAEA Statute imposed no restriction on the Agency undertaking such a role, or indeed on the use of PNEs by its member states, whether they were nuclear weapon states or not. The Statute's requirement that the IAEA promote the peaceful uses of nuclear energy could even be interpreted — and was later to be so interpreted by India — as obliging the IAEA to promote the development and use of PNEs by its members.[3] As David Fischer regretfully notes:[4]

> ... the barrage of publicity which PNEs received in the 1960s whetted the desire of many other countries to carry out their own PNE projects and bolstered the arguments of those who maintained that the language of the IAEA Statute does not preclude the use of IAEA-safeguarded material to carry out PNEs.

Canada, supported by Egypt, had been the first country to formally suggest a role for the IAEA in establishing and running an international PNE regime. But Canada's suggestion, made in the ENDC during the NPT negotiations, was not universally welcomed, even by Western states which normally supported the agency. The United States, for instance, did not agree initially that the IAEA should have sole responsibility for a PNE regime, largely because of the Federal Republic of Germany's reluctance to submit to IAEA monitoring of any of its nuclear activities. It was only at the end of the NPT negotiations, when the West German problem had been overcome, that the United States came around to supporting the IAEA option. On 25 July 1968 it urged the ENDC to:[5]

> ... agree upon the IAEA as the appropriate international organization through which non-nuclear parties may seek the benefits to be provided under Article V. And we can see no reason why we should not confidently turn to the IAEA for the development of procedures and agreements relating to the furnishing of the peaceful explosions service to be obtained through the IAEA.

Italy could see a reason. It declared that it would be premature to settle on the IAEA, at least as the forum for determining the organisational and procedural arrangements for PNEs. Italy counselled the ENDC not to anticipate the work of the Conference of Non-Nuclear Weapon States, to be held later that year.[6] Italy and other non-nuclear weapon states, mostly developing countries, saw the IAEA as unsuitable because its Board of Governors, known affectionately as the BOG, was biased heavily in favour of the states most advanced in nuclear technology.[7]

As a result of these differing opinions, the issue of whether or not the IAEA should organise the PNE regime remained unresolved at the conclusion of the NPT negotiations. Consequently Article V of the treaty made no mention of the IAEA, leaving open the possibility of a completely new international organisation being established to handle PNEs. According to George Quester, this immediately led to 'bureaucratic rivalries' between the IAEA and the United Nations Secretariat in New York over the form and location of a future PNE service. Quester alleges that 'when the NPT did not explicitly assign this task to the IAEA, subtle moves were made in New York and Geneva to draw it away from Vienna'.[8]

It was only with the convening of the Conference of Non-Nuclear Weapon States in September 1968, that the IAEA emerged as the favoured repository of the service. Pakistan had introduced a draft resolution, co-sponsored by Jamaica, Libya, Saudi Arabia and Somalia, at the 1966 session of the UN General Assembly, proposing the convening of such a conference to consider, *inter alia*, 'how nuclear devices might be used for exclusively peaceful purposes'.[9] The resolution was only 'halfheartedly' supported by the Assembly, but India alone had the temerity to vote against it.[10] The reason why Pakistan included the PNE issue in its Conference proposal, having never before evinced an interest in the subject, was clearly India's ambiguous attitude towards nuclear dynamite.[11] In July 1966 Pakistan addressed a letter to the UN Secretary-General warning that India was preparing to become a nuclear power and requesting that the matter be aired.[12] Pakistan obviously hoped that an international conference would help make sure that if India was genuinely interested in PNEs it could only use them under strict international control.

The Conference of Non-Nuclear Weapon States was convened in Geneva on 29 August 1968, only two-and-a-half months after the UN General Assembly approved the NPT. Ninety-six countries attended, including four of the nuclear weapon states (France, the Soviet Union, the United Kingdom and the United States). While the delegations of the nuclear weapon states had no vote and did not speak at the meeting, they were 'very active in the corridors'.[13] The PNE issue was considered under Item D on 'Programs for co-operation in the field of peaceful uses of nuclear energy'.[14] By this time, according to Elizabeth Young, the non-nuclear weapon states, particularly those which were also developing countries, were thoroughly excited by the prospect of 'untold nuclear goodies' which had been dangled before them to induce them to support the NPT:[15]

> Consequently most of the 'peaceful uses' recommendations of the conference displayed the uncritical, even greedy, enthusiasm with which the non-nuclears were responding to an apparent promise of economic miracle. They pointed out that they would require financial along with... technical assistance, suggested that special funds be set up for the purpose, and that representation on [the] IAEA be wider and poorer [sic].

Interest in PNEs was further stimulated by two papers commissioned for the Conference by the UN Secretary-General. One was by Swedish seismologist Dr Ulf Ericsson and bore the extraordinary title: *The Question of Nuclear Explosions for Peaceful Purposes by Non-Nuclear Weapon States and the Possibility of Misuse of Such Technology for the Production of Nuclear Weapons.* Ericsson's paper was notable for the suggestion that the PTBT be amended to allow cross-border release of radioactivity 'within agreed safe health limits' to ensure 'effective developmental work on peaceful nuclear explosion technique(s)'.[16] The paper was, however, sceptical about the possibility of establishing watertight verification provisions to ensure that weapons-related information was not obtained from PNEs.

The second paper was by Dr Theo Ginsburg, Professor at the Ecole Polytechnique Federale of Zurich. Entitled, more modestly, *The Question of Peaceful Explosions for the Benefit of Non-Nuclear Weapon States*, it was a paean to the benefits of PNEs. It described Plowshare excavation experiments as 'meaningful', and the use of PNEs in mining processes as 'very attractive from an economic point of view'.[17] Ginsburg concluded his paper, which could only have been encouraging to the non-nuclear weapon states interested in PNEs, with the reassurance that 'no new technology becoming available to man has ever been rejected'.[18] In Ginsburg's view:[19]

> The development of nuclear explosives for peaceful purposes will bring great benefit to mankind. It offers immense promise of adding to man's knowledge of his environment and improving his well-being.

India was one of the most vocal and active of the conference participants. While grudgingly acknowledging that the proliferation of nuclear weapons should be prevented, India asserted that this should be done without jeopardising the untrammelled development of all peaceful applications of nuclear energy, including PNEs, by all states. India argued that the PNE question should be linked to the conclusion of a nuclear test ban, when all nuclear testing by all states would be proscribed. A 'control' mechanism to govern PNEs should, in India's view, be limited to the conducting state giving advance notice to the IAEA, indicating the precise purpose of each explosion, and permitting 'full international inspection' in order to prevent stockpiling of nuclear devices and their use for military purposes.[20] No mention was made of a ban on non-nuclear weapon states developing or detonating their own PNEs.

Brazil's position was similar. It claimed that the existence of an international PNE service should not detract from the right of any state to manufacture and detonate PNEs, either by its own means or through agreements with other nations, whether nuclear or non-nuclear. The only qualification was that PNE detonations be for 'duly substantiated peaceful purposes under adequate control and international supervision'.[21]

One of the three draft resolutions tabled at the conference relating to PNEs was sponsored by Brazil and other Latin American states. It was an amalgamation of the proposals of two rival groups. On the one hand Mexico and eight of its supporters proposed that the IAEA, presumably through a decision of its Board of Governors, set up an autonomous PNE body within the Agency, to run an 'International Program for Nuclear Explosions for Peaceful Purposes'.[22] Argentina, Brazil and Peru, on the other hand, proposed the convening of a special session of the IAEA General Conference — where the developing countries would have more influence — to consider establishing a PNE Department within the IAEA by means of a new protocol to the IAEA Statute.[23] The compromise text ultimately tabled represented a victory for Argentina and Brazil. It recommended that the General Assembly 'consider the convening of a conference for the establishment, within the framework of the IAEA, of an international department' for PNEs. The text proposed that, through such a department, the nuclear weapon states would provide PNE services not just to NPT parties but to *all* states which had *renounced* nuclear weapons.[24] This was intended to permit parties to the Treaty of Tlatelolco and other nuclear weapon-free zones, or even states that had made unilateral declarations foreswearing nuclear weapons, to obtain the benefits of PNEs.[25]

Pakistan, with an eye to India's nuclear ambitions, reacted to the Latin American proposal by stressing that the renunciation of nuclear weapons would have to be by means of a binding multilateral agreement. Canada agreed, stating that a unilateral declaration would be insufficient.[26] The Latin American proposal also contrasted with the US and Soviet view that only parties to the NPT should be able to avail themselves of PNE benefits under the treaty. While the United States did envisage offering PNE services to non-NPT parties, these would be under bilateral arrangements and not at the expense of providing PNE services to NPT parties.

The draft Latin American resolution, which was opposed by Committee II of the Conference, failed to obtain the required two-thirds majority in Plenary.[27] Among US allies, only West Germany, Italy and Spain supported the proposal.[28] All other US allies, and the Soviet Union's, either voted against it or abstained, mostly on the grounds that establishment of a PNE service before progress had been achieved in the application of PNEs would be premature.[29] According to William Epstein, the nuclear weapon states had not been 'sitting idle in the lobbies'.[30]

A reason for American opposition to the Latin American proposal could be found in the recent report on the NPT by the US Senate Foreign Relations Committee. This report recorded the committee's 'concern at the open ended commitment' that Article V appeared to give regarding the provision of PNE services to non-nuclear weapon states. The committee rejected 'any suggestion that Article V constitutes an across-the-board pledge by the United States to support foreign and domestic commercial research and development projects'.[31] The committee's attitude was a direct challenge to AEC aspirations for an international Plowshare service. It was also a sign that at least one superpower was already preparing to back away from its NPT commitments.

While the Latin American resolution was defeated, two relatively innocuous draft resolutions on the subject of PNEs *were* adopted by the conference. Resolution H(C), sponsored by six of the smaller West European countries and Japan, requested the IAEA to initiate 'studies that are deemed advisable on its possible functions' in the PNE field and to 'compile and disseminate public information' on PNEs.[32] Resolution L, sponsored by Sweden and Nigeria, expressed the view that the PNE question was closely linked with a CTB, and urged the creation, by separate international instrument, of a regime to regulate and control all PNEs as exemptions from a CTB.[33] Italy abstained because it felt that linking the PNE issue to a CTB might delay conclusion of a PNE agreement.[34] Australia, concerned to protect its right to undertake joint Plowshare projects with the United States, abstained because of the lack of any reference to bilateral arrangements as provided for in Article V of the NPT.[35] According to Shaker, Australia was 'remarkably consistent' in its advocacy of bilateral arrangements.[36] Finally, the conference, in a separate resolution, also called upon all non-nuclear weapon states to accept the safeguards system of the IAEA, which was specifically described as preventing diversion of all nuclear material from peaceful applications to nuclear weapons or 'peaceful explosive uses'.[37]

Mexico, continuing its role as moderator of the aspirations of the radical non-nuclear weapon states, submitted to the conference a draft of an international PNE agreement, as envisaged in Article V of the NPT. Mexico considered that 'one of the most valuable contributions that the conference could make' would be to prepare a draft statute setting forth the 'structure, functions, powers, and procedure' of an international PNE body.[38] The Mexicans proposed that the international body, which they called the 'International Program for Nuclear Explosions for Peaceful Purposes' (IPNEPP), be open to all nuclear weapon states, as well as all non-nuclear weapon states which renounced the acquisition of such weapons. Mexico argued that since no existing body included all such states (China was at that stage not a member of the IAEA), a new body would have to be established. Although it would have complete autonomy, it would function within the framework of the IAEA, have the same headquarters, and hold its plenary meetings simultaneously with the IAEA General Conference.[39] The IAEA safeguards system would be used to monitor the use of PNEs.

Mexico proposed that IPNEPP be run by a governing council in which the non-nuclear weapon states would have a majority of seats. The council, which would comprise the five nuclear weapon states and a geographically balanced group of twenty-nine non-nuclear weapon states, would decide most matters by simple majority. In addition, Mexico suggested explicit safeguards to avoid PNE services being used as a means of interference in the internal affairs of non-nuclear weapon states and to guard against decisions on PNE services being based on 'political' considerations.[40] The Mexican paper also outlined verification procedures for carrying out PNE detonations. The state requesting a PNE service would be required to supply information on the:[41]

(a) date and exact site of the proposed explosion
(b) nature and origin of the explosive device
(c) specific purpose of the explosion
(d) anticipated power of the device
(e) possible radioactive contamination, and
(f) procedures for international observation.

IPNEPP personnel would observe preparations for, and the detonation of, the device and would have 'unrestricted access to any area near the site of the explosion', to verify that the device used and the procedures followed conformed with the information supplied beforehand.[42]

Finally, noting that the use of PNEs was 'destined to be of incalculable importance, especially to the developing countries', Mexico reiterated several proposals it had made during the General Assembly debate on the NPT, for alleviating the high cost of PNE services for such countries.[43] It suggested, for instance, that a Special Fund be established to provide financial assistance.[44] It also proposed that the nuclear weapon state providing a PNE service:[45]

> ... bear the entire cost of transporting the explosive devices to the territory of the benefiting Member or Members and the cost of the engineering work carried out in preparation for the explosion or of such safety measures as cannot be adopted or implemented by the benefiting Member or Members for lack of the necessary technical facilities or knowledge.

The supplier state would be obliged to defray the cost of damage insurance for PNE projects and absorb all research and development costs. Such an open-ended form of foreign aid was of course unlikely to be acceptable to any of the nuclear weapon states.

Italy also submitted a working paper, proposing an international conference to conclude an agreement to establish an international PNE organisation. Italy preferred this body to have the independent status of specialised UN agencies such as UNESCO, or alternatively to be located within the IAEA but as an autonomous entity, governed by its own statute and with its own secretariat.[46]

Due to lack of time, and to the great relief of the nuclear weapon states, neither the Italian nor Mexican proposals were considered by the conference.[47] In its Final Declaration, it simply stressed 'the importance of the potential use of nuclear explosive devices for peaceful purposes within appropriate and effective international safeguards which should be prepared as soon as possible and under strict international control'.[48] This was obviously carefully worded to avoid a dispute over whether non-nuclear weapon states should be permitted to acquire and use their own PNEs.

No sooner had the conference finished than the IAEA General Conference, in September 1968, asked the IAEA Director General to follow up the request of the non-nuclear weapon states by initiating 'studies of the procedures that the Agency should employ in performing the role envisaged for an international body under Article V of the NPT...' It also enunciated the 'willingness and availability' of the Agency to fulfil such a role.[49] This was all a bit presumptuous on the part of the IAEA, at least in the view of Mexico. While the Conference of Non-Nuclear Weapon States had spoken only of the IAEA's 'possible functions' in the PNE field, the IAEA now appeared to be assuming that such functions would automatically fall its way. At the 1968 session of the UN General Assembly, Mexico criticised the haste and manner in which the IAEA had adopted its resolution. It promptly tabled a resolution of its own, requesting the UN Secretary-General, rather than the IAEA, to prepare a report on the establishment of an international PNE service within the Agency. The resolution, co-sponsored by Austria, Iran, Italy, Ireland, Pakistan and 18 Latin American countries, was adopted by a vote of 75 to 9 with 30 abstentions — not a resounding success in UN terms. France and the United States abstained, while the Soviet Union and the United Kingdom voted against.[50]

The United States explained that while it supported an early start to studies on a PNE service, in its view the IAEA, not the UN Secretariat, was the appropriate body to carry out such a study. The United States had abstained, 'on the assumption' that the UN report would not impinge on the proper role of the IAEA and ultimately would be transmitted to the IAEA.[51] The Soviet Union and the United Kingdom on the other hand objected to the omission of any reference in the resolution to the NPT. The Soviets also complained that the resolution spoke of *all* PNEs being under international control, whereas the NPT was intended to apply only to PNEs conducted in the territory of non-nuclear weapon states.[52] Mexico defended the resolution on the grounds that the subject was of common interest and within the competence of both the IAEA and the UN.

Nearly all speakers at the 1968 General Assembly mentioned the need for international control over *all* PNEs, including those of the nuclear weapon states. Unlike the NPT negotiations, the Conference of Non-Nuclear Weapon States, as Elizabeth Young notes, had, if nothing else, 'at least been successful as a course of study in arms control'.[53] According to Young, the Soviets were alarmed by the newly-assertive and demanding behaviour of the non-nuclear weapon states towards such issues as the availability and regulation of PNEs. The Soviet delegate, Roshchin, was 'angry and withdrawn to an unusual degree' and ill-disposed towards most of the proposals of the non-nuclear weapon states. His attitude was that:[54]

> ... there would be no peaceful explosions at all for [NPT] non-signatories; no funds for anyone and no international service for peaceful explosions; certainly no periodic non-nuclear conferences.

The Soviet Union and its allies voted in solitude against the resolution endorsing the Declaration of the Conference of Non-Nuclear Weapon States.

Ironically, despite the procedural skirmishes about who should conduct a study into the PNE issue, the task ended up with the IAEA, as the United States had predicted. The UN Secretary-General simply handed over the job to it. In response, the IAEA Board of Governors in February 1969 established the *Ad Hoc* Committee on the Uses of Nuclear Explosives for Peaceful Purposes, open to all IAEA members. To stimulate discussion, the board circulated an analysis of the role that the Agency might play in relation to PNEs, as well as the status of the technology involved.[55] Not surprisingly, the committee recommended in its report to the board, which in turn reported to the UN Secretary-General, that 'the Agency's prospective responsibilities in the field of peaceful nuclear explosions fall within its statutory objectives and functions to accelerate and enlarge the contribution of atomic energy to peace, health and prosperity throughout the world'.[56]

The UN Secretary-General at the 1969 session of the UN General Assembly duly submitted his report containing the IAEA's recommendation as to its own competence. The report, prepared by a 'group of experts', concluded, in regard to the future of PNEs that:[57]

> The group believes these techniques will develop thanks to the pursuit of research and tests, but that developing countries will not necessarily gain immediate advantage from these technical innovations. There is hope nonetheless that completely underground explosions and earth-moving explosions will have practical applications at the end of the seventies.

The report recommended that the first requirement was for widespread dissemination of information about 'the potential of this new technology as well as its technical limitations and costs . . .'[58] The report also proposed that the IAEA keep developments in the PNE field under constant review in co-operation with those UN agencies that may be interested 'in their economic application and their effects upon the environment'.

The Secretary-General also presented a summary of the views of UN members on PNEs. Among forty countries, representing a wide cross-section of UN members, the predominant view was that PNE technology studies were required.[59] The general expectation was, however, that PNEs might be used for large-scale engineering projects or the recovery of gas, oil and minerals.[60] The report concluded that the technical expertise of the IAEA favoured its assuming the functions of an international PNE service.[61] The Secretary-General considered, however, that these functions would evolve gradually after continued international discussion in the IAEA, the United Nations and possibly other organisations.

Supposedly in response to the UN Secretary-General's recommendation as to the IAEA's competence in the PNE field, a draft resolution was submitted, sponsored by Austria, Canada, Denmark, Japan, the Netherlands and the United States. David Fischer confirms that the United States in fact 'connived' with the IAEA Secretariat in drafting the resolution.[62] AEC influence on US policy was by now notable for its absence. The only developing country to co-sponsor was Mexico. The resolution was a cautious one, noting that the IAEA's 'prospective responsibilities in this field will have to be defined on an evolutionary basis, taking into account the still experimental state of the technology'. The resolution invited the IAEA to keep the development of PNE technology 'under review' and to submit a special report on its studies and activities in the PNE field. The resolution also 'suggested', somewhat vaguely, that the IAEA continue to convene meetings on the scientific and technical aspects of PNE technology and 'initiate studies on the character of the international observation in which it might engage pursuant to Article V of the NPT'.[63] Finally, the resolution requested the inscription of a new item on the next General Assembly's agenda entitled, 'Establishment within the framework of the IAEA of an international service for nuclear explosions for peaceful purposes under appropriate international control'.[64]

This resolution fell far short of asking the IAEA to establish an international PNE service. 'International observation', of which a study had been requested, was only part of such a service. While such observation was of vital concern to the United States and the Soviet Union, it was the least important aspect as far as the developing countries were concerned. Small wonder that Brazil and India abstained on the resolution, arguing that it confused the 'international service' and the implementation of Article V of the NPT.[65] India, Brazil and Pakistan all argued, self-servingly, that PNE services should be available to all IAEA members 'irrespective of whether they had acceded to the NPT or not'.[66]

The resolution was a significant victory for the nuclear weapon states and their likeminded allies. It was the culmination of a series of manoeuvres, beginning with the defeat of the Latin Americans at the Conference of Non-Nuclear Weapon States, designed to commandeer the PNE issue and secure a home for the envisaged international PNE service in the IAEA. Within this Agency, where the developed countries wielded predominant influence, the non-nuclear weapon states could be weaned off the morsels which had been used to help lure them into the NPT.

# *Notes*

1 A. Kramish, *The Peaceful Atom in Foreign Policy*, Harper and Row, New York, 1963, p. 125.
2 Ibid., p. 126.
3 J.P. Jain, *Nuclear India*, Radiant Publishers, New Delhi, 1974, p. 109.
4 D. Fischer and P. Szasz, *Safeguarding the Atom*, Stockholm International Peace Research Institute, Stockholm, 1985, p. 76.
5 M. Kalkstein, *International Arrangements for the Peaceful Applications of Nuclear Explosives*, Stockholm International Peace Research Institute, Stockholm, 1970, p. 34.
6 Ibid., pp. 34–35.
7 Ibid., p. 35; V. Lamm, *The Utilisation of Nuclear Energy and International Law*, Budapest, 1984, p. 146.
8 G.H. Quester, *The Politics of Nuclear Proliferation*, Johns Hopkins University Press, Baltimore, Maryland, 1973, p. 241. David Fischer was unable to corroborate Quester's allegations in a conversation with the author in Canberra on 4 September 1986. Fischer recalls no such rivalry. While unable to offer explicit supporting evidence, Lawrence Scheinman, former Special Advisor to the IAEA, believes Quester to be correct. (From correspondence with the author, 21 March 1989.)
9 *The United Nations and Disarmament 1945–1970*, Department of Political and Security Affairs, United Nations, New York, 1970, p. 307.
10 E. Young, *A Farewell to Arms Control*?, Penguin, Harmondsworth, 1972, p. 118.
11 See Chapter 12.
12 W. Epstein, *The Last Chance*, The Free Press, New York, 1976, p. 126.
13 Young, p. 118.
14 *The United Nations and Disarmament 1945–1970*, p. 310.
15 Young, p. 119.
16 U. Ericsson, *The Question of Nuclear Explosions for Peaceful Purposes by Non-Nuclear Weapon States and the Possibility of Misuse of Such Technology for the Production*

*of Nuclear Weapons*, Document A/CONF. 35/DOC. 3, Conference of Non-Nuclear Weapon States, Geneva, July 1968, p. 19.
17 T. Ginsburg, *The Question of Peaceful Explosions for the Benefit of Non-Nuclear Weapon States*, Document A/CONF. 35/DOC. 2, Conference of Non-Nuclear Weapon States, Geneva, July 1968, p. 6.
18 Ibid., p. 19.
19 Ibid., p. 20.
20 Jain, p. 112.
21 M.I. Shaker, *The Nuclear Non-Proliferation Treaty: Origin and Implementation 1959–1979*, Vol. I, Oceana Publications, London, 1980, p. 413.
22 Ibid., p. 462.
23 Ibid., p. 463.
24 US Congress, Senate Foreign Relations Committee, Hearings on the NPT, 18 and 20 February 1969, US Government Printing Office, Washington DC, 1969, p. 458.
25 Shaker, p. 413.
26 Ibid., p. 411.
27 Ibid., p. 463.
28 Epstein, pp. 173–4.
29 Shaker, p. 463.
30 Epstein, p. 174.
31 Quoted in Ibid.
32 International Atomic Energy Agency (IAEA), Vienna, Document GOV/COM. 23/2.
33 Epstein, p. 131; *The United Nations and Disarmament 1945–1970*, p. 312.
34 Shaker, p. 444.
35 Ibid.
36 Ibid., p. 448.
37 Epstein, p. 130.
38 Conference of Non-Nuclear Weapon States, Geneva, 1968, Document A/CONF. 35/DOC. 15, p. 4.
39 Ibid., p. 5.
40 Ibid., Annex I, Article 8, p. 3.
41 Ibid., Annex II, Article 24, pp. 9–10.
42 Ibid., Annex II, Article 24, p. 10.
43 Ibid., p. 6; *The United Nations and Disarmament 1945–1970*, p. 427.
44 Conference of Non-Nuclear Weapon States, A/CONF. 35/DOC. 15, p. 6.
45 Ibid., Annex I, Article 7 (1), p. 3.
46 US Congress, Senate Foreign Relations Committee, Hearings on the NPT, p. 458.
47 Shaker, p. 453.
48 *The United Nations and Disarmament 1945–1970*, p. 315.
49 *Report of the Ad Hoc Advisory Group on Nuclear Explosions for Peaceful Purposes*, IAEA Document GOV/1854, 26 August 1977, IAEA, Vienna, 1977, p. 6.
50 *The United Nations and Disarmament 1945–1970*, p. 317.
51 US Congress, Senate Foreign Relations Committee, Hearings on the NPT, pp. 481–82.
52 Ibid.
53 Young, p. 121.
54 Ibid., p. 122.
55 IAEA Document GOV/COM. 23/2.
56 *The United Nations and Disarmament 1945–1970*, p. 324.
57 G. Fischer, *The Non-Proliferation of Nuclear Weapons*, Europa Publications, London, 1971, p. 114.
58 Epstein, p. 174.
59 Shaker, p. 404.
60 Ibid.
61 Epstein, p. 174.
62 From an interview with David Fischer, Canberra, 4 September 1986.
63 *The United Nations and Disarmament 1945–1970*, p. 325.
64 Ibid.
65 Shaker, p. 413.
66 Jain, p. 112.

# 9

# Australian Guinea Pigs: The Cape Keraudren Project

*Like a modern Moses' rod, the mighty subterranean blast will part the ocean, and the bed below it in an apocalypse that will be detected by seismologists the world over. Into the breach — months later — will sail the 1,000 foot leviathans now being built to ship Australian iron ore and later, pellets, to the four corners of the earth.*

The *Age*, Melbourne,
30 January 1969

WHILE the nuclear weapon states were squabbling with the developing countries over the establishment of a PNE 'regime', plans were being laid for what was heralded as the world's first actual use of nuclear dynamite. Arms control was the furthest thing from the minds of the planners. If carried out as planned, the project, the creation of a nuclear harbour in north-west Australia, would violate the PTBT. And far from being the first time a non-nuclear weapon state would obtain the benefits of PNEs under the auspices of Article V of the NPT, the project was to be carried out in Australia, a country which at the time was 'quietly opposed to the NPT'.[1] The timing of the project was in fact exquisite, landing it straight into controversy over US ratification and Australian signature of the treaty.

It was Edward Teller who first proposed the idea of a nuclear harbour in Western Australia, during a visit in the early 1960s.[2] Teller recalls telling West Australian mining magnate Lang Hancock about Project Chariot and arguing that public acceptance of a harbour project in Western Australia would be enhanced by a 'demonstration shot' at one of Hancock's iron-ore mines.[3] By 1964 the Plowshare 'team' at the Livermore Laboratory was considering the idea further (in addition to water diversion and conservation schemes in Australia), although 'only in the sense of paper studies of one sort or another'.[4] By 1966 the USAEC[5] was suggesting in its annual report that new harbours, 'particularly on the west coasts of Africa, Australia and South America', would 'greatly assist in the economic development of these regions'.[6] The Western Australian government, despite its experience with British nuclear weapon tests on the Monte Bello Islands off the West Australian coast in the early 1950s, was certainly interested. It raised the matter in 1966 with the then Australian Prime Minister, Robert Menzies.[7] In January 1967 Teller, writing in the US magazine *Technology Week*, repeated his proposal, along with the use of PNEs for deepening the Torres Strait and damming the Fortescue, Ord and Fitzroy rivers in Western Australia and the Darling and Diamantina rivers in eastern Australia.[8]

In January 1967 Glenn Seaborg visited Australia, along with USAEC Commissioner Tape and others from the USAEC.[9] He discussed with the Australian Atomic Energy Commission (AAEC) the possibility of a nuclear harbour in the Pilbara iron-ore region of Western Australia.[10] This was followed in May 1968 by a visit by Plowshare Director John Kelly and Dr G. C. Werth, head of Plowshare activities at Livermore, who toured a number of possible dam and port sites in the region, including Cape Keraudren.[11] According to the AAEC:[12]

> The visit aroused considerable interest, although it does not imply that Australia is about to undertake a Plowshare-type operation. Many technical, political and international problems would have to be resolved first. Nevertheless, the Commission is convinced that these difficulties can be overcome.

Australia's quickening interest in PNEs soon had an impact on its attitude towards the NPT. In the same month that the Plowshare team was in Australia, the Australian Ambassador to the United Nations, Patrick Shaw, enunciated Australia's policy on PNEs to the First Committee of the United Nations General Assembly in New York. Australia, he said, 'as a continent with a low rainfall, a poorly indented coastline and little topographical relief', had a 'special interest in the possible use of nuclear explosions for major engineering projects'.[13] While Australia recognised that 'at this stage of technological development' an effective NPT could not permit the production of nuclear explosive devices by non-nuclear weapon states, nevertheless it strongly believed that the NPT 'must not impede progress in the development and application of the technology of peaceful applications of nuclear explosives'.[14] Australia felt that *all* countries should have access to PNEs — a view at variance with the idea that Article V of the NPT was an incentive for states to sign the treaty. Australia also opposed the Swedish view that technical and economic judgements on projects should be the responsibility of an international body and not of the state directly concerned. In Australia's view international arrangements under the NPT should go no further than providing 'appropriate assurances' on safety and an 'adequate demonstration' that the explosives would not be used for nuclear weapons development.

This somewhat confused PNE policy reflected divisions within the government over the NPT. While the External Affairs Department favoured signature and the Departments of Defence and Prime Minister took 'a mixed and more complex view', the AAEC was strongly opposed.[15] According to Ian Bellany, it is hard to avoid the conclusion that Australia's early NPT policy was 'more the product of introspection by the Atomic Energy Commission than the outcome of balanced discussion among other relevant government departments', a clearly unsatisfactory situation given that the NPT had a 'wide political importance quite distinct from its work-a-day relevance to a country's technological progress'.[16] The chairman of the Commission, Professor (later Sir) Philip Baxter, who was widely believed to support the acquisition of nuclear weapons by Australia, was opposed to any action that would curtail Australia's choices in the nuclear field.[17] For the anti-NPT forces within the government and bureaucracy, such as Baxter, nuclear dynamite symbolised the type of technology they feared Australia could miss out on if it signed the NPT.

Confusion over Australian policy was heightened when, just three months after Australia's statement to the General Assembly, the Minister for National Development, David Fairbairn, received a report which dashed hopes that PNEs could be used to create water storages in Australia. Fairbairn responded pessimistically that:[18]

> The idea is current in some circles that the many problems associated with civil engineering works of this nature have been solved. This is not so. There are very real practical difficulties and international obligations precluded the use of these explosions in Australia. In any case the limited test ban treaty prohibits excavation by nuclear blasting when this released radioactivity into the atmosphere. This effectively prohibits the use of nuclear explosions of this type anywhere in Australia, irrespective of the non-proliferation treaty.

Despite these strong Ministerial reservations and the AAEC's assurances that no Plowshare project was about to be undertaken in Australia, by 17 September 1968 talks were being held on a nuclear harbour for Western Australia. No Australians took part and the Australian government was not aware of their existence. They were held at the Livermore Laboratory in California and involved Plowshare scientists and officials, including Edward Teller, and National Bulk Carriers (NBC), an American corporation.[19] Discussion focused on the general aspects of nuclear excavation and the possibilities, advantages and problems of a nuclear harbour at Cape Keraudren in Western Australia.

A subsidiary of NBC, Sentinel Mining Company, had high-grade iron-ore deposits at Nimingarra, about 60 kilometres inland from Cape Keraudren. Its agreement with the West Australian government obliged it to build a harbour at the Cape to avoid overcrowding nearby Port Hedland.[20] There were no alternative natural harbours and the continental shelf was too shallow and the tides too extreme to allow ships to approach the shore. Possessed of a hinterland that, according to Bruce Bolt, was 'isolated and barren, with little ecological value', the Cape was 'ideal' for a nuclear

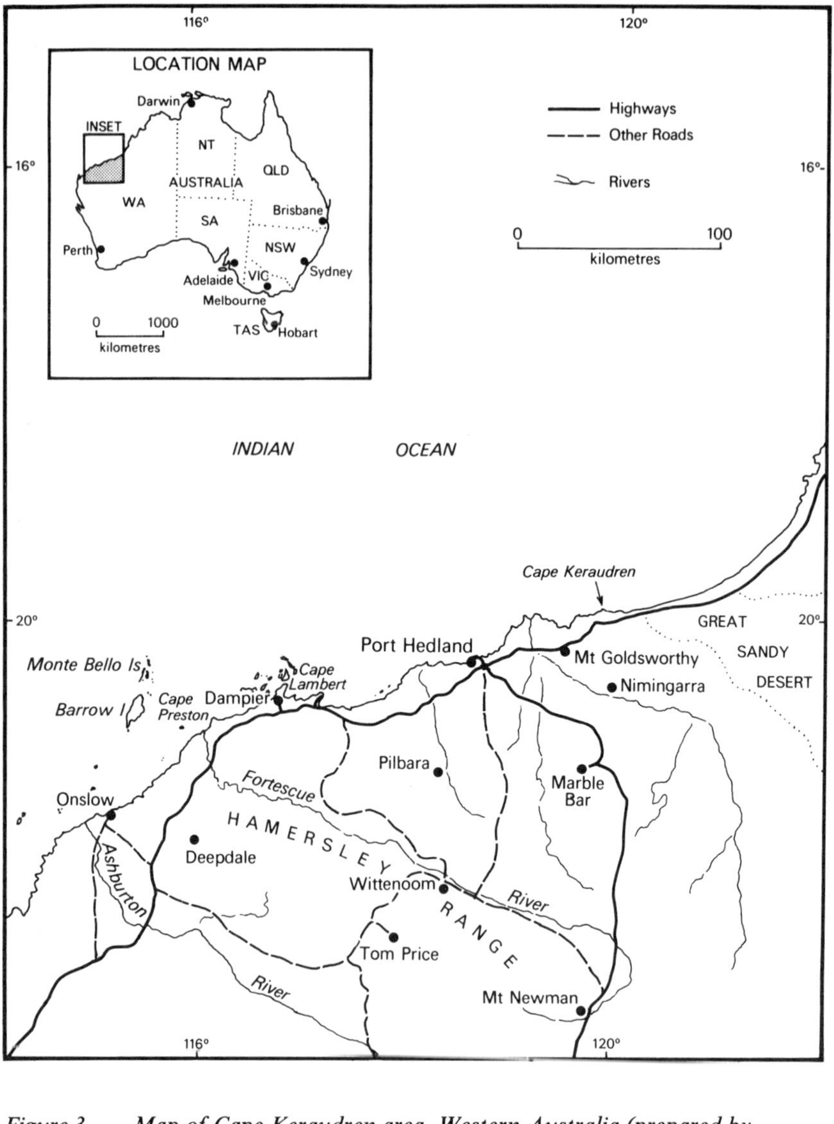


*Figure 3 Map of Cape Keraudren area, Western Australia (prepared by Cartography Department, Research School of Pacific Studies, Australian National University, Canberra).*

excavation project.[21] Glenn Seaborg says that the project could not be accomplished at all by conventional means.[22] Edward Teller had informally raised the idea of a nuclear harbour at Cape Keraudren with NBC in the late northern summer of 1968.[23] There is no evidence that NBC or Sentinel had thought of the idea before then.

Livermore's plan was to use five underwater 200 kiloton nuclear explosions to blast a trench in the sea bed 1300–1600 feet (400–500 m) wide, 6000 feet (1800 m) long and 200–400 feet (60–120 m) deep, surrounded by broken rock. The crater would either abut the shore or be connected to it by a causeway. The entire project would be conducted from a ship anchored off the Cape.[24] The nuclear explosives would be prepared at Livermore and transported to Australia on this ship. At the site, personnel would be housed and fed on board. Emplacement of the explosives would be also done from there, using barge-mounted cranes. The devices would be placed 1100 feet (330 m) apart in the sea bed at a depth of 800 feet (240 m). The vessel would then move to a safe distance and the 'row charge' of explosives would be detonated simultaneously, by radio.

One issue that soon emerged was Sentinel's concern that the project be completed within eighteen months. Its agreement with the Western Australian government specified that mining could commence only when a harbour was in place. In October 1968, NBC, on Sentinel's behalf, laid down three pre-conditions for considering the project further:

(1) agreement of the governments involved should be obtained by 1 March 1969
(2) conventional earthmoving needed to convert the nuclear trench into a harbour should be started by 1 March 1970
(3) costs to NBC should be limited.

Livermore replied that it would be impossible to give a commitment on government approvals, but the Laboratory was optimistic. On the second condition, Livermore thought such a schedule was possible. As the Cape Keraudren project would have great relevance to the Panama Canal project, it would clearly suit the USAEC to have it carried out well before 1 December 1970, when the Canal Commission was due to submit its report. The third condition would be a matter for discussion between the USAEC and NBC, but the Laboratory felt that data from Cape Keraudren would make a 'substantial contribution to the Plowshare program and the transisthmian canal studies', the implication being that the bulk of the costs would be borne by Plowshare.

In December 1968, as agreed by Livermore and NBC, Sentinel Mining sought the support of the West Australian government for a feasibility study of the project. The State Premier, Mr Brand, agreed, subsequently informing the Australian government of his government's interest in the project, providing it cost the state nothing and safety measures satisfied both governments.[25] Meanwhile, on 27 December, US Congressman Hosmer

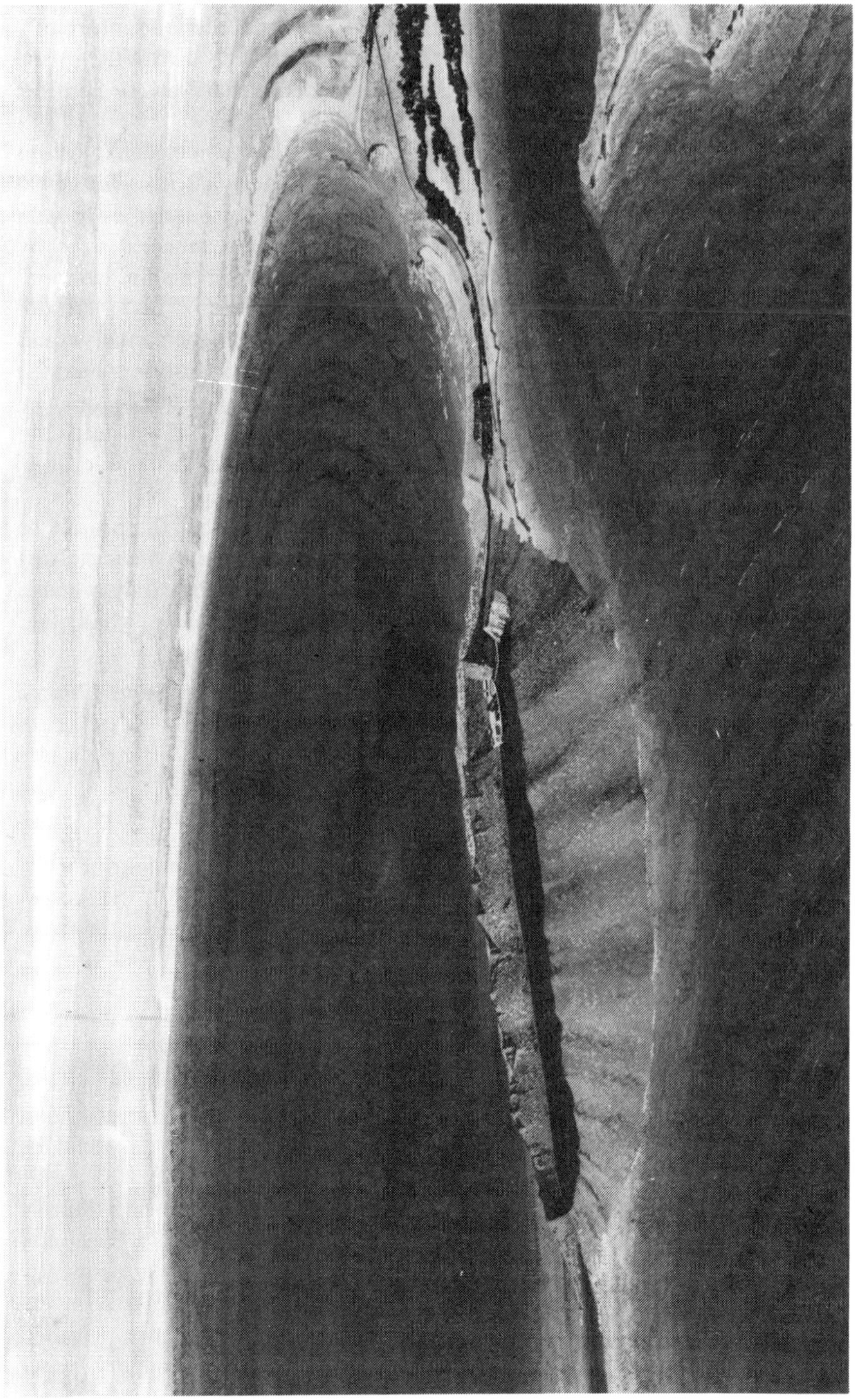

wrote to President Johnson, urging that $267,000 be immediately added to the USAEC budget to fund the Cape Keraudren feasibility study, with an additional $2.2 million to be added in five months for carrying out the actual project.[26] Hosmer estimated that the total cost of the project would be $8 9 million.

Ignorant of the extent of plans for an unprecedented nuclear project on Australian territory, and apparently not yet having received Brand's request, the Australian government was caught 'flatfooted' by the public release of Hosmer's letter in Washington on New Year's Day 1969. Fairbairn reacted cautiously, saying:[27]

> Such techniques may eventually play a part in our development. The only thing is everyone wants everyone else to try the first one.

Noting that 'the many technical problems' associated with nuclear dynamite would be minimised by 'the remoteness of Cape Keraudren from any major centres of population', Fairbairn warned nonetheless that 'complications would be raised by the nuclear test ban treaty with possible diplomatic repercussions'.

But with Cape Keraudren now front page news, the US President involved, and Western Australia keen, Canberra found itself 'press-ganged' into the project. On 22 January the Australian Embassy in Washington delivered a note to the State Department formally requesting a joint AAEC/USAEC feasibility study into the Cape Keraudren proposal.[28] In apparent anticipation of the request, the chairman of the Texas Radiological Advisory Board, Dr J. R. Maxfield, had inspected Cape Keraudren a week before, opining: 'I think Western Australia should be given the opportunity to construct a harbour with the aid of nuclear energy'.[29]

In announcing the request, Fairbairn said that the feasibility study, involving detailed investigation of the technical and economic aspects of the project, plus a comprehensive safety assessment, would be merely the first step. If the results were favourable, further studies into aspects such as geology and meteorology would be needed. If all aspects of the project, 'including fulfilment of Australia's international obligations', were satisfactory, the nuclear detonations would take place 'in about twelve months time'. The project was being treated as urgent, Fairbairn said, because of Sentinel's program for iron-ore shipments, and 'because the spread of population is reducing the number of sites at which construction by nuclear blast can be carried out'.[30] This latter reason seems rather implausible given that the vast, empty coastline of Western Australia was unlikely to become populated overnight.

Fairbairn admitted that problems could arise because 'underwater blasting' had not yet been applied in practice and because the United States

*Figure 4 An artist's conception of the proposed nuclear harbour for Cape Keraudren, Western Australia. (Courtesy of Lawrence Livermore National Laboratory.)*

had not previously used 'civil nuclear explosives' outside its territory. He misleadingly reassured the Australian people, however, that 'Plowshare techniques so far have proved markedly successful in other fields', a marked contrast to his statement only months earlier about the problems associated with using PNEs to create water storages.

President Nixon took just five days to agree to Australia's 'request', directing the USAEC to 'actively and promptly study the feasibility of the project'.[31] A joint US/Australian press release announcing the decision was nevertheless delayed for forty-eight hours, reportedly while Australia successfully sought assurances that the announcement would not result in pressure on it to sign the NPT.[32] Responding to the joint announcement, the Australian Prime Minister, John Gorton, waxed lyrical, describing the Cape Keraudren project as 'the first practical application of the atom as a mighty engineering tool for civil construction, helping to open up new frontiers'.[33] Despite the fact that the project had been initiated by the Americans for their own purposes, Gorton said that the Australian government 'warmly appreciated' the US President's 'prompt, positive response'. The feasibility study would, he said, be followed by 'an examination in the light of the results of that study of the applicability or otherwise of the limited test ban treaty' and then a 'new decision by each Government concerned as to whether the execution of the project will take place'.[34]

Later Fairbairn detailed the administrative arrangements for the feasibility study.[35] To oversee the project, an Australian Plowshare Committee, headed by Doug McKay of the AAEC, would be established. In addition, a Cape Keraudren Project Committee, headed by Alan Wilson, Deputy Director of the AAEC's Lucas Heights research establishment, would co-ordinate technical and economic investigations, act as the working body for the feasibility study and liaise with the USAEC. This committee, which included representatives of the Bureau of Mineral Resources, the Bureau of Meteorology, the West Australian government and Sentinel, was by 14 February meeting Plowshare officials at the Nevada Test Site and inspecting Plowshare excavation craters.[36]

Press reaction to the Cape Keraudren proposal was almost universally favourable in both Australia and the United States. The *Sydney Morning Herald* exclaimed that there would be 'some honour and excitement' in trying out for the first time the nuclear excavation of a harbour.[37] The Sydney *Sun-Herald* reported that the project, 'because it is an experiment never before attempted by man ... is being likened to the historic Apollo 8 mission around the moon last Christmas'.[38] The *Washington Star*, in an article entitled 'The Atom As Excavator', was ecstatic:[39]

> ... the Australian proposal will be a trailblazer in the application of the non-proliferation treaty's provisions for co-operation between the atomic 'haves' and 'have-nots'. And beyond that, of course, it will demonstrate that the atom is a revolutionary sort of excavator to be used for the speedy and economic building of harbours and waterways around the world, including a new canal across Central America.

The Melbourne *Age*, after claiming that the blast would be 'five times Hiroshima' (in fact it would be almost seventy times) predicted that the project would be a world 'first' and 'will spotlight the nation as few other events could'.[40] The *Age* described the project, more accurately than it knew, as a 'dummy run' for the Panama Canal project. As a result, it said, the United States government would be providing much of the cost. The *Age* also recorded that seismologists expected the blast would register 4 on the Richter scale, while the earthquake that levelled the Western Australian town of Meckering in 1968 registered 7, a fact that would not have reassured residents of Port Hedland, less than 150 kilometres south of Cape Keraudren.

The *Age* reported American sources as saying that the *experiment* would be 'completely safe', with 'negligible fallout' and that explosions of similar force had been carried out closer to Las Vegas than Cape Keraudren was to its nearest centre of population, Port Hedland.[41] This was misleading. Unlike nuclear tests at the Nevada Test Site, which were mostly fully contained underground explosions, the Cape Keraudren detonations would necessarily vent radioactivity into the atmosphere. And Las Vegas had never experienced five 200 kiloton nuclear explosions detonated simultaneously. In the light of the 1985 report of the Royal Commission on British atomic testing in Australia in the 1950s,[42] which was critical of safety measures taken at the time, it is instructive to note Fairbairn's assurances that:[43]

> Before any nuclear excavation project would be authorized in Australia, safety assessments would be made embracing all relevant aspects and the findings would be assessed independently by leading scientists in Australia as had been the case with other nuclear experiments conducted in this country.

It is also notable that, as in the case of the British nuclear tests in Australia, there was no mention, in all the talk of the Cape Keraudren region being uninhabited, of the Aborigines. As nomadic people the Aborigines were of course more susceptible to radioactive fallout than whites. The white inhabitants, at least those in Port Hedland were — so the *Sydney Morning Herald* reported — quite blasé about the project, living as they did in a region where 'men are moving mountains, and developers talk in hundreds of millions of dollars'.[44] Port Hedland cab drivers and workmen were reported as shrugging the matter off with comments like:[45]

> It's just another big bang. We had one in the Monte Bellos off the coast a few years ago, you know, when they tested the atom bomb.

The only sour note in the press coverage was an article in the *Sydney Morning Herald* which implied that Australia was being used as a nuclear 'guinea pig' by the Americans, a charge described by AAEC executive commissioner Timbs as 'unreasonable'.[46] In Canberra the US Ambassador, Mr Crook, declared that there would be 'nothing experimental' about the project, as everything proposed for Cape Keraudren had been tried experimentally already.[47] Despite these denials, the issue soon embroiled the

Cape Keraudren project in hearings of the US Senate Foreign Relations Committee on US ratification of the NPT, which began in Washington on 19 February 1969.

The Committee Chairman, Democrat Senator J. W. Fulbright, expressed concern that the first ever use of nuclear dynamite was to take place in a country that had not yet signed the NPT and appeared reluctant to do so. This appeared to sharply diminish the value of Article V as an inducement to states to adhere to the treaty to obtain the benefits of PNE technology.[48] Fulbright quizzed Glenn Seaborg:[49]

> I must say as long as we are entering the Nonproliferation Treaty to choose a country that has not signed it and is not interested enough to sign it seems a little odd. There must be some other country you could favor with such an experiment, assuming it is a favor. I am not sure it is.

Seaborg replied, completely contradicting the US Ambassador to Australia, that the USAEC saw Cape Keraudren as an 'experiment', not an actual use of PNEs.[50] It did not, therefore, fall under NPT auspices. The Chairman, not entirely convinced, pressed further:[51]

> *Chairman:* Article V is a central element, as you say, in our ability to encourage other prospective signers [of the NPT] to relinquish their options. It doesn't seem to apply to Australia. You are using it in a case apparently where they don't want to sign it.
>
> *Dr Seaborg:* Yes; I was referring there to the furnishing of nuclear explosion services for developed uses. The Australian project is a research and development project which conforms to our own programmatic interests, and replaces an experiment that we would otherwise do elsewhere.
>
> *Chairman:* I think we are bothered by semantics. You told me that you know so much already about the project, that there is no danger in conducting it. If that is true, I think the project is developed. If it is purely experimental, how do you know it wouldn't destroy the ecology of that area, which would be a great disaster?
>
> *Dr Seaborg:* Well, we need to conduct more excavation experiments in order to develop the technology so that it can then be used — nuclear explosives can be used — for these engineering projects in the future such as the construction of canals and harbors and so forth. We clearly need to do such experiments . . .

Seaborg's casuistry reflected a genuine dilemma on the part of the USAEC. By describing the Cape Keraudren project as an 'experiment', the Commission risked giving the impression that the Australians were indeed being used as 'guinea pigs'. Yet such a description was needed to secure funds for the project from Congress, which had to be convinced that Cape Keraudren would replace one of the Nevada experiments — identified by John Kelly as probably the Galley cratering shot — designed to assist the Panama Canal Commission.[52] More important, under its charter the USAEC was not permitted to provide actual PNE services, either domestically or abroad. The Commission was therefore obliged to describe Cape Keraudren as an experiment to avoid acting illegally.[53] The paradox behind this semantic charade was that even if the Commission had been legally entitled to claim that it was offering an actual PNE 'service' to Australia, the Cape Keraudren harbour project, being the first nuclear excavation project ever undertaken, would still have been very much an experiment. As a Senior Lecturer in Physics at Monash University in Melbourne was to point out, the project would involve a multiple explosion ten times larger than any previous PNE and would therefore be 'as much an experiment as a work of development'.[54] Clearly, the major motivation behind the Cape Keraudren project was not that it would assist Australia or Sentinel Mining, but that it would provide invaluable experience and data for the Panama Canal project. As Glenn Seaborg told the JCAE:[55]

> If such a project were to materialize, it would have the bonus effect of providing direct engineering experience difficult to achieve in experiments on the Nevada test site . . .

Meanwhile in Australia, suspicions were being aroused that the US Senate was deliberately using Cape Keraudren to pressure Australia to sign the NPT.[56] This was denied by both the Australian government and by the US Ambassador.[57] While the Senate hearings, by publicising Australia's hesitancy about the NPT, certainly increased pressure on the country to sign the treaty, this was almost certainly incidental to the committee's main concern, which was to probe US credibility in regard to Article V. In any case the USAEC had cast some doubt on the relevance of the NPT to the Cape Keraudren project.

Questions were also asked in the US Congress and the Australian Parliament about the implications of Cape Keraudren for the PTBT. Both governments tried to fob the questions off by claiming that the issue did not arise since the project was only at the feasibility stage.[58] When pressed by Senator Fulbright, Glenn Seaborg said that if it were estimated that radioactivity at Cape Keraudren could not be 'controlled', he would seek an amendment or 'understanding' to the PTBT. An 'understanding', Seaborg explained, would be short of an amendment, which he agreed would be a 'time-consuming and laborious process'. Fulbright seemed relieved, noting that 'clearly you are not intending to just go ahead and violate the treaty just to satisfy some experimental urge'.[59]

Environmental concerns were also raised in both Congress and Parliament. Senator Fulbright warned Seaborg that: 'You might be doing them [the Australians] a great disfavor by destroying their beaches'.[60] A Labor Party Senator quoted to the Australian Senate a warning by Dr Marvin Kalkstein of the State University of New York that:[61]

> There has not been sufficient testing of large nuclear explosives for earth moving to warrant proceeding with an application; and what testing there has been has indicated difficulties in containing the radioactive debris. Results of cratering tests in Nevada alluvium to date might also not be directly

> applicable to a deepening operation beneath the ocean floor in Australia. It is highly doubtful that the technical feasibility could be established in so short a time as a year, particularly in view of the problems and delays that have so far presented themselves in the [US] Atomic Energy Commission's own program.

On the same day, in the Australian House of Representatives, a West Australian Labor Party MP, Mr Collard, asked whether or not the United States had found a way of eliminating radioactivity from PNEs or of accurately estimating beforehand their likely fallout. Fairbairn replied that 'the latest forms of nuclear device do reduce very considerably the amount of radioactivity that is released'.[62]

In fact, as far as is known, the United States had never detonated a nuclear device under the seabed and, as Teller and his colleagues admitted in their 1968 work, *The Constructive Uses of Nuclear Explosives*, it could only be assumed that the radioactivity vented into the atmosphere from such an explosion would be less than that from a shot on dry land.[63] The United States *had* by 1969 detonated at least five nuclear devices underwater.[64] The first of these, in the Bikini Island lagoon in 1946, threw a huge volume of water into the air, where it formed a column contaminated with the products of the explosion. When the column fell back to earth it spread out rapidly, contaminating a wide area, a phenomenon known as 'base surge'. British nuclear scientist William Penney later described it as 'for all the world like a thin pancake mixture spreading as it is poured into a frying pan'.[65]

Finally, questions were asked in the Australian Parliament as to why the project was being rushed. Glenn Seaborg said he expected the Cape Keraudren study to take just two months.[66] As Professor Kalkstein put it:[67]

> We have here a case where the commercial plans of a private enterprise appear to take precedence over treaty obligations and present limitations of the technology.

Opposition Leader Gough Whitlam sought confirmation of Alan Wilson's announcement, at a Sydney press conference in February, that the feasibility study was expected to be completed by June 1969 and, if it was favourable, the blast could take place in January 1970. Whitlam quoted the 1964 Australian report on Plowshare to the effect that, 'since the occurrence of plants and animals may vary widely between seasons it is essential that the ecological study [for PNE projects] should cover at least the complete seasonal cycle', that is, at least one year.[68] This compared with the environmental impact study for Project Chariot which, according to its authors, 'set some pattern for future studies', was carried out over three seasonal cycles, and had taken four years to complete.[69] That report had warned of the speed and scope of nuclear excavation projects and:[70]

> . . . the ever-present danger of achievement simply for achievement's sake, with inadequate considerations for effects on related ambient phenomena, con-

> ditions, and bioenvironmental processes ... Man's oversights can be costly, and ignorance of his bioenvironment can only result in irretrievable losses, extended disaster, and possible diminishment of his dominion over the earth.

Australian scientists were beginning to have qualms about the Cape Keraudren project on such grounds. The Director of the Western Australian Museum, Dr Ride, criticised the lack of research on the possible effects of nuclear detonations on marine life around Cape Keraudren.[71] A Senior Lecturer in Physics at Monash University, warning that the Cape Keraudren project should be 'regarded with some suspicion', expressed the hope that the technical committee would decide that more time was needed to study the project.[72] On the other hand, the Professor of Thermonuclear Physics at Sydney University, Dr Watson-Munro, predicted that there would be no danger to people or animals from the Cape Keraudren detonations.[73]

The growing controversy was cut short by Fairbairn's announcement in Parliament on 25 March 1969 that a meeting was being held in Washington that day between Sentinel, the AAEC and the USAEC. Clearly there were difficulties with the feasibility study. If proceeded with, said Fairbairn, it would take 'rather longer than we had originally expected'.[74] Four days later, on 29 March, less than two months after the study had been agreed upon, the AAEC and the USAEC announced a halt to the project.[75] The official reason was Sentinel's decision, after a 're-evaluation of its opportunities in the mining and marketing of iron-ore', to 'limit its participation in', or more specifically 'limit its financial commitment to', the feasibility study.[76]

Suspicions about the real reason for the cancellation were rife on both sides of the Pacific. US Congressman Hosmer asked John Kelly, during JCAE hearings, to state 'in truth and in fact' the reason why the Cape Keraudren project 'fell apart'. Kelly replied:[77]

> ... the industrial sponsors felt that they could not participate in depth financially because of the limitations on the opportunities in their particular iron-ore mining venture in that area.

Hosmer volunteered, 'More specifically, they failed to sign a contract with the Japanese for the ore they were going to ship from the point, as I understand it'. Kelly replied, 'I think that was part of it, yes, sir'. Hosmer repeated this explanation at a symposium in Las Vegas on 8 April 1969, when he stated that the project collapsed 'for totally non-nuclear reasons'.[78] Bolt speculates that the reason for the failure of the iron-ore sale to Japan may have been the Japanese attitude towards nuclear testing.[79]

Kreith and Wrenn suggest the project was cancelled because 'it was feared there would be severe ecological damage as well as radiation hazards'.[80] George Quester asserts that it was the US government which changed its mind, either because Australia had not signed the NPT or because Australian plans for the project were 'badly thought out'.[81] A West Australian Senator alleged that the project was cancelled 'following opposition in the United States of America, because Australia had not signed the Nonproliferation

Treaty'.[82] Glenn Seaborg has claimed that '... there were serious political obstacles' and 'the economic picture also changed'.[83]

The truth, as revealed by the Livermore Laboratory's own documents, is that Sentinel Mining Company got cold feet about the project for a variety of reasons. As early as 19 February 1969, the USAEC noted that the company was showing reluctance to commit itself to the feasibility study.[84] To start with, Sentinel had begun to lose confidence in the quality of the iron-ore deposits at Nimingarra.[85] The company felt it had an 'absolute obligation' to the Western Australian government to satisfy itself of the economics of the venture before it went ahead. Sentinel was also worried by the amount of publicity the project was attracting. The USAEC was, for instance, planning to inform the IAEA Board of Governors of the details of the project. Like many companies, Sentinel was wary of getting involved in politics.

The most important reason for Sentinel's withdrawal was, however, confusion over who would pay for the nuclear-excavated harbour. Indeed, all the parties involved seemed to have assumed that others would bear the major share of the costs. The USAEC believed, as Seaborg told the Senate Foreign Relations Committee, that the USAEC would pay only for the cost of the nuclear explosives, while Sentinel and 'other participants' (presumably the Australian and Western Australian governments) would bear part of the remaining costs.[86] For their part, the two governments believed they would be liable for no costs at all.

As for Sentinel, its manager, Mr Peter Stork, stated that the company had originally understood that Plowshare would pay for the entire cost of the feasibility study and the blast, while the company would be responsible only for converting the crater into a harbour. Sentinel claimed it told Plowshare officials at the outset that its financial contribution would be 'quite limited'. Stork confirmed that disagreement over cost sharing for the feasibility study had caused the project to be 'deferred indefinitely'.[87] He said there was a large amount of the cost that none of the parties involved was prepared to pay. Alan Wilson confirms that it was indeed disagreement over Sentinel's share of the costs that killed the feasibility study.[88] He and Plowshare's John Kelly had agreed before the final meeting with Sentinel that the feasibility study could not proceed any further without a cost sharing agreement.

NBC claims that Sentinel was pressured into participating in the project by the Livermore Laboratory and the Plowshare Division of the USAEC. According to an NBC official, the proposal for the feasibility study was the result of '2–3 years promotion by the USAEC Plowshare Division', whose objectives were to 'get some company to do something in the Plowshare field outside the United States and to say something publicly', to help pave the way for the Panama Canal project. The evidence suggests that the Cape Keraudren project was indeed cooked up by Livermore, at Edward Teller's instigation, sold to a reluctant Sentinel and then foisted onto an unsuspecting Australian government. Unlike the Australians, the USAEC and Livermore clearly regarded the project as primarily an experiment which would give Plowshare its long-awaited field test of a nuclear harbour project, without the environmental and political difficulties that had killed Project Chariot. Australia was to be a proving ground and Australians the guinea pigs in a 'planetary engineering' experiment.

Yet, even if the feasibility study had not collapsed in March 1969, the project itself would probably have done so sooner or later, if only for financial reasons. The USAEC did not have the funds to carry out the project in 1970 even if the feasibility study was favourable, as pointed out by a concerned Chet Holifield to Seaborg in January 1969.[89] In the end, despite President Nixon's approval of the project and lobbying by Hosmer, the Commission failed to obtain from the Bureau of the Budget even the special allocation requested for the feasibility study. Hosmer, in a speech to Congress on 14 April 1969, bitterly accused the Bureau, the State Department and ACDA of opposing funds for Cape Keraudren as part of their anti-Plowshare campaign.[90]

Despite the political and financial difficulties that had torpedoed the Cape Keraudren project, the Australian and US atomic energy commissions remained unbowed. In their joint press release announcing the termination of Cape Keraudren, they announced they would be seeking an alternative site for a nuclear excavation project in north-west Australia.[91] This corroborates the thesis that it was for financial reasons rather than any problem over Australia's attitude towards the NPT, that Cape Keraudren was cancelled. Speaking to reporters before leaving Sydney for a visit to the United States, Prime Minister Gorton confirmed that while the project was 'probably' unlikely to go ahead, other sites would be examined.[92] Having had their fingers burned over Cape Keraudren, however, the two commissions moved cautiously. On 21 April 1969 the AAEC wrote to Seaborg inquiring whether the USAEC would be prepared to participate in a 'general, preliminary assessment of the suitability of various sites in Australia for the construction of harbours with nuclear explosions'. Seaborg agreed, but noted that if any suitable sites were identified they would require 'more detailed study if specific interest in developing a port actually arises'.[93] The funding arrangements for the survey were also carefully delineated. Representative Hosmer revealed to Congress that Cape Hedland and Cape Preston were emerging as alternative sites for alternative companies. He exclaimed, excitedly:[94]

> An Australian Plowshare harbor is still a possibility. You will be hearing about it quite soon. My lips are sealed for now.

In 1969 the Australian Plowshare Committee produced a report on the possibility of a nuclear harbour at Cape Preston, south-west of Cape Keraudren. Australia's largest company, Broken Hill Proprietary Limited (BHP) had at one stage been interested in such a project. Although the report recommended that the project was feasible, nothing ever came of it, there being no Australian individual or organisation enthusiastic enough to push the idea forward. After 1969 neither the USAEC nor the AAEC ever again mentioned an Australian harbour project in their annual reports.

The role of the Australian government in the abortive Cape Keraudren project can only be described as incredibly naïve. As Bellany put it:[95]

> ... the Australian Government would do well to gather around itself a better balance of technical advisers to meet future situations than appears to have been the case with the NPT: the apparent naïveté of the advice received by the government on the Cape Keraudren business alone should be taken as a warning.

In the first place, Australia was a party to the PTBT, which US officials had long admitted would preclude the use of nuclear dynamite in excavation projects. Glenn Seaborg told the US Senate Foreign Relations Committee in 1963 that a harbour in Alaska 'could not now be built until the treaty is amended'.[96] As Marvin Kalkstein, who described the USAEC's attitude towards the PTBT as 'careless',[97] pointed out:[98]

> The detection in Canada of radioactive debris attributed to the Project Schooner cratering test of December 8, 1968, suggests the strong possibility that radioactive debris from Cape Keraudren could reach Indonesia or other Pacific islands.

Once such debris was detected outside Australia's three-mile offshore territorial zone, a violation of the PTBT would be recorded. An amendment to accommodate a project like Cape Keraudren was remote in the best of circumstances, but extremely unlikely within the time frame set by Sentinel. The USAEC lamely told the JCAE in 1969 that a study of the PTBT problem in relation to Cape Keraudren had been undertaken by the US government but that no decision could be taken because 'the utility of the various alternatives' could not be discussed until the feasibility study for the project had been completed.[99]

The Soviets were certainly aware of the implications of the Cape Keraudren project for the PTBT. A senior Soviet Foreign Ministry official recalls discussions at Geneva with the British, presumably during the NPT negotiations, about whether the project was prohibited by the PTBT because it involved 'underwater' explosions (which were banned outright by the treaty) or whether, because the explosives would be buried in the seabed, it would run afoul of the treaty's provisions concerning venting of underground explosions.[100]

A second reason why the Cape Keraudren proposal was ill-conceived, was Australia's opposition to French atmospheric testing in the South Pacific. As early as April 1963 Australia had expressed deep regret over France's intention to move its nuclear testing from Algeria to the Pacific.[101] In an *aide memoire* to the French government on 9 September 1963, Australia claimed that 'scientific knowledge of the effects of radioactive fallout is incomplete and... the results of even a small overall increase in the general level of radioactivity cannot be predicted with certainty'.[102] Now, six years later, the same Australian government was seriously contemplating permitting five nuclear devices to be detonated off Western Australia.

In April 1970, only two months after announcing the Cape Keraudren study, the government hypocritically protested to France that 'the resumption of nuclear testing in the atmosphere without regard to the deep concern that this action arouses amongst the peoples of the area and without regard to the wishes of these peoples is a matter of deep regret to the Government of Australia'.[103] Further, Australia noted, it approached this question:[104]

> ... as a party to the Partial Test Ban Treaty of 1963, which it would like to see universally applied and accepted. Continued atmospheric testing weakens the force of the treaty and makes more difficult the negotiation of an effective comprehensive nuclear test ban. This trend could adversely affect the state of world security.

Presumably the same would have applied to Cape Keraudren's five 200 kiloton explosions under the seabed. Of the thirteen nuclear explosions detonated by France in the South Pacific to February 1970, only five were over 200 kilotons. Another five were below 30 kilotons.[105]

Finally, the Australian government was naïve to accept assurances that the project was feasible within the time frame sought by Sentinel and the USAEC. The USAEC had never built a nuclear harbour before, never detonated a nuclear device on the territory of another sovereign country and never detonated a nuclear device under the seabed. As Seaborg admitted to Senator Fulbright in 1969:[106]

> This would be the first explosion — the first experiment utilizing a nuclear explosive — that would be conducted under the bottom of the ocean.

None other than Edward Teller and his collaborators in 1968 cast doubt on the effectiveness of nuclear dynamite underwater because it would create a wider, shallower crater than the same shot on dry land. To offset this phenomenon, 'unnecessarily high explosive yields' would be needed to achieve the necessary depth. Teller *et al.* specifically noted that, 'It is this property of wide, deep trenches that rendered the use of nuclear explosives unsuitable for an offshore shipping channel in north-western Australia'.[107] In 1971 Alan Wilson noted that there was a 'scarcity of directly relevant experience on which to assess the effect of the overlying water in an offshore cratering project'.[108]

Moreover, the USAEC had not yet perfected the nuclear devices for the type of precise 'nuclear engineering' required at Cape Keraudren. In 1968 Glenn Seaborg testified to the US Senate Foreign Relations Committee that, even under optimum conditions, the type of explosive needed for digging canals would not be ready for another five to ten years.[109] Assuming the Cape Keraudren project would have used the same type of cratering devices as canals, then clearly such devices would not have been ready by 1970. If the devices were different, then surely they would have needed testing underwater prior to five of them being detonated at Cape Keraudren. There is no evidence that the Americans planned such tests.

Last but not least, the timetable set by Sentinel was far too short to carry out the necessary bioenvironmental studies for a 'planetary engineering' project of this size. Fairbairn claimed that the Cape Keraudren project would not require the same thorough study as Project Chariot, because a significant amount of 'basic data' had been accumulated since then. However, this would have been of little assistance in determining the effect of nuclear detonations on the unique flora and fauna of Cape Keraudren. Fairbairn further claimed, incorrectly, that the Cape Thompson report's primary purpose had been to determine whether the project would be hazardous to the local inhabitants, the implication being, presumably, that Australia had no equivalent of the Eskimos over whom to be concerned. In fact, the Chariot study covered the likely impact on the entire biosphere.[110] As Gerald Johnson pointed out after the cancellation of Chariot:[111]

> . . . to establish public confidence, care must be taken in any future experiments and demonstrations that all appropriate backgrounds are acquired before the experiments are conducted and that all appropriate biological-ecological work is encouraged. Only with such information for the record might it be possible to answer the critics in the public forum. The preparatory work for the Chariot experiment provides a good example of such necessary exhaustive preparation.

Despite the lessons that should have been learned from the Cape Keraudren fiasco, new PNE proposals for Australia continued to surface after the project's demise. Lang Hancock in late 1969 announced that he had sought permission to use a nuclear device to mine iron-ore at the Wittenoom Gorge in the Hamersley Range in Western Australia, a suggestion first made by John Kelly during his 1968 visit.[112] The AAEC also studied, at the behest of BHP, the possibility of using PNEs to break up iron-ore at its Deepdale deposits at Robe River in Western Australia.[113] The reaction of Senator Cant of Western Australia to Hancock's idea was equally applicable to Cape Keraudren.[114]

> Remember that this area is 7,000 miles from the coast of America. The Americans are not concerned about whether or not the broken ore is contaminated by radiation. Their people will not be affected. This would be a good experiment to enable them to find out whether or not there is any contamination because it would be 7,000 miles away from their shores.

According to Gerald Johnson, Hancock's project was not proceeded with because the Japanese steel industry indicated that the ore would be unacceptable because nuclear explosions would be involved.[115]

In 1970, at the Fourth Plowshare Symposium, a University of California scientist resurrected Teller's proposal for deepening the Torres Straits with nuclear dynamite. He proposed two sets of nuclear 'dashes' with a yield of roughly 8–10 megatons per dash, with a maximum total device yield of 150–175 megatons, to create a 120 mile (180 kilometre) channel. He said, however, that:[116]

> Further study of the Torres Strait concept is required in order to assess the effects of the water-wave generated by the explosion, the beach run-up of the wave produced, possible interaction of the water wave with solid ejecta, and oceanic transport and diffusion of radioactivity deposited in the ocean.

In 1971 Glenn Seaborg suggested the creation of a large earth dam on the Fortescue River in Western Australia using PNEs to cave in the sides of a gorge near the Hamersley Range.[117] He suggested injecting water into the dam and freezing it, creating a permafrost barrier to seepage.[118] The energy generated by a contained, underground nuclear explosion, presumably an additional one to that which created the dam, could be tapped to power such a freezing process! In theory, according to Seaborg, after the initial freezing 'only a little power from the new hydroelectric plant would suffice to keep the earth-insulated ice frozen'. That such an undertaking should be proposed for the Australian desert is truly mind-boggling.

Finally, Edward Teller in 1982 proposed using PNEs to turn back Australia's great easterly-flowing rivers to provide water for the dry interior of the country:[119]

> In Australia, too much water runs the wrong way... We can pierce the mountain ranges and excavate a system of necessary reservoirs most easily and efficiently with nuclear explosives. The savings over conventional methods of moving large quantities of earth would be enormous. This project is a dream only because it is so great. Courage and enterprise could make it a reality.

Australia eventually signed the Nuclear Nonproliferation Treaty in February 1970 after Japan and West Germany decided to do so and after strong representations from the United States.[120] Despite or perhaps because of the Cape Keraudren experience, and no doubt to placate the AAEC, the Australian government made a declaration when signing, which noted that the NPT would:[121]

> ...in no way inhibit and is in fact designed to assist non-nuclear-weapon States in their research, development and use of nuclear energy and nuclear explosives for peaceful purposes either individually or collectively.

In addition Australia also welcomed the fact that the Treaty provided for 'international co-operation for the development of the... peaceful application of nuclear explosives'. Of the other NPT parties only the Federal Republic of Germany, Italy, Japan, Mexico and Yugoslavia made specific statements on PNEs at the time of signing or ratifying the treaty. None were as pro-PNE as Australia's.

Since Australia ratified the NPT in 1972 no Australian government has seriously contemplated a project involving nuclear dynamite. On the contrary, Australia has adopted a fiercely critical attitude towards PNEs. Such an about-face may be traced to the election of the Whitlam government in 1972 and the Indian 'peaceful' nuclear explosion in 1974. Popular opposition to nuclear testing in general and to testing by France in the South Pacific, combined with strong support for a CTB, have also contributed to Australia's position. In 1985, in order to place on record its change of heart, Australia informed the three NPT depositary states, the United States, the Soviet Union and the United Kingdom, and the UN Secretary-General, that those parts of the declaration Australia made when signing the NPT which referred to PNEs were no longer to be taken as accurately reflecting Australia's position.[122]

# *Notes*

1 G. H. Quester, *The Politics of Nuclear Proliferation*, Johns Hopkins University Press, Baltimore, Maryland, 1973, p. 165.
2 J. Koop, 'Plowshare and the Nonproliferation Treaty', *Orbis*, XII, 3, Fall 1968, p. 804.
3 T. C. Findlay, 'Twenty Minutes with Edward Teller', *Newsletter*, Peace Research Centre, Australia National University, Canberra, 2, 1, 1987, p. 2.
4 US Congress, JCAE, Hearings on AEC Authorizing Legislation for Fiscal Year 1965, US Government Printing Office, Washington DC, 1964, p. 1220.
5 To avoid confusion with the Australian Atomic Energy Commission (AAEC), the acronym USAEC will be used in this chapter to designate the US Atomic Energy Commission.
6 G. T. Seaborg *et al.*, *Plowshare*, Division of Technical Information, US Atomic Energy Commission, Oak Ridge, Tennessee, 1966, p. 43.
7 Western Australian Parliamentary Hansard, Thursday 27 March 1969, p. 2857.
8 The *Sydney Morning Herald*, 8 February 1969, p. 6.
9 G. T. Seaborg and B. S. Loeb, *Stemming the Tide: Arms Control in the Johnson Years*, Lexington Books, Lexington, Mass., 1987, p. 252.
10 Seaborg also learned from Sir Leslie Martin, AAEC Commissioner and Scientific Advisor to the Department of Defence, that Australia, in Seaborg's words, was 'struggling with the decision of whether to produce a nuclear weapon', a fact of which President Johnson had not been informed during his visit to Australia (Seaborg and Loeb, *Stemming the Tide*, p. 252).
11 Press release, Minister for National Development, Canberra, 23 January 1969.
12 Australian Parliamentary Hansard, Senate Estimates Committee, 15 October 1968, p. 1975.
13 Australian Department of External Affairs, *Current Notes*, May 1968, p. 210.
14 Australia, like Japan, apparently believed that nuclear dynamite without military applications would some day be developed.
15 T. B. Millar, 'Australia: A Recent Ratification' in R. M. Lawrence and J. Larus (eds), *Nuclear Proliferation Phase II*, University of Kansas Press, Witchita, 1974, p. 78.
16 I. Bellany, *Australia in the Nuclear Age*, Sydney University Press, Sydney, 1972, p. 107.
17 Ibid., p. 77. On 15 April 1968 the *Age*, Melbourne, reported Baxter as saying that to

protect itself from an overpopulated outside world, Australia should equip itself with 'the most sophisticated weapons that man can devise — with no type excluded'. Alan Wilson recalls, however, that Baxter had a cautious attitude towards PNEs.

18 The *Age*, 15 August 1968, p. 7.
19 US Congress, Senate Foreign Relations Committee, Hearings on the NPT, 18 and 20 February 1969, US Government Printing Office, Washington DC, 1969, p. 329.
20 AAEC, *Annual Report 1968–69*, Australian Government Publishing Service, Canberra, 1969, p. 77.
21 B. A. Bolt, *Nuclear Explosions and Earthquakes: The Parted Veil*, W. H. Freeman and Co., San Francisco, 1976, p. 197. It is notable that the same was said of the Cape Thompson region in Alaska before the Project Chariot environmental study revealed its incredible natural abundance.
22 Seaborg and Loeb, *Stemming the Tide*, p. 347.
23 The following account of dealings between Sentinel Mining Company, National Bulk Carriers, the Livermore Laboratory, the USAEC and the West Australian government is taken from an unclassified telex from Livermore to the USAEC in Washington DC, 12 March 1969.
24 L. J. Vortman, 'Nuclear Excavation' in L.E. Weaver (ed.), *Education for Peaceful Uses of Nuclear Explosives*, University of Arizona Press, Tucson, Arizona, 1970, p. 67.
25 Western Australian Parliamentary Hansard, Thursday 27 March 1969, p. 2857.
26 US Congress, JCAE, Hearings on PNE Services for Domestic and Foreign Users, 8–9 May and 17 July 1969, US Government Printing Office, Washington DC, 1969, p. 62; The *Sydney Morning Herald*, 2 January 1969, p. 1.
27 The *Sydney Morning Herald*, 2 January 1969, p. 1.
28 The *Sydney Morning Herald*, 25 January 1969, p. 13.
29 The *Sydney Morning Herald*, 15 January 1969, p. 10.
30 Press Release, Minister for National Development, Canberra, 23 January 1969.
31 Hosmer later touted this, unjustifiably, as the budgetary process would reveal, as evidence of Nixon's strong support for Plowshare.
32 The *Sydney Morning Herald*, 8 February 1969, p. 6.
33 Australian Department of External Affairs, *Current Notes*, February 1969, p. 57.
34 Ibid.
35 Ibid., p. 58.
36 The *Sydney Morning Herald*, 7 February 1969, p. 1; *Current Notes*, February 1969, p. 59. The craters inspected were those created by the Buggy, Cabriolet and Sedan detonations; see Chapter 10. At its headquarters in Sydney the AAEC reportedly had a diorama of the Cape Keraudren project installed for public viewing. At the press of a button, five lights lit up to represent the five nuclear explosions to be conducted.
37 The *Sydney Morning Herald*, 7 January 1969, p. 2.
38 The *Sun-Herald*, Sydney, 2 February 1969, p. 2.
39 US Congressional Record, 28 January 1969, US Government Printing Office, Washington DC, 1969, p. 2022.
40 'Five Times Hiroshima', the *Age*, 30 January 1969, p. 6.
41 The *Age*, 24 January 1969, p. 1.
42 The Royal Commission found, *inter alia*, that, 'The decision to use the mainland for atomic tests was made without specific consideration by Australian scientists or others of whether weapons could be safely fired' (R. Milliken, *No Conceivable Injury*, Penguin, Ringwood, 1986, p. 348).
43 Australian Parliamentary Hansard, 13 May 1969, p. 1731.
44 The *Sydney Morning Herald*, 6 February 1969, p. 4.
45 Ibid.
46 The *Sydney Morning Herald*, 2 January 1969, p. 1 and 13 February 1969, p. 8.
47 The *Sydney Morning Herald*, 6 March 1969, p. 5.
48 M. Kalkstein, *International Arrangements for the Peaceful Applications of Nuclear Explosives*, Stockholm International Peace Research Institute, Stockholm, 1970, p. 24.
49 US Congress, Senate Foreign Relations Committee, Hearings on the NPT, p. 315.
50 Ibid., p. 316.
51 Ibid., pp. 328–29.
52 US Congress, JCAE, Hearings on AEC Authorizing Legislation for Fiscal Year 1970, US Government Printing Office, Washington DC, 1969, p. 505.

53 Representative Hosmer tried to rectify this by repeatedly introducing a Commercial Plowshare Services Bill into the Congress after 1968 (US Congressional Record, 14 April 1969, p. 8945). The so-called Hosmer Bill was reintroduced in each succeeding Congress until the demise of Plowshare, but was never passed, even with the bipartisan support of all the House members of the JCAE (*PNE Activity Projections for Arms Control Planning* (the *GURC Report*), Gulf Universities Research Consortium, Galveston, Texas, US Arms Control and Disarmament Agency Document, ACDA/PAB-253, Washington DC, 1975, I, pp. 20–21; F. Kreith and C. B. Wrenn, *The Nuclear Impact*, Westview Press, Boulder, Colorado, 1976, p. 11).
54 The *Sydney Morning Herald*, 24 March 1969, p. 8.
55 US Congress, JCAE, Hearings on AEC Authorizing Legislation for Fiscal Year 1970, p. 303.
56 See the *Age*, 3 March 1969, p. 7 and the *Sydney Morning Herald*, 1 March 1969, p. 4.
57 See Australian Senate Record, 25 February 1969, p. 11 and the *Sydney Morning Herald*, 6 March 1969, p. 5.
58 See Australian Senate Record, 25 February 1969, p. 8 and US Congress, Senate Committee on Foreign Relations, Hearings on the NPT, p. 319.
59 US Congress, Senate Committee on Foreign Relations, Hearings on the NPT, p. 319.
60 The *Sydney Morning Herald*, 20 February 1969, p. 4.
61 The *New York Times*, 23 February 1969, p. 13E.
62 Australian Parliamentary Hansard, 25 February 1969, p. 1.
63 E. Teller *et al.*, *The Constructive Uses of Nuclear Explosives*, McGraw-Hill, New York, 1968, p. 231.
64 J. Smith, *Clouds of Deceit: The Deadly Legacy of Britain's Bomb Tests*, Faber and Faber, London, 1985, p. 62; *SIPRI Yearbook 1987*, Oxford University Press, Oxford, 1987, p. 54.
65 Smith, p. 62.
66 The *Sydney Morning Herald*, 30 January 1969, p. 3.
67 The *New York Times*, 23 February 1969, p. 13E.
68 Australian Parliamentary Hansard, 25 March 1969, p. 823.
69 *Environment of the Cape Thompson Region, Alaska*, USAEC, US Government Printing Office, Washington DC, 1966, p. vi. Only after cancellation of the project did Australian Labor Party MP, Dr Jim Cairns, raise this point in the Australian Parliament (see Australian Parliamentary Hansard, 13 May 1969, p. 1731).
70 *Environment of the Cape Thompson Region, Alaska*, p. viii.
71 Australian Senate Record, 19 September 1969, p. 1156.
72 The *Sydney Morning Herald*, 24 March 1969, p. 8.
73 The *Sydney Morning Herald*, 3 January 1969, p. 4.
74 Australian Parliamentary Hansard, 25 March 1969, p. 823.
75 Bolt, p. 197.
76 The *Age*, 31 March 1969, p. 1; AAEC, *Annual Report 1968–69*, p. 77.
77 US Congress, JCAE, Hearings on AEC Authorizing Legislation for Fiscal Year 1970, p. 303.
78 US Congressional Record, 14 April 1969, p. 8944.
79 Bolt, p. 197.
80 Kreith and Wrenn, p. 10.
81 Quester, p. 160.
82 Australian Parliamentary Hansard, 19 September 1969, p. 1156.
83 G. T. Seaborg and W. R. Corliss, *Man and Atom: Building a New World Through Nuclear Technology*, E. P. Dutton and Co., New York, 1971, p. 189.
84 Letter of 20 February 1969 from John Kelly, Director, PNE Division, USAEC to Glenn Seaborg, Chairman, USAEC.
85 USAEC file note, 3 March 1969.
86 US Congress, Senate Foreign Relations Committee, Hearings on the NPT, p. 316.
87 The *Age*, 31 March 1969, p. 1.
88 From an interview with Alan Wilson, Sydney, 7 August 1987.
89 US Congress, JCAE, Hearings on AEC Authorizing Legislation for Fiscal Year 1970, p. 307.
90 US Congressional Record, 14 April 1969, p. 8945.

91 Vortman in Weaver (ed), p. 67.
92 The *Age*, 31 March 1969, p. 1.
93 US Congress, JCAE, Hearings on AEC Authorizing Legislation for Fiscal Year 1970, p. 337.
94 US Congressional Record, 14 April 1969, p. 8944.
95 Bellany, p. 111.
96 US Congress, Senate Foreign Relations Committee, Hearings on the Nuclear Test Ban Treaty, 12–15 August, 19–23 August, 26–27 August 1963, US Government Printing Office, Washington DC, 1963, p. 240.
97 Kalkstein, p. 22.
98 The *New York Times*, 23 February 1969, p. 13E.
99 US Congress, JCAE, Hearings on PNE Services for Domestic and Foreign Users, 8–9 May and 17 July 1969, US Government Printing Office, Washington DC, 1969, p. 62.
100 From a discussion with Mr V. V. Shoustov, Director, Research Co-ordination Centre, Soviet Ministry of Foreign Affairs, at Sochi, USSR, 25 March 1988.
101 International Court of Justice, *Application Instituting Proceedings on behalf of Australia against the French Republic*, Australian Department of Foreign Affairs, Canberra, 1972, p. 3.
102 Ibid., Annex 3.
103 Ibid., Annex 6.
104 Ibid.
105 *SIPRI Yearbook 1974*, Almqvist and Wiksell, Stockholm, 1974, p. 505.
106 US Congress, Senate Foreign Relations Committee, Hearings on the NPT, p. 317.
107 Teller *et al.*, p. 231.
108 A. R. W. Wilson, 'Current Status of Civil Engineering and Mineral Resources Development Applications of Peaceful Nuclear Explosions' in Proceedings of the Fourth International Conference on the Peaceful Uses of Nuclear Energy, 6–16 September 1971, IAEA, New York and Vienna, 1972, p. 221.
109 Kalkstein, p. 16.
110 According to Ralph Sanders, the study had attempted to assess the 'total probable effects of the Chariot experiment'. (R. Sanders, *Project Plowshare*, Public Affairs Press, Washington DC, 1962, p. 78.)
111 Johnson in Weaver (ed.), p. 15.
112 The *Age*, 13 May 1968, p. 6.
113 From an interview with Alan Wilson, Sydney, 7 August 1987.
114 Australian Senate Record, 19 September 1969, p. 1157.
115 G. W. Johnson, 'Plowshare at the Crossroads', *Bulletin of the Atomic Scientists*, June 1970, p. 88.
116 *Fourth Plowshare Symposium Summaries*, American Nuclear Society, Hinsdale, Illinois, 1970, pp. 16–17.
117 Seaborg and Corliss, p. 115.
118 Ibid., p. 116.
119 Edward Teller, *The Legacy of Hiroshima*, Doubleday and Co., New York, 1982, p. 90.
120 Millar, p. 80.
121 *Status of Multilateral Arms Regulation and Disarmament Agreements*, United Nations, Second Edition, UN Department for Disarmament Affairs, New York, 1983, pp. 84–85.
122 *Disarmament Newsletter*, 11, 24 October 1985, Australian Department of Foreign Affairs, Canberra, p. 23.

# 10

# Collapsing Craters: The End of Nuclear Excavation

*The only meaningful support for the excavation program was never technical, it was political from the start, which might account for its ultimate failure.*[1]

Peter Metzger, 1972

THE year 1968 was an eventful one for the future of nuclear dynamite. The NPT negotiations were concluded, planning for the Cape Keraudren project was initiated, the Conference of Non-Nuclear Weapon States instigated movement towards a PNE 'regime' and the Plowshare budget, at $18 million, was at its highest level ever. And, after a three-year hiatus, authorisation was finally given by President Johnson for Plowshare nuclear excavation experiments for the Panama Canal project to resume. The first of these was Cabriolet. Glenn Seaborg, in a letter to the Canal Commission of 28 August 1968, observed that 'previous inability to obtain requisite approval for the nuclear cratering experiments' seemed to have been 'alleviated'. He did not elaborate on the reasons.[2] The successful detonation of Cabriolet on 26 January 1968, he notes, 'set the stage' for two more excavation experiments in 1968, Buggy and Schooner.[3] According to Seaborg, in neither case was there major opposition from within the government.[4]

## *Buggy*

In the view of the Livermore Laboratory, the Buggy experiment was especially significant. Conducted in basalt, it involved the simultaneous detonation of five 1.1 kiloton nuclear explosives buried 41 metres (135 feet) deep and spaced 46 metres (150 feet) apart. This 'row charge' produced an enormous trench, 260 metres (860 feet) long, 77 metres (280 feet) wide and 20 metres (68 feet) deep which, the Laboratory speculated, would be 'well suited to the excavation of canals, harbors, or highway cuts'.[5] The AEC announced, on the basis of preliminary information, that the amount of radioactivity released by Buggy was 'very low', 'beyond the immediate crater area', while for Cabriolet (a single 2.3 kiloton shot) it had been 'small', 'beyond the throwout area'.[6] After viewing a film of Buggy, Seaborg noted 'somewhat testily' in his diary that 'this and Cabriolet should have been approved for execution long ago'.[7]

On the basis of these experiments the AEC unexpectedly announced in May 1968 that it had developed a virtually 'clean' nuclear explosive.[8] According to Theo Ginsburg, it was now possible to produce nuclear explosives up to megaton size with no more than a few kilotons of fission.[9] In Ed Fleming's view, Plowshare came 'extremely close' to developing a 'clean' nuclear explosive.[10] Progress had also been made in shielding and entrapment techniques to trap radioactivity underground in cratering explosions. Ginsburg noted that, while the 100 kiloton Sedan event in 1962 produced a total dose of 0.5 roentgen 100 kilometres from the explosion, the devices and techniques of 1968 would have shortened this to 50 kilometres. In the future, Ginsburg estimated, the distance could be reduced to 20 kilometres or less.[11]

Some observers were sceptical. First, there was no indication as to whether the explosives were more costly when fallout was minimised, a factor which could affect the economics of nuclear dynamite. Second, only radiation from fallout, not that induced by 'neutron activation', was reduced. Third, the low fallout was obtained by using very small detonations. As the AEC admitted, 'the key... has been the small size of the explosives used... a very small thermonuclear explosive triggered by a really tiny fission device'.[12] This also had critical implications for the economics and practicability of PNEs. As Congressman Hosmer quipped to Glenn Seaborg in 1965, 'You are not talking about a clean device but a *cleaner* device, then, are you not?'[13] According to Brooks and Krutilla, even the 'cleanest' device will distribute *some* radioactivity, and:[14]

> ... if the explosive device is large enough to be economically feasible in excavation applications, it is very likely (despite improvements in explosives technology) to cause fallout over an area of several hundred square miles. The fallout problem is particularly serious because certain of the most worrisome isotopes (strontium-90 and cesium-137) have gaseous precursors and are therefore widely distributed.

Despite the alleged development of a 'clean' explosive, the nuclear excavation program, along with the Panama Canal project itself, was headed for extinction within two years. This had more to do with escalating budgetary difficulties facing the Johnson Administration than with the question-

able performance of Plowshare. Without waiting for the Canal Commission's report, the Administration, under pressure to curb inflation fuelled by the Vietnam War, quietly allowed the Plowshare budget to be cut. This lent further credence to the view that the Panama Canal project and the nuclear excavation program were political ploys rather than practical undertakings.

In early 1968 the Bureau of the Budget, the powerful fiscal arm of the White House, rejected the AEC request for $28 million for Plowshare for 1969, conceding only $14 million. This was $4 million down from the all-time high of 1968. Plowshare had always been erratically funded, since it had to compete for money with the vastly more significant nuclear weapons program. The 1968 cut was not therefore generally perceived as threatening the existence of the excavation program. Indeed, the Panama Canal Commission, in a letter to Seaborg of 22 August 1968, still insisted that the AEC carry out five experiments (Schooner, Yawl, Galley, Gondola and Phaeton) by the end of 1970.[15] The Commission indicated, however, that it intended to submit its report by 1 December 1970, 'with or without a final determination of the feasibility of nuclear excavation'.[16]

## *Schooner*

The third excavation shot for 1968, the 35 kiloton Schooner experiment, was conducted on 8 December 1968, creating a crater 850 feet (260 m) in diameter and over 240 feet (70 m) deep,[17] and venting a significant amount of radioactivity into the atmosphere. Ironically, having been postponed because of the PTBT, the shot violated the treaty. The radioactive cloud drifted north-east across the United States, tracked by the US Air Force. While the aircraft stopped at the Canadian border, the cloud, oblivious to such geopolitical niceties, crossed into eastern Canada, where abnormally high levels of radiation were detected between 13 and 15 December in Hamilton, Ottawa, Montreal and Toronto.[18] Tungsten isotopes were later detected in the air over Europe.[19] While not enough to result in measurable exposures to humans, the radioactivity was above normal background levels.[20] The State Department announced that 'extensive precautions were taken to insure that [the test] was carried out in a manner consistent with the limited test ban treaty'.[21] It did not comment on whether these precautions had been successful. In testimony before the Senate Foreign Relations Committee in 1969, Glenn Seaborg was cagey, saying only that:[22]

> ... there were reports that fluctuations in the background radiation in Canada were higher than normal. Our own monitoring stations in that region... did not show fluctuations greater than normal.

*Figure 5 Crater produced by the Buggy experiment, a demonstration of nuclear excavation technology. The small object at the lower right end of the crater is a truck. (Courtesy of Lawrence Livermore National Laboratory.)*

Contradicting Seaborg, the US Public Health Service reported that Schooner had resulted in the highest airborne radioactivity levels in eight western American cities since atmospheric testing ended in 1962. Levels were 30–40 times normal background levels in Idaho Falls, Spokane, Salt Lake City and Denver.[23] On 21 January 1969, the first full day of the Nixon Administration, the Soviet *chargé d'affaires* in Washington delivered an *aide-memoire* stating that Schooner had caused a 'two to fivefold increase in fallout in the regions along the Baltic, Volga, Northern Caucasus, and Crimea'.[24] Seaborg later explained to a White House aide, Robert Ellsworth, that this corresponded to an 'absurdly small amount of radioactivity', since detection techniques were probably as refined in the Soviet Union as in the United States. Seaborg also pleaded Plowshare's cause to Ellsworth, in what he called 'an attempt on my part to gain the new administration's sympathetic support'.[25] Schooner was in fact to be the last nuclear excavation experiment conducted by the United States.

**Table 10.1** Major Plowshare Cratering Experiments*

| Code Name | Date | Size of Explosion | Remarks |
|---|---|---|---|
| Danny Boy** | 5 March 1962 | 0.43 kt. | Low-yield shot; crater 214 feet (63 m) in diameter, 68 feet (21 m) deep |
| Sedan | 6 July 1962 | 100 kt. | 1200 feet (360 m) diameter crater, 320 feet (100 m) deep |
| Cabriolet | 26 Jan. 1968 | 2.5 kt. | Point shot; crater 360 feet (110 m) in diameter, 120 feet (35 m) deep |
| Buggy | 12 March 1968 | 5 × 1.1 kt. | First nuclear row-charge experiment; row 860 feet (260 m) long, 280 feet (77 m) wide, 68 feet (20 m) deep |
| Schooner | 8 Dec. 1968 | 35 kt. | Point shot; crater 850 feet (260 m) in diameter, 240 feet (70 m) deep |

* All shots fired at the Nevada Test Site

** Department of Defense experiment with Plowshare participation

*Source:* G.T. Seaborg and N.R. Corliss, *Man and Atom: Building a New World Through Nuclear Technology*, E.P. Dutton and Co., New York, 1971, p. 187

In early 1969 the Nixon Administration rejected the AEC funding request of $29.2 million for Plowshare for fiscal year 1970, pegging it at $14.5 million, roughly the same level as fiscal year 1969. The new Administration obviously felt no obligation to provide increasing funding for a politically-motivated excavation program initiated by its predecessor. This was despite the President's speedy approval of Australia's request regarding the Cape Keraudren project and what Seaborg describes as Nixon's 'special prejudice' in favour of Plowshare.[26] George Quester speculates that nonproliferation considerations were involved, the Administration apparently believing that a slowing down of Plowshare would mean a slowing of international publicity about the alleged benefits of PNEs.[27]

This time, however, the budgetary threat to Plowshare attracted the ire of the JCAE, now under the firm chairmanship of Chet Holifield, once described as 'the most pivotal individual controlling the destiny of atomic energy in the United States'.[28] The committee was especially exercised by the AEC's admission that it would only be able to conduct the second excavation shot (Yawl) of the four remaining shots requested for the Panama Canal study, by the time the Canal Commission's extended deadline expired on 1 December 1970. Galley (which had been the most likely candidate for replacement by the Cape Keraudren project), Gondola and Phaeton would have to be postponed.[29] Craig Hosmer and Chairman Holifield were outraged, perceiving not only a threat to the Panama Canal project but to the ability of the United States to provide the benefits of PNEs to non-nuclear weapon states under the NPT. Hosmer declared bitterly that:[30]

> ... for the last several years during the negotiations of the Nonproliferation Treaty in the Eighteen-Nation Disarmament Committee in Geneva, subsequently during discussions on the ratification on two occasions in the US Senate, there was constant laudatory reference to Article V and the obligation of the nuclear powers to provide peaceful nuclear explosion services for the non-nuclear powers. That was a *quid pro quo* for their agreement to forego the development of nuclear explosives themselves. Yet, under both administrations, the previous one and the present one, this Plowshare program has not been given the funding and has not been provided the impetus and has not been given the direction to go ahead and live up to these treaty obligations of Article V of the NPT, let alone the promise that the program has constantly made to the domestic users and that the Atomic Energy Commission has indicated could be achieved if the proper amount of work was done to establish the commercial Plowshare services.

In its report to Congress the committee noted that Plowshare had 'clearly been funded at half the level for which it was technologically ready'.[31] It could not understand why the Administration proposed to have the program 'limp along at the same hand-to-mouth funding level as it has suffered in years past'.[32] Citing the 'important' domestic uses of Plowshare technology and US obligations under the NPT, the Committee stated that it was 'high time that the Administration begin to match word with deed'.[33]

Surprisingly however, despite the JCAE's influence, neither the Administration nor Congress responded to its demands for extra funds, an indication of the lack of support Plowshare now commanded outside the JCAE and the AEC. It was also a portent of the long-term fiscal threat to Plowshare.

Hosmer ranted against those he considered responsible during a speech in Las Vegas on 8 April 1969:[34]

> In addition to the assorted professors, scientists, lawyers and literati who whine over Plowshare for philosophical reasons, a hard core of Plowshare opponents seem to have developed within the Executive Branch of the government itself — particularly within the Budget Bureau, the State Department and the Arms Control and Disarmament Agency. Behind the scenes, this group strenuously fights to obstruct every attempt at upgrading the program. These people seem to have a paranoic distrust and abhorrence for Plowshare, which they cannot divorce in their minds from the weapons program. I am sure Article V of the NPT, which gives Plowshare international respectability, must have broken their bleeding hearts.

Alongside Plowshare's growing funding difficulties, there were also growing doubts among scientists and engineers about the technical feasibility and environmental impact of the proposed Panama Canal project. An article in *Science* magazine in January 1969 noted that 'the protests of scientists concerned about US plans to build a new interoceanic Atlantic-Pacific sea-level canal seem, like television commercials, to grow louder and longer'.[35]

In 1968, an Italian nuclear physicist, Carlo Schaerf, writing in the *UNESCO Courier*, claimed that radioactive waste from the Panama Canal project 'would contaminate a vast area, so that some 25 000 people would have to leave and settle elsewhere for a few years'.[36] In 1969 an ecologist at Cornell University, Lamont Cole, warned that existing expertise was not sufficient to prevent radioactivity from the Panama Canal explosions contaminating water and land, eventually upsetting the food chain. He described the project as 'the most irresponsible suggestion that I can remember since Admiral Byrd's senile proposal to blow the ice caps off Antarctica'.[37] Herbert Scoville described the project as 'ill-starred'.[38]

In September 1968 the American Institute of Biological Sciences held a symposium at Ohio State University on the bioenvironmental studies being carried out for the canal project. One paper, entitled 'Potential Biological Catastrophe', by Professor John Briggs of the Department of Zoology at the University of South Florida, argued that joining the Pacific and Atlantic Oceans by a sea-water canal would reduce the marine diversity of both oceans disastrously.[39] This would, of course, have applied whether the canal was built with nuclear dynamite or not. Other papers argued that the radiation problem connected with the transisthmian proposal was 'manageable' in spite of 'the present uncertainties in fallout prediction'.[40]

In February 1969 the Corps of Engineers Nuclear Cratering Group released an interim report questioning the technical feasibility of a nuclear-excavated Panama Canal. The report recorded that a major 'slope stability' problem would exist along 20 miles (30 kilometres) of Route 17, in Panama, where a weak, saturated clay shale had been discovered.[41] In April 1969 an article critical of the nuclear excavation program was published in *Environment* magazine.[42] According to its author, Dr Edward Martell of the Cambridge Research Laboratories, the article had been successfully blocked from publication in *Science* magazine by Philip Abelson, 'an old AEC stalwart'. In 1970 SIPRI delivered what ACDA later described as 'a closely argued assault' on the Panama Canal project in terms of its economic and social costs.[43] SIPRI warned that the project could lead to a major upheaval in the 'economic and social make-up of the [Panamanian] nation' if nuclear explosives resulted in the relocation of thousands of people.

The Panamanian government was also becoming concerned. Having little expertise of its own to draw on, it approached the IAEA for advice on the safety and environmental aspects of the use of nuclear explosives. The IAEA turned to Alan Wilson of the Australian Atomic Energy Commission, one of the few non-Americans with any involvement in Plowshare.[44] Wilson visited Panama to inspect the alternative canal routes and 'educate' the Panamanians. He could do little more, however, than provide general advice on the potential hazards of the project, as no decision had yet been made on the route or the number of nuclear explosions likely to be involved.[45]

In 1970, several months before the Canal Commission report was due, the final budgetary blow was delivered to Plowshare's nuclear excavation experiments. While the PNE Division of the AEC had requested $36.1 million for 1971, the Commission, aware that such a request was unrealistic, chopped it back to $22.9 million, a rise of $6.4 million on 1970. The Bureau of the Budget slashed this to a mere $8 million.[46] The Commission told the JCAE that as a result of such savage budgetary cuts the Plowshare team that was designing a special low-fallout, low-radioactivity excavation explosive would have to be disbanded, a program for investigating superheavy elements would have to be cancelled, and cratering experiments would definitely not be completed in time for the Canal Commission deadline.[47] Livermore's Dr May also testified that the United States could not meet its NPT obligations with a Plowshare budget of only $8 million.

A horrified Holifield wrote to Secretary of State Rogers and Secretary of Defense Laird to enlist their support for Plowshare as an essential part of US obligations under the NPT. The State Department replied that Article V of the NPT did not remove US discretion regarding the level of Plowshare funding. It quoted Seaborg's statement during the 1969 Senate hearings on the NPT that 'we will continue to conduct, *within the limitations of available funds* [emphasis added], an active program to develop peaceful nuclear explosive devices'.[48] Hosmer concluded, in response to this missive, that Article V of the NPT was 'only a snare, a delusion and an illusion as far as the non-nuclears are concerned'.[49] The Department of Defense was as blasé as the State Department. While noting the valuable co-operation between the AEC and the US Army Corps of Engineers on Plowshare projects, the Department said that 'while we would have benefitted from the continuation of the nuclear excavation program, cessation of these activities appears consistent with other recent budgetary actions'.[50]

Faced with this indifference to the future of Plowshare, the JCAE again failed to convince Congress or the Administration to provide more funds. In addition, despite Nixon's 'favourable prejudice', the AEC was unable to obtain Administration approval for even one more excavation shot after Schooner.[51] On 7 July 1970 the AEC was thus obliged to inform the Panama Canal Commission that it:[52]

> ... had not been able to do all of the experiments required to make a determination of the feasibility of using nuclear explosions for the excavation of canals ... and that, consequently, any decision in the near future to construct a sea-level canal would have to be made without reliance on nuclear excavation.

In December 1970 the Atlantic Pacific Interoceanic Canal Commission presented its report to the President.[53] The most detailed study ever carried out for a project involving the possible use of nuclear dynamite, it had taken five years to complete at a cost of $22 million.[54] Because the AEC had not been able to complete its nuclear excavation shots in time, the Canal Commission was forced to conclude that 'no current decision on US canal policy should be made in the expectation that nuclear excavation technology will be available for canal construction'. The Commission recommended that the existing Panama Canal be widened by conventional means. As to the much vaunted economics of nuclear dynamite, the Commission concluded that the cost of using it along the most favourable route for its application would actually exceed by $200 million the updated cost estimate of the most favourable conventional route.[55] According to Glenn Seaborg:[56]

> The difference in the two sets of estimates lay almost entirely in the costs of the nuclear alternative — they had skyrocketed, basically because cost factors overlooked in the first estimates began to be taken into consideration.

These included the political and social costs as well as the purely economic.

As to the extent of nuclear excavation required, the Commission estimated that for the three routes studied, the number of nuclear explosives needed was 103, 399 and 106 respectively.[57] The explosive energy to be used would be twenty times the cumulative amount of explosives used in all wars prior to Vietnam. According to Brooks and Krutilla, these figures begin to 'suggest the massive assault on the physical environment' involved in the project.[58] Glenn Seaborg now says, surprisedly, that 'the potential for blast damage and radiation from explosions of such magnitude was sobering indeed'.[59]

The Commission found that the social impact of nuclear dynamite was likely to be devastating. In relation to one route through Panama, for example, the Commission predicted that as a result of the expected fallout, 43 000 people would have to be evacuated from 16 000 square kilometres of territory during the nuclear operation and for several months thereafter.[60] Resistance from Indian tribes in the region would be strong due to their 'cultural attachments' to the land. Ecological studies indicated that there would be a mixed impact on the physical environment, with certain 'positive' aspects such as interior swampy lowlands becoming useful. As to seismic ground motion, it was estimated that some damage to weak masonry structures would extend outside the evacuation area as far as Panama City, 160 kilometres away.[61]

The Commission concluded that technical uncertainties associated with the use of nuclear explosions militated against using such technology to construct a new canal. According to Iris Borg, these technical uncertainties were twofold: watersaturated rock in the region, '... a unique situation which to that point had not been fully investigated', and the stability, 'in the presence of wave action...', of canal walls produced by nuclear row explosions.[62] According to Gerald Johnson the greatest problems with nuclear excavation are slope stability and radioactivity.[63] Johnson concluded, in an article entitled 'Plowshare at the Crossroads', published six months before the Canal Commission report, that:[64]

> Despite great accomplishments, the program to reduce radioactivity has not progressed at a rate to keep up with the growth of constraints. As a consequence nuclear excavation overall is probably less acceptable now than it was ten years ago.

Glenn Seaborg also says that 'in the final analysis, radiation raised the greatest concerns'.[65] Although the AEC, he says, was confident that this hazard could be controlled, it 'weighed so heavily on many minds that there was grave doubt a majority of the parties to the Limited Test Ban Treaty would ever have approved amending the treaty to permit the canal project to go forward'.[66] Gerald Johnson says that 'we all knew this was impossible for the canal'.[67] Such factors 'cast... a pall of uncertainty over the nuclear excavation program'.[68]

In the view of Professor Long of Cornell University, perhaps the greatest problem with nuclear excavation of canals arises from the 'high risk associated with the consequences of a miscalculation or failure in the deep cuts', a characteristic drawback of nuclear dynamite.[69] Nuclear explosives are not subtle engineering tools. Their 'take it or leave it' quality allows little opportunity for subsequent tailoring or adjustment. According to Long, an expensive full-scale test program would have been required to reduce the risks associated with this single problem. Because of the uniqueness of the site geology in Central America, such a program would almost by necessity have to have been conducted in the planned canal area. In Long's view, 'this fact in itself would be a substantial inhibition to the consideration of nuclear excavation as a viable alternative for such a construction project'.[70]

Such uncertainties also plagued the economics of nuclear excavation. As Brooks and Krutilla put it, most nuclear excavation proposals:[71]

> ... are so large in scale that they involve extrapolation far beyond the range where current experience permits any degree of confidence. The uncertainties in this regard are compounded by a wide range of unforeseeable as well as unintended consequences which seriously impair the use of cost estimates as of this stage in knowledge... it is premature to repose any confidence in this application of nuclear explosives.

The Canal Commission strongly implied that there was never any possibility that it would have recommended using PNEs for a new Panama Canal even if the AEC had been able successfully to complete its experiments on time:[72]

> The technical feasibility of the use of nuclear explosives for sea-level canal excavation has not been established. Whether the technology can be perfected and the international treaty obstacles to its use removed are not now predictable. Removal of the technical and treaty obstacles to employment of nuclear excavation would still leave major political and economic obstacles to a sea-level canal remote from Panama's population centres.

This did not, however, stop the Commission stating, outside its terms of reference, that it was 'confident that someday nuclear explosions will be used in a wide variety of massive earth moving projects'. It recommended that the United States pursue the development of such technology.[73]

Despite this final recommendation, the Canal Commission had delivered the *coup de grace* to Project Plowshare's nuclear excavation program. First, it demolished the technical and economic justification for the pursuit of nuclear excavation. Although it only considered a single project, albeit the largest PNE project ever envisaged, the report illustrated the enormous technical, environmental and political difficulties that would be faced by *any* 'planetary engineering' project. For the first time in Plowshare's history an independent body had assessed the implications of using PNEs for a particular project and concluded that they could not and should not be used. The Canal Commission thereby confirmed the gut reaction of many lay people to the use of nuclear dynamite: namely that the imponderables involved in 'correcting nature's mistakes' with large numbers of nuclear explosives are too terrifyingly great to contemplate.

Second, the Canal Commission report removed the political *raison d'être* for nuclear excavation. Panamanians could be certain that if a sea-level canal were built it would not be with nuclear dynamite, and it would be located in Panama, close to major population centres.[74] In Glenn Seaborg's words, the CSC report 'effectively closed the book' on a nuclear-excavated canal.[75] If the Canal Commission and the nuclear excavation program had been intended as political ploys to pressure Panama during the negotiations on a new canal treaty, then they had failed dismally. The Panamanians simply refused to conclude the negotiations during the Commission's six-year term. Not until September 1970, three months before the Commission handed down its report, did Panama formally reject the three draft treaties the United States had proposed in 1967. Even when negotiations resumed in June 1971, it still took another six years before the new Panama Canal Treaty was finalised.[76] In it the United States agreed that it would not use nuclear explosives to build a new canal without Panama's permission and would not construct a canal outside Panama, without Panamanian agreement, before 1999, when the treaty expires.

With Chariot, Cape Keraudren and the new Panama Canal now defunct, and having lost its political utility, Plowshare's nuclear excavation program simply disappeared after 1970. For Fred Holzer, there could be no Plowshare excavation program without the impetus of the Canal project.[77] And there were no other projects of similar dimension on the horizon. As Peter Metzger has pointed out 'in the whole world there are only two sites for sea-level canals [the Isthmus of Kra and the Isthmus of Panama] and about a dozen sites for harbors which fulfil the criteria of remoteness and practical value'.[78]

In 1973 two US consulting firms undertook an engineering study of the possibility of using PNEs to excavate an international waterway across the Kra Isthmus to link the Gulf of Thailand with the Andaman Sea.[79] The detonation of 139 explosives, ranging from 100 to 1000 kilotons, totalling 41.35 megatons, was contemplated. According to Iris Borg, although it was estimated that nuclear excavation would save several billion dollars and up to four years' construction time, the Thai government was uncertain whether the nuclear option was practical 'in light of current world perspectives'.[80] Nothing has been heard of the idea since.

Besides inter-oceanic canals and harbours, several other planetary engineering projects had been investigated under Plowshare, including a canal through one of the Aleutian islands, the deepening of the Bering Straits and the Tennessee-Tombigbee Waterway Project, involving a canal to connect two river systems in Alabama and Mississippi.[81] A feasibility study was also undertaken on a 'fresh water management' project (Project Aquarius) sponsored by the Arizona Atomic Energy Commission and the University of Arizona.[82] Also briefly considered was a proposal by the Parsons Company for the use of nuclear explosives to help build a nation-wide system of canals and river diversions for both irrigation and power generation purposes.[83] Known as the North American Water and Power Alliance (NAPA) project, this idea was supported by JCAE member Hosmer[84] and promoted by Glenn Seaborg.[85]

All these projects were shortlived. For instance, an AEC study concluded that the Tennessee-Tombigbee project would be unduly complicated and costly because of the population density of the area.[86] Another proposal, Project Carryall, contemplated a cut through the Bristol Mountains north of Amboy, California, to accommodate the Atchison, Topeka and Santa Fe Railroad and Interstate Highway 40. While a joint study by the California State Division of Highways, the railroad company and the AEC found the project technically feasible, it also concluded that nuclear excavation could not be accomplished in time for completion of that section of the interstate highway system.[87] Alan Wilson of the Australian Atomic Energy Commission noted in 1971 that 'the need for a smooth stable bed makes nuclear excavation *less* [emphasis added] attractive for cuttings than for canals'.[88]

The demise of nuclear excavation was, in Glenn Seaborg's words, 'a heavy blow to the Plowshare program, whose hopes for the future rested so heavily on the foreseen opportunities to perform excavation projects as a service for other nations'.[89] Herbert Scoville has remarked that only the Soviet Union and perhaps China, have the 'geographic and technological potential' for independent programs employing nuclear explosions for excavation.[90] In point of fact no country, including the United States, has to this day ever carried out a 'planetary engineering' project using nuclear dynamite.

The most incredible aspect of the entire nuclear excavation saga had been the air of unreality about the chances of the PTBT being amended — even to permit the larger excavation experiments required, much less an actual project. To start with, no-one seemed to have any idea whether the Soviet Union would agree or not:[91]

> *Senator Anderson:* Do you feel that Tsarapkin[92] would change his mind?
> *Mr Kelly:* That is possible, sir.
> *Chairman Holifield:* He has on one or two occasions.
> *Senator Anderson:* Very much?
> *Chairman Holifield:* Not much.
> *Senator Anderson:* I don't think so. I have never seen him change a particle.

When Representative Hosmer asked AEC Commissioner Palfrey what steps were being taken to amend the PTBT, Palfrey replied vaguely:[93]

> I think steps are pretty preliminary now. I think it was reported last year there were expressions of considerable interest at the Geneva Conference from several different sources. The steps that will have to be taken are looking down the pike about five years, I guess. I suspect it is too early to say now what specific steps will be taken.

When Hosmer asked whether it was not time that 'somebody in the Administration started to get moving on this', Palfrey could do little but agree that the sooner the United States talked to the Russians about the problem the better.

While there were occasional vague mutterings from the State Department and others about talks with the Soviets — Glenn Seaborg claimed that 'we had been trying to get such an agreement for two years without success'[94] — there is little evidence that the United States ever seriously approached the Soviet Union over the issue. The barriers were just too formidable. The AEC was in any case unwilling to push for such a move, since an amendment would probably take years to achieve and in the meantime only delay Plowshare experiments. The AEC hoped that if Plowshare continued to make progress in minimising the release of radioactivity from cratering shots, an amendment to the treaty might become unnecessary. In any event the government agencies responsible for seeking an amendment, the State Department and ACDA, would oppose such a move. The AEC was unlikely to be able to command the political and bureaucratic resources to overcome such opposition. Perhaps the greatest obstacle, however, was the fact that President Johnson regarded the Panama Canal project's purpose as political rather than practical. He could therefore safely ignore the international legal implications of actually using nuclear excavation techniques in Central America, just as he could afford to permit his 1964 directive concerning nuclear excavation to be quietly overturned by a strict interpretation of the PTBT. President Nixon was of course committed to neither the Canal project nor nuclear excavation in the first place.

Yet even if the entire US government had been behind a move to amend the PTBT and had secured Soviet agreement, it was unlikely that the United States could have obtained the support of a majority of the treaty's parties — mostly developing, non-aligned countries — as required by the treaty's amendment provisions. India, for instance, despite making a statement on signing the PTBT that appeared to favour the use of PNEs, declared in 1968 that the treaty should not be amended to permit development of nuclear excavation technology which would result in radioactivity venting into the atmosphere.[95] While probably intended as a deliberate snub to the United States, India's stand was indicative of the problems the Americans would face in seeking to amend the PTBT.

And if it was impossible to amend the PTBT, what was the point of pursuing nuclear excavation technology, the *raison d'être* of Plowshare? The JCAE's Senator Anderson quizzed Plowshare director John Kelly on this in 1966 in relation to the $5 million Cabriolet experiment and the Panama Canal project:[96]

> *Senator Anderson:* What good is the test if you still can't do it [the Panama Canal Project]? You can't do it without changing the test ban treaty, can you?
> *Mr Kelly:* I don't think you can dig the canal without changing the treaty.
> *Senator Anderson:* Why go through the preliminaries if you can't dig the canal without changing the treaty?
> *Mr Kelly:* I think there is some considerable hope and perhaps even optimism that by the time one could [sic] want to dig the canal it might be possible to amend the Test Ban Treaty.

# Notes

1 H. P. Metzger, *The Atomic Establishment*, Simon and Schuster, New York, 1972, p. 233.

2 US Congress, JCAE, Hearings on AEC Authorizing Legislation for Fiscal Year 1970, US Government Printing Office, Washington DC, 1969, p. 307.

3 G. T. Seaborg and B. S. Loeb, *Stemming the Tide: Arms Control in the Johnson Years*, Lexington Books, Lexington, Mass., 1987, p. 344.

4 Ibid.

5 *Energy and Technology Review*, Lawrence Livermore National Laboratory, September 1982, p. 16.

6 T. S. Lough, 'Peaceful Nuclear Explosions and Disarmament', *Peace Research Reviews*, Canadian Peace Research Institute, Ottawa, 1968, p. 21.
7 Seaborg and Loeb, *Stemming the Tide*, p. 344.
8 D. B. Brooks and J. V. Krutilla, *Peaceful Use of Nuclear Explosives: Some Economic Aspects*, Resources for the Future, Inc., Washington DC, 1969, p. 14.
9 T. Ginsburg, *The Question of Peaceful Explosions for the Benefit of Non-Nuclear Weapon States*, Document A/CONF. 35/DOC. 2, Conference of Non-Nuclear Weapon States, Geneva, July 1968, p. 17.
10 From an interview with Ed Fleming, Livermore Laboratory, California, 3 December 1986.
11 Ginsburg, p. 17.
12 Brooks and Krutilla, p. 14.
13 US Congress, JCAE, Hearings on AEC Authorizing Legislation for Fiscal Year 1966, US Government Printing Office, Washington DC, 1965, p. 29.
14 Brooks and Krutilla, p. 12.
15 US Congress, JCAE, Hearings on AEC Authorizing Legislation for Fiscal Year 1970, p. 304.
16 Seaborg and Loeb, *Stemming the Tide*, p. 345.
17 Ibid., p. 344.
18 J. P. Jain, *Nuclear India*, Radiant Publishers, New Delhi, 1974, p. 123.
19 A. C. McEwan, 'Environmental Effects of Underground Nuclear Explosions' in J. Goldblat and D. Cox (eds), *Nuclear Weapon Tests: Prohibition or Limitation?*, SIPRI, Oxford University Press, Oxford, 1988, p. 85.
20 *International Herald Tribune*, 10 January 1969, p. 5.
21 T. Ehrlich, 'Legal Problems (of PNEs)' in L. E. Weaver (ed.), *Education For Peaceful Uses of Nuclear Explosives*, University of Arizona Press, Tucson, Arizona, 1970, p. 228.
22 US Congress, Senate Foreign Relations Committee, Hearings on the NPT, 18 and 20 February 1969, US Government Printing Office, Washington DC, 1969, p. 319.
23 *International Herald Tribune*, 10 January 1969, p. 5.
24 Seaborg and Loeb, *Stemming the Tide*, p. 345.
25 Ibid.
26 Ibid.
27 G. H. Quester, *The Politics of Nuclear Proliferation*, Johns Hopkins University Press, Baltimore, Maryland, 1973, p. 27.
28 Metzger, p. 15.
29 US Congress, JCAE, Hearings on AEC Authorizing Legislation for Fiscal Year 1970, pp. 62 and 304.
30 Ibid., pp. 308–09.
31 US Congress, JCAE, Hearings on AEC Authorizing Legislation for Fiscal Year 1972, US Government Printing Office, Washington DC, 1971, p. 2375.
32 Ibid.
33 Ibid.
34 US Congressional Record, 14 April 1969, p. 8945.
35 *Science*, 10 January 1969, quoted in US Congress, Senate Foreign Relations Committee, Hearings on the NPT, Annex, p. 444.
36 *The American Experience with Peaceful Nuclear Explosions*, US Arms Control and Disarmament Agency, US Government Printing Office, Washington DC, 1978, p. 3.
37 *Science*, p. 444.
38 H. Scoville Jr., 'Peaceful Nuclear Explosions — An Invitation to Proliferation' in A. W. Marks (ed.), *NPT: Paradoxes and Problems*, Arms Control Association, Washington DC, 1975, p. 48.
39 J. C. Briggs Jr., 'The Sea-Level Panama Canal: Potential Biological Catastrophe', *Bioscience*, January 1969.
40 G. J. Ferber and R. J. List, 'Prediction of External Gamma Dose From Nuclear Excavation of a Sea-level Canal', *Bioscience*, March 1969, p. 236.
41 G. W. Johnson, 'Plowshare at the Crossroads', *Bulletin of the Atomic Scientists*, June 1970, p. 84.
42 Metzger, p. 276.
43 *The American Experience*, p. 3.
44 Australian Department of External Affairs, *Current Notes*, May 1968, p. 210.
45 From an interview with Alan Wilson, Sydney, 7 August 1987.

46 US Congress, JCAE, Hearings on AEC Authorizing Legislation for Fiscal Year 1971, US Government Printing Office, Washington DC, 1970, p. 600.
47 Ibid., pp. 601–03.
48 Ibid., p. 596.
49 Ibid., p. 635.
50 Ibid., p. 597
51 Seaborg and Loeb, *Stemming the Tide*, p. 346.
52 US Atomic Energy Commission, *Annual Report to Congress, 1970*, US Government Printing Office, Washington DC, 1970, p. 202.
53 B. A. Bolt, *Nuclear Explosions and Earthquakes: The Parted Veil*, W. H. Freeman and Co., San Francisco, 1976, p. 183.
54 US Congress, Senate Foreign Relations Committee, Hearings on the TTBT/PNET, 28 July, 3 August, 8 and 15 September 1977, US Government Printing Office, Washington DC, 1977, p. 136.
55 Ibid.
56 Seaborg and Loeb, *Stemming the Tide*, p. 348.
57 US Congress, Senate Foreign Relations Committee, Hearings on the TTBT/PNET, 1977, p. 90.
58 Brooks and Krutilla, p. 41.
59 Seaborg and Loeb, *Stemming the Tide*, p. 348.
60 Bolt, p. 195.
61 Ibid.
62 I. Y. Borg, *Nuclear Explosions for Peaceful Purposes*, Lawrence Livermore National Laboratory Paper No. UCRL-95026, Rev. II, November 1986, p. 5.
63 Johnson, 'Plowshare at the Crossroads', p. 85.
64 Ibid., p. 87.
65 Seaborg and Loeb, *Stemming the Tide*, pp. 348–49.
66 Ibid.
67 From correspondence with Gerald Johnson, La Jolla, California, 28 September 1987.
68 Seaborg and Loeb, *Stemming the Tide*, p. 349.
69 *An Analysis of the Economic Feasibility, Technical Significance and Time Scale for Application of Peaceful Nuclear Explosions in the US, with Special Reference to the GURC Report Thereon (The Long Report)*, Program on Science, Technology and Society, Cornell University, US Arms Control and Disarmament Agency, Washington DC, 1975, p. 26.
70 Ibid.
71 Brooks and Krutilla, p. 43.
72 *The Long Report*, p. 28.
73 US Atomic Energy Commission, *Annual Report to Congress 1970*, p. 201.
74 As further reassurance the United States negotiated the abrogation of the 1916 treaty which had given it exclusive rights to build a canal in Nicaragua (see D. N. Farnsworth and J. W. McKenney, *US/Panama Relations 1903–1978: A Study in Linkage Politics*, Westview Press, Boulder, Colorado, 1983, p. 112).
75 Seaborg and Loeb, *Stemming the Tide*, p. 346.
76 Farnsworth and McKenney, p. 113.
77 From an interview with Fred Holzer, Livermore Laboratory, California, 3 December 1986.
78 Metzger, p. 233.
79 Bolt, p. 196.
80 Borg, p. 10.
81 G. W. Johnson and H. Brown, 'Non-military Uses of Nuclear Explosives' in H. F. York (ed.), *Arms Control*, W. H. Freeman and Co., San Francisco, 1973, p. 410; *Report of the Ad Hoc Advisory Group on Nuclear Explosions for Peaceful Purposes*, IAEA Document GOV.1854, 26 August 1977, IAEA, Vienna, 1977, p. 81; L. J. Vortman, 'Nuclear Excavation' in L. E. Weaver (ed.), p. 67.
82 *A Legal Framework for Industrial Plowshare Applications*, Atomic Industrial Forum Inc., New York, 1969, pp. 1–2.
83 From correspondence with Gerald Johnson, La Jolla, California, 28 September 1987.
84 US Congress, JCAE, Hearings on AEC Authorizing Legislation for Fiscal Year 1967, US Government Printing Office, Washington DC, 1966, pp. 1366–67.
85 G. T. Seaborg and W. R. Corliss, *Man and Atom: Building a New World Through Nuclear*

*Technology*, E. P. Dutton and Co., New York, 1971, p. 115.
86 Vortman, p. 67.
87 Ibid., p. 66; Brooks and Krutilla, p. 36.
88 A. R. W. Wilson, 'Current Status of Civil Engineering and Mineral Resources Development Applications of Peaceful Nuclear Explosions' in Proceedings of the Fourth International Conference on the Peaceful Uses of Nuclear Energy, 6–16 September 1971, IAEA, New York and Vienna, 1972, p. 221.
89 Seaborg and Loeb, *Stemming the Tide*, p. 346.
90 Scoville, p. 49.
91 US Congress, JCAE, Hearings on AEC Authorizing Legislation for Fiscal Year 1967, p. 1349.
92 Chief Soviet delegate during the PTBT negotiations.
93 US Congress, JCAE, Hearings on AEC Authorizing Legislation for Fiscal Year 1967, p. 1366.
94 Seaborg and Loeb, *Stemming the Tide*, p. 337.
95 *The United Nations and Disarmament 1945–1970*, Department of Political and Security Affairs, United Nations, New York, 1970, p. 245.
96 US Congress, JCAE, Hearings on AEC Authorizing Legislation for Fiscal Year 1967, p. 1349.

# 11

# Plowshare Goes Underground

*Why is the AEC so keen on the gas stimulation program, in spite of the opposition to it? One reason is that the program is all that is left of Plowshare. Once a grandiose scheme for using nuclear explosions for all sorts of weird and wonderful projects such as excavating harbours and canals, Plowshare has been forced underground because of the radiological consequences of setting off nuclear blasts in the atmosphere — not to mention the test ban treaty.*[1]

*Nature*, 23 February 1973

BESIDES its excavation program Plowshare had, from its inception, also encompassed several 'underground engineering' applications of nuclear dynamite, including the recovery of oil, the leaching of copper, the creation of underground storage cavities, the stimulation of natural gas and the creation of 'heavy' elements. This work moved slowly. As the AEC said in a statement of 6 May 1964, almost seven years after Plowshare had begun, 'more research and development is needed before any routine commercial applications are practical'.[2] Part of the problem was securing industrial interest in such techniques. According to Gerald Johnson, it was clear that 'unless industry saw a way to use the explosives to economic advantage, there would be no future to the program'.[3] The idea was that while the AEC would design and construct explosive devices and be responsible for nuclear safety and security, the field work would be planned and executed largely by industrial partners, who would supply the expertise pertinent to the particular uses contemplated.[4] The AEC also envisaged that industry would provide most of the funding.

For a time this looked like being successful. Several underground Plowshare projects attracted industrial participation, especially gas stimulation. Three firms, the Continental Oil Company, Edgerton, Germeshausen and Grier (EG & G), and the Reynolds Electric Company, even created a new organisation called CER Geonuclear to specialise in the provision of commercial 'Plowshare services'.[5] In 1968 three companies, El Paso Natural Gas Company, of the United States, Nobel Bosch of France and Belgium's *Poudrerie Réunies de Belgique*, the last two experienced manufacturers of chemical explosives, formed Nobel Paso Geonuclear, an international consultant and servicing firm for users of nuclear dynamite.[6]

There seemed, however, to be confusion over whether the AEC was or should be soliciting industrial participation. Plowshare director Kelly told the JCAE in 1965 that '... we try to respond to each company or group that comes to us with an idea or a project that they want to study... We do not go out and solicit or have never gone out to solicit this kind of interest'.[7] Gerald Johnson and Glenn Seaborg said the opposite:[8]

> *Dr Johnson:* We have done everything we can to indicate interest. We have put substantial effort into it. We are anxious to join in co-operative endeavors to study this problem, and have stated so publicly.
> *Dr Seaborg:* I think the way I would state our policy is that we indicate our interest to any individual group, point out the possibilities in these various areas, and then determine whether or not we will proceed with any given project by how much the company is willing to participate with us. I think it has to be a joint enterprise, jointly financed...

With the collapse of nuclear excavation in 1970, Plowshare was left with its 'underground engineering' programs. Some of these were already in tatters. Of the remainder, only the gas stimulation program would amount to anything. It would outlast the excavation program by three years and then only because of the 'energy crisis' gripping the United States. Herbert Scoville concluded in 1974 that 'but for the emergence of the energy crisis the once enthusiastic US peaceful nuclear explosives program would already have been confined to oblivion'.[9] *Nature* magazine described the so-called energy crisis as Plowshare's 'trump card'.[10] Like other underground PNE applications, gas stimulation, too, would fall victim to the fatal combination of budgetary, environmental and political pressures and lack of support from industry.

## Oil Recovery

One of the earliest proposed uses of nuclear dynamite was for oil recovery. Different Plowshare techniques were proposed to retrieve oil from different types of deposits.[11] For depleted oil wells, a nuclear explosion would be used to increase the permeability or level of saturation of an oil-bearing strata. The oil would then be pumped out normally. In the case of oil shale, the nuclear blast would be used to heat the shale, which would decompose into gaseous and liquid hydrocarbons plus some free carbon. These products would then be ignited in the underground cavity, with air being pumped in to keep the combustion process alive, a process known as *in situ* retorting. In theory this would produce recoverable oil. In the case of oil sands, a nuclear explosion would be used to create a cavity into which the oil would seep. The oil could then be retorted *in situ* as in the case of oil shale.

One of the earliest Plowshare proposals was for the recovery of oil from oil sands in Colorado. In 1958 several oil companies rejected out of hand a joint AEC-Bureau of Mines proposal to conduct such an experiment.[12] Shortly afterwards, the Canadians rejected a similar proposal for the Athabasca tar sands.[13] At this stage there was scepticism on the part of the oil companies and the US Bureau of Mines as to whether *in situ* retorting was even feasible.[14]

In the mid-1960s the Bureau of Mines apparently changed its mind. An experiment, code-named Handcar, was conducted at the Nevada Test Site on 5 November 1964, in part to investigate the amount of underground breaking that would occur, a phenomenon relevant to the retorting technique.[15] In January 1967 Secretary of the Interior Udall announced an oil shale program using nuclear explosions. This, he declared, could produce over 400 billion barrels of oil, with a value of over one trillion dollars, from the oil shales of the Green River formation in Utah, Colorado and Wyoming.[16] In 1968 a contract was negotiated between CER-Geonuclear Corporation and the AEC, with the participation of the Bureau of Mines and the Livermore Laboratory, for an oil shale study called Project Bronco. However, according to Livermore, 'financial and patent disagreements among the participants aborted the project'.[17] By 1970 the oil shale development program had still not got underway; in fact, it never did. Gerald Johnson concluded that 'all in all, economic recovery of oil from oil shale using nuclear explosives seems a highly speculative prospect at this time'.[18]

In the mid-1970s the idea of recovering oil from Canada's Athabasca tar sands was revived when a Canadian took out a patent on the PNE method of *in situ* retorting. The Canadian government again responded with a study, entitled *The Practicality and Economics of Peaceful Nuclear Explosions in Canada*. The report, drawn up by a panel of experts including seismologists and representatives of the Atomic Energy Control Board, concluded that PNEs were neither practical nor economic for Canada.[19]

In discussing oil shale recovery before the JCAE in 1972, Livermore Director Dr Roger Batzel conceded: 'We are still in the situation for underground engineering that we don't understand the underground phenomenology very well — the effect of the nuclear explosion'. Representative Hosmer replied: 'No wonder these oil shale people are worried'.[20] In 1975 the *GURC Report* estimated that even with the best planning, the PNE method of oil stimulation would result in only 25 per cent effectiveness, mainly because of the danger of nuclear explosions blowing into existing nuclear cavities.[21] By the mid-1970s the US oil industry was convinced that nuclear dynamite was inappropriate for oil stimulation because it could result in serious oil reservoir damage, to the detriment of long-term production.[22] No oil stimulation experiment was ever conducted under the Plowshare program.

## *Production of Heavy Elements*

During Glenn Seaborg's ten-year term, several Plowshare experiments were carried out in the attempt to discover new 'heavy' elements. One such experiment, code-named Hutch, conducted on 16 July 1969, created the largest neutron flux known to man, 'generating extraordinarily large quantities of heavy isotopes'.[23] Overall, however, the program was a failure,[24] and was discontinued after Seaborg left the Commission. According to the Livermore Laboratory, no new elements were found, 'presumably because the spontaneous fission lifetimes of such isotopes are much shorter than the time it took to retrieve the debris samples'.[25]

## *Leaching of Copper*

In October 1967 the AEC inaugurated Project SLOOP, designed to examine the feasibility of fracturing, with a PNE, and then leaching, with an acid solution, a large deposit of copper, near Safford, Arizona.[26] The economic feasibility of the project could not, however, be demonstrated and it was dropped. The idea was revived in 1973 when the AEC and the Kennecott Copper Corporation began a joint study of a concept developed at Livermore. Again several problems were identified, among them the fact that radioactive material, primarily ruthenium-106, with a half-life of one year, would be present in the copper. While the amounts would be 'almost unmeasurable', they would nonetheless contaminate the national copper supply since copper tends to be recycled.[27] Many uses of copper, such as photographic and medical equipment, require zero radioactivity. According to the *GURC Report*, 'the hazards of such contamination would require extremely careful study'.[28]

## *Underground Storage Cavities*

The concept of PNE-created underground cavities for storing hydrocarbons or waste material had been demonstrated by the 1964 Salmon detonation in the Tatum salt dome in Mississippi. In contrast to the rubble-filled chimneys and cavities associated with nuclear explosions in brittle rock, the Salmon salt cavity was nearly spherical and devoid of rubble. Almost all the radioactivity was trapped in a salt puddle at the bottom of the cavity.[29]

In 1967, in co-operation with the Colombia Gas Systems Service Corporation, the AEC proposed the creation of a natural gas storage cavity in Pennsylvania. The venture, called Project Ketch, envisaged the detonation of a 24 kiloton nuclear explosive 3000 feet (1000 m) underground near the town of Renovo.[30] Physicist Ernest Sternglass recalls his reaction:[31]

> My eyes popped out when I saw that any possible accident might release as much as 4 million curies of radioactive iodine into central Pennsylvania, actually not very far from where a big release happened later at Three Mile Island...

As he told an editor of the *Pittsburgh Post Gazette*:[32]

> These people are crazy. This is the heart of the dairy country. Millions of curies of radioactive iodine would poison the milk all the way up to New England, all the way to New York, Washington and down to Philadelphia. This is madness...

Sternglass, an authority at the Pittsburgh Medical School on the dangers of low-level radiation, subsequently publicised the potential radiation hazards of Project Ketch at public meetings, including one held by the AEC at Carnegie-Mellon University involving Plowshare staff from Livermore. It was especially disturbing to Sternglass that the State Governor had already signed the papers approving the detonation before the AEC held its meeting. Preliminary holes had also been drilled and approvals obtained from the various state authorities. As a result of what Sternglass calls the 'first citizens' rebellion of the nuclear age in this part of the world', public opposition led to the cancellation of Project Ketch.[33] In Gerald Johnson's words, 'the project was indefinitely deferred due to public opposition to nuclear explosions in their backyards'.[34]

After the Ketch disaster the future of nuclear storage looked bleak. As Brooks and Myers put it: [35]

> Increasing citizen involvement in environmental matters suggests that it is unlikely that any significant number of storage projects will be allowed at sites that are in close proximity to market areas.

As Ketch demonstrated, the main problem with nuclear cavity storage is that while 'it is desirable that storage be close to large population centres... these areas are precisely the places where public opposition will be greatest because of the fear of radiation and seismic damage'.[36]

According to Peter Metzger, one of the most promising applications for underground nuclear cavities, the storage of radioactive wastes, was blocked off by the AEC itself.[37] Livermore scientists proposed the storage of such wastes in nuclear cavities 6000 feet (1800 m) underground. The rock in such a nuclear cavity would gradually solidify, permanently trapping radioactive wastes and preventing groundwater contamination.[38] In this application of nuclear dynamite, as in no other, the fact that radioactivity would be produced by the explosion would be of no consequence. Alan Wilson has noted that despite some possible problems, 'the potential advantages of the concept are such that it merits urgent study'.[39] He still believes, like Peter Metzger, that it is one of the better ideas for using PNEs.[40]

The AEC, by contrast, pressured Livermore to adopt a 'low profile' on the idea, because 'waste policy is established and this proposal is contrary to that policy and therefore will only confuse and impede the orderly implementation of our policy of radioactive-waste disposal in salt'.[41] Apparently the Commission had already invested too much money and effort in the idea of storing radioactive waste in salt formations to be able to change course. As Metzger noted in 1982:[42]

> It will be an interesting paradox (but historically consistent) if the very technique which finally solves the radioactive-waste problem has to do so over the objections of the AEC itself.

In the mid-1970s, as a result of the oil crisis, the creation of oil storage cavities by nuclear means was once more mooted. It was now proposed that such cavities be created offshore under the seabed, thereby attenuating seismic shock waves resulting from the PNE detonations involved. This would enable such cavities to be created near population centres. But as the *GURC Report* concluded, 'in the United States, public reaction may upset this calculation by creating an insistence on practically *no* seismic damage'.[43] The report also pointed out that the 1971 Seabed Treaty 'may impose a barrier to moving offshore beyond national waters and may hinder the flushing of PNE chimneys into international waters'.[44] The Seabed Treaty is in fact ambiguous in regard to PNEs. While it bans the emplacement on the seabed of nuclear weapons or other weapons of mass destruction, it does not define such weapons or speak of 'other nuclear devices' as do the PTBT and NPT.[45]

## Geothermal Power Generation

Otherwise known as 'pacer' technology, this concept, which can be traced back to the very beginnings of Plowshare and the ill-fated 1961 Gnome detonation, involves the explosion of nuclear dynamite in a deep PNE cavity partially filled with water. The high-pressure steam generated would be used to run conventional steam turbines. In addition, neutrons from the explosions could be used to breed fuel for nuclear power reactors.[46]

In March 1971 Edward Teller proposed to the JCAE, somewhat inappropriately in view of the budgetary stringencies being suffered by Plowshare, that a new research program be established to investigate such technology.[47] The committee recommended an appropriation of $500 000. The AEC did not share Teller's enthusiasm and the funds remained unspent. Alan Wilson described the geothermal concept as 'highly speculative'.[48]

A detailed study of the idea was nevertheless undertaken some years later, not at Livermore, but at Los Alamos. Code-named Project Pacer, the study concluded that Teller's concept was flawed. It was estimated that the generation of 2000 megawatts of electricity would require at least two 50 kiloton nuclear explosions *per day*! According to the *GURC Report*:[49]

> The public problems in exploding nuclear devices in such an installation, regularly every eleven hours, seem most forbidding in today's political environment.

Furthermore, the economics did not appear favourable, being dependent on the manufacture of approximately 10 000 nuclear explosive devices per year. An ACDA report in 1975 commented that technical problems with pacer technology included the radiation hazard, the ability of the cavern to tolerate numerous explosions, the cost of the nuclear explosives involved and the capital and operating costs.[50] Professor Long has pointed out that for 'pacer' technology to be economic, the salt cavity used would need to be stable for a normal operating life of perhaps 40 years or 30 000 explosions. Early failure of the cavity would have 'catastrophic effects on the economics'.[51] The idea did not proceed.

## *Gas Stimulation*

Only in the case of one underground engineering application, the stimulation of natural gas, was Plowshare able to attract sufficient industrial interest to actually reach the field testing stage. According to Brooks and Myers, too much should not be made of this. Risk taking is normal for gas producers. In the case of Plowshare, in which costs were shared with the AEC, the potential return was very large compared with completely commercial ventures.[52] In any case about half the proposed experiments never eventuated, including Project Dragon Tail, sponsored by the Continental Oil Company, Project Pinedale sponsored by the El Paso Natural Gas Company and Project WASP sponsored by the Wyoming Atomic Stimulation Project.

Unfortunately for the AEC, the experiments which did eventuate coincided with several important political developments. The first was increasing public dissatisfaction with all the operations of the AEC, especially its civil nuclear power program. The second was the rise of the environmental movement, resulting in such legislation as the 1969 National Environmental Protection Act (NEPA). The third was increasing opposition to all forms of underground nuclear testing.

Although the nuclear excavation program had aroused some public concern because of the venting of radioactivity from tests at the Nevada Test Site, the problems and dangers of nuclear excavation remained largely hypothetical because no tests were ever conducted 'in the field'. The gas stimulation program, on the other hand, whose experiments had to be carried out in actual gas-producing areas, directly increased public awareness of the dangers of PNEs. The first gas stimulation explosion, Gasbuggy, attracted little interest. It was the second experiment, Rulison, that aroused the feisty opposition of US citizenry. By the time the third experiment, Rio Blanco, was conducted, public opposition had become 'vigorous and directed'.[53] The principal concerns were radioactive contamination, seismic shock and the dislocation caused by precautionary safety measures such as evacuation. Public opposition continued when gas from the experiments was flared, releasing radioactivity into the atmosphere.[54]

*Figure 6* *Lowering the nuclear explosive into the test hole for the Gasbuggy experiment designed to liberate natural gas from a sandstone formation. (Courtesy of Lawrence Livermore National Laboratory.)*

## Gasbuggy

By the mid-1960s the United States was facing the prospect of a severe shortage of natural gas if economic new wells were not found. Plowshare theorists had long speculated that gas in 'tight' formations could be stimulated with nuclear explosives. In co-operation with several industrial partners, the AEC began planning a series of field tests of such a technique in the Rocky Mountains. The first experiment, Gasbuggy, involved the detonation of a single 26 kiloton device in New Mexico on 10 December 1967, under the technical direction of the Livermore Laboratory and with the joint participation of the El Paso Natural Gas Company and the Department of the Interior.[55] According to Livermore the results were 'encouraging'. Gas production increased six to eight times over previous rates.[56] According to Glenn Seaborg, Gasbuggy was 'highly successful'.[57] In the view of Fred Holzer of the Livermore Laboratory, Gasbuggy succeeded in the sense that compared with conventional wells nearby there was a very large increase in gas flow.[58]

The explosion also demonstrated the need for a 'cleaner' nuclear explosive, as the Gasbuggy device produced unexpectedly large amounts of radioactive tritium.[59] This had obvious implications for the successful marketing of nuclear-stimulated gas, as the gas itself would be radioactive. The experiment provided the impetus for development of a small (down to 7½" (19 cm) in diameter), cheap, tritium-free nuclear device, able to withstand high temperatures and pressures deep underground.

## Rulison

Project Rulison, near Rifle, Colorado, was sponsored by the Austral Oil Company and the AEC. It was under the technical direction, not of Livermore, but of Los Alamos.[60] A 40 kiloton fission device was to be used, rather than a fusion bomb, in an attempt to reduce the concentration of tritium in the resulting gas. Project Rulison encountered heated public debate, enquiries by the Colorado Water Pollution Control Commission into possible underground water contamination, questioning by the Oil Shale Commission about the safety of radioactive gas, and coal miners' protests about seismic shock waves. The Governor of Colorado sought postponement of the experiment to allow further study. In August 1969 three separate suits were filed in the Denver District Court seeking injunctions against Rulison. One of these, by the Colorado Open Space Coordinating Council, was brought on behalf of 'all persons entitled to protection from possible harmful effects resulting from release of artificially created radiation into the biosphere and into natural resources'.[61] None of the injunctions was granted.

The Rulison shot was conducted on 10 September 1969. Even then it was delayed by protestors. Edward Teller, one of the supervising scientists, recalls that when five demonstrators staged a sit-down at the site, he offered to sit with them during the blast. According to Teller it was not dangerous because the bomb would be exploded deep below the surface. But Teller was not permitted to join them. Instead a helicopter was sent in, but two of the protestors refused to leave. As Teller recalls:[62]

> Now we told these two people that when the shock from the nuclear explosion arrives it will be felt on the soles of their feet and it might conceivably hurt their backbone, but if they bend their knees, so that there is a bit of tether between the soles of their feet and their backbone, then there won't be any damage. Well nothing happened. Or at least we don't know what happened. It may be that the Devil took the two environmentalists but if he took them, he left no hide nor hoof nor cleft. They just vanished... I am very sure they did not suffer any hardship, because if they had... we would have heard about it.

Rulison was the deepest nuclear explosion ever conducted in the United States, at 2568 metres.[63] Preliminary results indicated that gas stimulation did occur, although less than expected. Tritium levels were down, although still present in 'substantial quantities'.[64] As in the case of Gasbuggy, Glenn Seaborg pronounced the experiment 'highly successful'.[65] According to Fred Holzer, the experiment was 'not nearly as successful' as Gasbuggy.[66] Gas production was disappointing. For Holzer, part of the problem was that geology, 'as much an art as a science', could not always pinpoint the location of gas accurately enough. Investors risked spending millions of dollars on a nuclear explosion that might fail to affect the deposit. To lessen this risk, Plowshare scientists conceived the idea of multiple explosions in one drill hole. Because drilling of the emplacement hole is a major expense, this would minimise drilling costs and ensure maximum impact on the gas. Planning began for Project Rio Blanco, which would use three nuclear devices, and Wagon Wheel, which would use five.

Some Plowshare scientists were, however, by this stage becoming sceptical about the future of nuclear gas stimulation because of the radioactivity problem. In June 1970 Gerald Johnson, in an article entitled 'Plowshare at the Crossroads', published in the *Bulletin of the Atomic Scientists*, argued that radioactivity was 'the principal and most serious limitation' confronting gas stimulation.[67] Noting that nuclear-stimulated gas contains both tritium and krypton–85, Johnson argued that 'through appropriate design of explosive, method of emplacement and perhaps through encouragement of exchange processes to keep the tritium in water', the tritium concentration in the gas could be reduced to a small fraction of the concentration of krypton–85. However the krypton problem would remain.[68] Another problem was the recommendation by two Livermore scientists that the 'guideline' exposure levels for radioactivity be immediately reduced at least ten-fold, which would make the use of nuclear-stimulated natural gas completely unacceptable unless it were used in electricity generating plants. As Johnson, many years later, put it, the image of a woman in the kitchen with a baby on her knee and radioactive gas burning on the stove was a powerful deterrent to the commercial use of nuclear-stimulated gas in the home.[69]

In the same month that Rulison was detonated, a Federal Power Commission report on National Gas Supply and Demand took a dismal view of the prospects of nuclear-stimulated natural gas:[70]

> ... important economic and social considerations must be resolved before such techniques experience large-scale commercial use. We do not anticipate that nuclear stimulation techniques will have a measurable effect on domestic supply availability in the foreseeable future.

The report noted that nuclear stimulation required thousands of nuclear detonations to produce a substantial increase in natural gas. Full field development of the Rulison area alone would require over 100 nuclear detonations of approximately 200 kilotons each. One scientist involved in Project Rulison, no doubt in an attempt to be reassuring, countered that 'less energy than has been used in the highest Russian bomb test could open up enough gas reserves in this peaceful application to supply total US needs for four months'.[71]

Following the Rulison detonation a new injunction was taken out to prevent flaring of radioactive gas into the atmosphere. The District Court in Denver ruled that the proposed flaring was safe and could proceed.[72] According to the Atomic Industrial Forum:[73]

> ... some individuals wearing luminous dial wrist watches received much greater radiation exposures due to tritium from their watches than anyone received from the flaring of gas in production testing of the re-entry well. And the exposure from Rulison was only a very small fraction of the additional exposure received from living in the mountains versus living at the seashore, or from living in a brick house versus living in a wood frame house. And mankind has been making this latter sort of decision without apparent harmful effect throughout the entire time the human race has existed.

In 1972 the Austral Oil Company, the AEC's partner in Rulison, accused the Commission of getting its sums wrong on the economics of nuclear-stimulated natural gas. Austral complained of the escalating cost of the project, including the cost of lawsuits brought against Rulison. Austral claimed it had spent $11 million on Rulison without recovering 'a penny', including $3 million on engineering and environmental work, before there was any 'device or knowledge of any device that would allow us to do the job at Rulison'. The company noted that it was going to take '20 or 30 wells and devices fired at Rulison to produce any significant amount of gas and make it economic at the price we are looking at'.[74] A study by the University of California, financed by the National Science Foundation, later estimated that $11 million was spent on Rulison to gain access to gas worth $1.5 million. That gas, moreover, was contaminated and of lower quality than that obtained from conventional wells in the area.[75] The AEC's response was that Rulison was designed to refine the technique of nuclear stimulation, not to demonstrate its economics. In 1974, in a somewhat belated move, Austral sued the Commission over the poor results of the test.[76] The only good that seems to have come out of Project Rulison was its use for seismic research purposes, the results being reported to the 1970 session of the Conference of the Committee on Disarmament (CCD) in Geneva.[77]

In early 1970, less than six months after Rulison, during JCAE hearings on the Plowshare budget for 1971, the gas stimulation program came under critical scrutiny. John Kelly was quizzed by Representative Aspinall:[78]

> *Representative Aspinall:*... You have two detractors, of course. One of them is money.
> *Mr Kelly:* That is right. Yes, sir. We are hopeful, of course, that that will gradually fade as the inflationary trend is halted.
> *Representative Aspinall:* The other is public acceptance.
> *Mr Kelly:* That is correct.
> *Representative Aspinall:* If you think the public acceptance of Rulison is any greater today than it was six months ago or eight months ago, I think you are mistaken.

Aspinall complained that he had trouble promoting Plowshare because of:[79]

> ... the politicians and the theorists and the environmentalists, and a few self-proclaimed scientists and engineers that apparently don't get any other employment than comes to them from being against something. I can't overcome this... it is impossible. If that is one of your detracting agents, we had better go ahead and do one of these jobs well and get it done and prove that there are no dangers involved.

A further exchange with Kelly was revealing:[80]

> *Mr Kelly:* This is exactly my point... I think even our detractors would agree that, and I am talking about environmental detractors here, would agree that we really should be allowed to go forward far enough to be sure that the benefits as well as any possible detrimental effects are clearly and well understood by everyone.
> *Representative Aspinall:* I would agree with you that your goal is fine. But when you begin to talk about Gasbuggy, Rulison, Dragon Trail, Wagon Wheel, WASP, Ketch, Sloop, Bronco, and all the rest of them before you have one of them that you can prove anything with, I think that we are whistling in the dark as far as the public is concerned.

Kelly was forced to acknowledge that under current funding levels, nuclear-stimulated commercial gas would not be available for another three to five years.[81] While the AEC would continue its efforts to develop specialised explosives for gas stimulation, the Commission was itself waiting for 'stimulation' from industry before undertaking commercial gas stimulation ventures.[82]

Such exchanges illustrated a growing dilemma for Plowshare. Successive administrations were reluctant to fund the project in the absence of demonstrated industrial support, while industrial support was not forthcoming because funding cutbacks prevented Plowshare from demonstrating the feasibility of industrial applications. Livermore Director Dr Michael May defended Plowshare, claiming that it had 'in certain areas made a great deal of progress since it started thirteen years ago... Most programs like Plowshare proceed in a somewhat uneven fashion, especially programs which have a great deal of unknowns attached to them'.[83] He then listed several factors that had slowed Plowshare: technical problems, the secrecy requirements, the cautious attitude of industry and fiscal restrictions.

The committee also heard a report on the progress of the Soviet Union's PNE program, details of which had been revealed during US/Soviet talks on PNEs in Vienna in April 1969 and in Moscow in February 1970. An article in the *New York Times* of 5 March had attempted to whip up 'missile gap' fever over PNEs with the banner headline, 'Civilian Atom Blasts in Soviet Found to Outpace US Project'.[84] Noting that JCAE members felt that Plowshare had been 'unduly cut back by the Nixon Administration', the article predicted that the Soviet program was certain to be used by the JCAE to pressure Congress to increase the Plowshare budget. In fact, the JCAE was again unsuccessful in seeking increased funds for Plowshare.

In 1971 the Administration had no need to cut back the Plowshare funding request. The AEC sought only $8 million, despite the PNE Division's request for $23 million.[85] The Office of Management and the Budget, successor to the Bureau of the Budget, opted for $5 million and urged the AEC to obtain maximum funding from industry. Incredibly, the Administration and the AEC, in a joint statement, told the JCAE that, 'despite the declining dollar support for Plowshare in recent years', it was not the intention of the Administration to 'close out the natural gas stimulation part of the program'.[86] The committee was not so sanguine. The Chairman, Representative Price, noted that some of the private companies involved in the gas stimulation program were 'rapidly approaching the point of no return — either they must be able to go forward with a defined program or it is time for them to get out of this business'.[87]

Senator Hansen of Wyoming, whose state was scheduled to host Project Wagon Wheel, pointed to the critical shortage of natural gas facing the United States. He claimed that the $5 million in the President's budget for Plowshare was a bare subsistence level that would require tripling to permit the carrying out of the AEC's announced five-year program for providing nuclear-stimulated natural gas on a commercial basis.[88] Representative Aspinall queried Senator Hansen's assumption that nuclear stimulated gas was economically feasible:[89]

> We know... that one detonation apparently is not going to do it, because that we have done. We have had two previous experiences [Gasbuggy and Rulison]. Now it is a question of multiple detonations, the question of fracturing more and lengthening the chimney that is involved, if I understand their language. This will be the next step. But as far as economics, they

haven't been able to show me yet that they are anywhere close to producing gas economically.

PNE Division Director Kelly emphasised that despite the budget cuts 'there has been no decision to suspend the program nor to suspend peaceful nuclear explosion tests'. He confirmed, however, that there were probably not sufficient funds in the proposed budget for an entirely government-funded detonation in 1972, or for the Rio Blanco shot in co-operation with CER-Geonuclear.[90] Representative Hosmer grilled Kelly in his usual fashion:[91]

> You have indicated that the industry will continue to support these joint projects, but they are not going to support these joint projects unless these joint projects are starting to move at a much, much faster rate than this $5 million a year, which practically is nothing but peanut, gum and overhead money.

Kelly admitted that due to budget cuts the PNE Division now had the minimum staff necessary to maintain 'a viable competence', but would be kept available so that the program could be picked up in future years.[92]

El Paso Natural Gas Company also testified that budget cuts would delay the gas stimulation program, while Austral Oil Company warned that if the cutback in the fiscal year 1972 budget was sustained it would be forced to withdraw from the program.[93] Austral continued with the following telling comments:[94]

> We cannot justify continued financial outlays when the government, as our partner in this pioneering development program, has apparently decided to back off in its support of the essential technological effort. And it isn't just Austral we are talking about. There should be no mystery as to why other companies haven't been coming forward with projects. Based on the experience we have had with declining government funding, and the unreasonable financial burden that has consequently been placed on Austral, it would be difficult to recommend that others participate in the program... In summary, if the government has in fact lost interest in the Plowshare program, it is time for the Administration to say so, forthrightly, so that all of us, both in industry and government, can turn our R & D energies and resources to other pursuits which could be more rewarding.

Edward Teller, in addition to floating his proposal for a Plowshare geothermal project, staunchly defended the gas stimulation program. He concluded by reiterating his unbounded optimism about the future of PNEs, unaffected by fourteen years of fruitless research:[95]

> I think progress cannot and will not be stopped, and I know that Plowshare will proceed. Whether it will proceed as rapidly as it should, whether it will proceed in the United States or in some other part of the world, these are most important questions of detail that we should consider with some care. Today those conservationists who have become reactionary, who are opposed to all progress, who seem to believe that everything that is good lies in the past, may bring about another dark age. I hope that they will not succeed.

This time the JCAE managed to convince Congress to increase the 1972 appropriation for Plowshare by $2 million to $7 million, including $500,000 for Teller's geothermal experiment.

In July 1971 James Schlesinger replaced Glenn Seaborg as chairman of the AEC, with a directive to give the agency a new image. The Commission was by now the centre of bitter controversy, largely over its nuclear power reactor program but also many other aspects of its activities. According to Daniel Ford, Schlesinger could see that the environmental movement, which was leading the attack on the AEC, was growing and public sympathy with its objectives increasing.[96] 'Environmentalists have raised many legitimate questions', Schlesinger said. In future the AEC would respond to these questions 'as a referee serving the public interest', not as an exclusive agent of the nuclear industry.[97] As for Plowshare, Schlesinger described it as 'iffy'.[98] Industry, he said, was 'on its own' as far as the gas stimulation program was concerned and should be made to pay its way.[99] The Plowshare Division was promptly absorbed into the Division of Applied Technology. Incredibly, President Nixon, in spite of these developments, declared on 4 June 1971, that one of the four means by which the government proposed mitigating a potential critical shortage of natural gas was the use of PNEs.[100]

In requesting funds for Plowshare for fiscal year 1973, the Division of Applied Technology sought $18 million, which the Commission reduced to $7 million, the same as appropriated in 1972. The OMB reduced this to $6.8 million. Most of the cuts had therefore again been made by the Commission itself. Commissioner Abbadessa testified to the JCAE that the AEC had priorities other than Plowshare. These included electric power for uranium enrichment; the fast breeder program and related nuclear safety; waste management; industry regulation; controlled thermonuclear research; and, finally, nuclear weapons production — where the Commission had 'a built-in mortgage'.[101] According to Ed Fleming, Abbadessa, a very influential Commissioner, had by this stage turned against Plowshare and was instrumental in producing lower funding requests.[102]

John Kelly, now Assistant Director for PNEs in the Applied Technology Division, testified that almost all the funds requested for 1973 would be devoted to the gas stimulation program, in particular the Rio Blanco shot in Colorado. Detailed planning would also begin for Wagon Wheel, involving five 100 kiloton explosions. Hosmer concluded that, with the level of funding envisaged, the five-year gas stimulation program had in effect been 'junked'. Kelly admitted that the program was now more likely to take ten or more years. Still, his faith in Plowshare remained unbounded:[103]

> At last year's hearings I stated that I believed that within a few years and at modest cost to the Government, Plowshare technology could make a positive contribution in assisting in alleviating the natural gas shortage. I continue to believe this; and I trust that with continued support for the development of technologies, such as natural gas stimulation, Plowshare can contribute substantially in the recovery and better use of various natural resources...

Our goal will be to determine how Plowshare can best assist in meeting the needs of society.

## Rio Blanco

The third gas stimulation experiment, Rio Blanco, was a technical failure, generating more controversy than gas, and killing off Plowshare for good. The project was sponsored by CER-Geonuclear and the AEC. The Department of the Interior, which had been 'dragging its feet' on the gas stimulation program because it feared that nuclear explosions risked damaging contiguous oil shale deposits, declined to co-sponsor Rio Blanco, although it was obliged to be involved.[104] Rio Blanco envisaged the detonation of three 30 kiloton nuclear devices simultaneously in the same drill hole. These were expected to create three interconnecting cavities to form one large vertical 'chimney'. A special fission explosive (Miniata) had been designed for the experiment, to minimise the size of the drill hole required, the tritium yield and the cost.[105]

To the great surprise of Plowshare and AEC officials, major public opposition to Rio Blanco surfaced. This was partly the result of the availability of a number of new avenues of protest. Governor Love of Colorado formed an advisory body on the project. The AEC, under Schlesinger's new rules, was obliged to hold public hearings and, under the National Environmental Protection Act, to prepare for the first time an environmental impact statement. Senator Haskell of Colorado also held hearings on Rio Blanco in the Subcommittee on Public Lands of the Senate Committee on Interior and Insular Affairs. Finally, legal suits were filed against the detonations. Most public debate centred on the potential dangers of radioactivity in the atmosphere, underground and in ground water, resulting not just from the planned triple explosion, but from future commercial development of the gas field. Consumer advocate Ralph Nader became involved. Oil companies also voiced opposition, claiming, like the Department of the Interior, that underground oil shale deposits could be damaged by repeated nuclear explosions. Ironically though, the local counties around the Rio Blanco site supported the experiment because of the potential economic value of nuclear-stimulated gas.[106]

Opposition to Rio Blanco did come from local communities in neighbouring Wyoming, where a grassroots movement had arisen against the planned Wagon Wheel project. This movement, which had decided that the best way to stop Wagon Wheel was to stop Rio Blanco, lobbied actively in Washington, including Capitol Hill, where they targeted Wyoming Senators, and AEC headquarters. At the AEC, the lobbyists gave a presentation, including slides, that lasted all afternoon and which, according to Ed Fleming, was 'totally emotional'. No rational arguments, including economic justifications, were entertained. According to Fleming, as a result of this experience, many of the AEC employees present began to see the personal side of 'going into someone's backyard and letting one of those things off'.[107] This seemed to increase antipathy towards Plowshare within the AEC.

AEC Chairman Schlesinger had promised that Wagon Wheel would be cancelled if the citizens of the area, Sublette County, objected. When Senator McGee of Wyoming wrote to Schlesinger, expressing doubts about Wagon Wheel, Edward Teller described him as a 'nincompoop' and 'feeble-minded'.[108] Schlesinger's successor, Dixie Lee Ray, formerly Professor of Zoology at the University of Washington, Seattle,[109] was more sympathetic to McGee, assuring him that the interests of the people of Sublette County would have a 'very important value bearing on any ultimate decision' on Wagon Wheel.[110] When a straw poll of 78 per cent of the county's voters revealed them to be opposed to Wagon Wheel by a margin of 4:1,[111] the project was 'indefinitely postponed'. Dixie Lee Ray declared it 'dead as a doornail'.[112] While Ray was reportedly a Plowshare supporter, she was both unwilling and unable to sustain the program in the face of such opposition. According to Gerald Johnson, by the time Dixie Lee Ray was appointed it was too late to save Plowshare: 'the war was over by then'.[113]

In the meantime, the battle over Rio Blanco was still raging. Just ten days before the experiment was to be conducted, Senator Haskell began his hearings in Washington, by which stage the three explosive devices had been lowered into the shaft. According to Haskell, however, this was 'by no means irreversible'. He had written twice to the Secretary of the Interior seeking cancellation of Rio Blanco because:[114]

> Full field development at Rio Blanco is a dubious proposition. This view is shared by virtually every competent scientific, economic and governmental authority with whom I have discussed the proposal... Why accept any risk of radioactive contamination of air or water, or create a radioactive cavern demanding perpetual monitoring, if the program will produce nothing of value?

During his hearings Haskell heard evidence that full field development of Rio Blanco could contaminate the Colorado River, that seismic damage would amount to $250 million over the eighty years of the field's projected lifetime, and that in a period of relative uranium scarcity Plowshare was an inefficient use of fissionable material.[115] There was also concern about the domestic use of radioactive gas. One elderly lady from Aspen, Colorado, was advised by Ed Fleming that, because natural radiation levels are lower at lower altitudes, she could counteract the minor effects from using such gas by moving her house eight feet (2.5 m) down the mountain.[116]

On 4 April 1973 the Federation of American Scientists wrote to Haskell claiming that it would be 'foolhardy' to proceed with Rio Blanco when its value would be zero if full field development was not undertaken. The Federation noted that according to the AEC's own Environmental Statement, the Rio Blanco experiment would be followed by Phase II, involving 4–6 stimulations using 3–5 explosives each, and Phase III involving 20–

60 stimulations. In total, 300 nuclear explosions with yields of 30 kilotons each were proposed, just to determine the feasibility of nuclear stimulation. This was more than the AEC had detonated in its entire nuclear weapon development program since the PTBT was signed.[117] Up to 1000 detonations would be required for full field development.

The Federation identified several potential dangers in full field development. These included accidental venting of radioactivity; flaring of the gas which would release tritium, carbon–14, argon and krypton into the atmosphere; and the possibility of groundwater contamination. Pointing out that accidental venting had occurred in more than twenty nuclear tests at Nevada, the Federation warned, with some exaggeration, that:[110]

> We should not repeat the mistakes made in the Plowshare excavation program in which premature experiments were carried out, spreading large amounts of radioactive material over extensive areas of the US.

The Federation was particularly scathing about the AEC's decision to ignore the groundwater contamination problem because, in the Commission's own words, the AEC 'could not hypothesise any mechanism whereby [radioactivity] could get into ground water'. Federation Secretary Herbert Scoville testified that the Longshot test in Amchitka had resulted in surface radioactivity of unknown origin.[119]

Scoville also berated the AEC for installing the nuclear explosives at the Rio Blanco site while attempts were still being made to have the shots postponed. He said it was like testifying with 'a nuclear explosive at your head', especially as the AEC believed that the problem of removing the explosives was 'very difficult'.[120] Scoville testified that:[121]

> There has been a previous case in which President Johnson refused to go along with a test which had been planned and which they had the explosives already in the ground and there was a great hoola about them moving these explosives, but President Johson refused to give his approval at the last moment and the explosives had to come out and they did come out. This was a bloody nuisance, if I may say so, in rather bald terms, and it is certainly not something one wants to do, but I think if one forced the AEC to do this a few times they would be willing to listen to the public first, perhaps, rather than after they have put the explosives in the ground.

When all efforts to stop Rio Blanco failed, including the legal injunctions, eleven conservationist groups appealed to President Nixon to halt the blast. Nixon was not moved by such appeals. A White House staffer apparently rang the Department of the Interior and told them that the President wanted the shots to proceed, so 'do it'![122] Rio Blanco's three shots were detonated on 17 May 1973 'like clockwork'.[123] The results: several 'unexpected and little understood problems'.[124] The three blast cavities failed to join; high levels of cesium–137 were unexpectedly found in the gas; the amount of radioactive water produced was greater than expected; and deadly strontium–90 escaped from the chimney. The multiblast technology had failed. Only small amounts of gas were obtained from the upper cavity.[125] Ironically, according to Fred Holzer, the nuclear explosion itself may have prevented gas from seeping into the cavity by sealing its sides through heat and blast.

The reasons for the lack of connection, according to Iris Borg, remain conjectural. In her view, the result was 'more of an inconvenience' than a demonstration of failure to stimulate natural gas.[126] Incredibly, there had been no testing of the gas reservoir prior to the Rio Blanco experiment. The amount and quality of gas shown to exist by post-shot drilling proved to be considerably less than predicted.[127] A principal participant in the project summed it up with the statement: 'if we had known in 1972 what we know now about this site, the project would not have been executed there'.[128]

Rio Blanco was the last nuclear stimulation experiment and the last PNE conducted by the United States. As Herbert Scoville remarked soon after the blast, 'there is a distinct possibility that the test may have spelled the death knell of this particular technology'.[129] The *GURC Report* in 1975 exclaimed: 'It is fair to say that less confidence is felt regarding PNE technology since the Rio Blanco event than before it'.[130] None of the other planned gas stimulation projects ever eventuated.

The Livermore Laboratory has described the failure of the gas stimulation program in the following terms:[131]

> Although the Gasbuggy, Rulison and Rio Blanco experiments provided invaluable experience and established the technical feasibility of gas stimulation techniques using nuclear explosives, its economic viability was never demonstrated. Because Federal policy required private industry to fund 90 per cent of the costs, and because sites were limited to low-production fields, incentives to undertake further gas-stimulation experiments were weak. No gas from the experimental wells has been marketed.

But there were other factors just as significant. Franklin Long told the Senate Committee on Foreign Relations that the abandonment of gas stimulation tests was due partly to the existence of conventional alternatives but 'at least as important... is a strongly negative reaction of segments of the US public to the last two field tests [Rulison and Rio Blanco]'.[132]

In 1973 the Western Interstate Nuclear Board recommended that the Rocky Mountain states press Congress to assure them of an appropriate role in the regulation of PNE technology before passing any commercial Plowshare legislation.[133] In 1974 the voters of Colorado passed by referendum a constitutional amendment preventing further underground nuclear explosions in the state without a prior affirmative vote of the people. This was particularly impressive given the existence of the so-called 'energy crisis'.[134] It was apparently the judgement of Coloradans that the 'possibility of long-term physical damage caused by continued underground nuclear blasting and the radiation hazards associated with this technology were too high a price to pay for a short term increase in the supply of natural gas'.[135] In response to such concerns, Congress in 1974 added a restriction to the Energy Research and Development Act forbidding any funds being used for 'field testing of nuclear explosives in the recovery of oil and gas'.[136]

# Notes

1 'AEC Plans Peaceful Explosions But Raises Anger', *Nature*, Vol. 241, 23 February 1973, p. 495.
2 Australian Atomic Energy Commission, *Peaceful Uses of Nuclear Explosions — An Evaluation, for Australian Purposes, of Proposed Civil Engineering and Mining Applications*, AAEC, Sydney, 1964, p. 219.
3 G. W. Johnson, 'Plowshare at the Crossroads', *Bulletin of the Atomic Scientists*, June 1970, p. 84.
4 *US Perspectives on Peaceful Nuclear Explosions*, unpublished ms, ACDA, 7 February 1978, p. 16.
5 J. Koop, 'Plowshare and the Nonproliferation Treaty', *Orbis*, XII, 3, Fall, p. 803.
6 M. I. Shaker, *The Nuclear Non-Proliferation Treaty: Origin and Implementation 1959–1979*, Vol. I, Oceana Publications, London, 1980, p. 391.
7 US Congress, JCAE, Hearings on AEC Authorizing Legislation for Fiscal Year 1966, US Government Printing Office, Washington DC, 1965, p. 20.
8 Ibid., p. 22.
9 H. Scoville Jr., 'Peaceful Nuclear Explosions — An Invitation to Proliferation' in A. W. Marks (ed.), *NPT: Paradoxes and Problems*, Arms Control Association, Washington DC, 1975, p. 51.
10 *Nature*, Vol. 241, 23 February 1973, p. 496.
11 G. T. Seaborg and W. R. Corliss, *Man and Atom: Building a New World Through Nuclear Technology*, E. P. Dutton and Co., New York, 1971, p. 181.
12 US Congressional Record, 14 April 1969, US Government Printing Office, Washington DC, p. 8945.
13 See Chapter 2.
14 US Congress, JCAE, Hearings on AEC Authorizing Legislation for Fiscal Year 1966, p. 21.
15 Ibid.
16 G. F. Tape, 'National Policy on Peaceful Uses of Nuclear Explosives' in L. E. Weaver (ed.), *Education for Peaceful Uses of Nuclear Explosives*, University of Arizona Press, Tucson, Arizona, 1970, pp. 5–6.
17 *Energy and Technology Review*, Lawrence Livermore National Laboratory, September 1982, p. 18.
18 Johnson, 'Plowshare at the Crossroads', p. 91.
19 From a conversation with Dr Peter Basham, Department of Energy, Mines and Resources, Ottawa, 11 December 1986.
20 US Congress, JCAE, Hearings on AEC Authorizing Legislation for Fiscal Year 1973, US Government Printing Office, Washington DC, 1972, p. 128.
21 US Congress, Senate Foreign Relations Committee, Hearings on the TTBT/PNET, 28 July, 3 August, 8 and 15 September 1977, US Government Printing Office, Washington DC, 1977, p. 135.
22 *PNE Activity Projections For Arms Control Planning* (the *GURC Report*), Gulf Universities Research Consortium, Galveston, Texas, US Arms Control and Disarmament Agency Document, ACDA/PAB-253, Washington DC, 1975, p. 131.
23 *Energy and Technology Review*, pp. 16–17.
24 I. Y. Borg, *Nuclear Explosions for Peaceful Purposes*, Lawrence Livermore National Laboratory Paper No. UCRL-95026, Rev. II, November 1986, p. 8.
25 *Energy and Technology Review*, p. 17.
26 The *GURC Report*, p. 126.
27 Ibid., pp. 128–29.
28 Ibid., p. 129.
29 Borg, *Nuclear Explosions for Peaceful Purposes*, p. 9.
30 L. J. Freeman, *Nuclear Witnesses: Insiders Speak Out*, W. W. Norton and Co., New York, 1981, p. 63.
31 Ibid., p. 64.
32 Ibid.
33 Seaborg and Corliss, p. 180.
34 Johnson, 'Plowshare at the Crossroads', p. 89.
35 D. B. Brooks and H. R. Myers, 'Plowshare Evaluation' in B. Boskey and M. Willrich

(eds), *Nuclear Proliferation: Prospects for Control*, Dunellen Publishing Co., New York, 1970, p. 94.
36 The *GURC Report*, p. 105.
37 H. P. Metzger, *The Atomic Establishment*, Simon and Schuster, New York, 1972, p. 255.
38 The *GURC Report*, p. 121.
39 A. R. W. Wilson, 'Current Status of Civil Engineering and Mineral Resources Development Applications of Peaceful Nuclear Explosions' in Proceedings of the Fourth International Conference on the Peaceful Uses of Nuclear Energy, 6–16 September 1971, IAEA, New York and Vienna, 1972, p. 228.
40 From an interview with Alan Wilson, Sydney, 7 August 1987.
41 Metzger, p. 256.
42 Ibid.
43 Ibid., p. 118.
44 Ibid.
45 *Arms Control and Disarmament Agreements*, United States Arms Control and Disarmament Agency, Washington DC, 1982, p. 102.
46 The *GURC Report*, p. 135.
47 US Congress, JCAE, Hearings on AEC Authorizing Legislation for Fiscal Year 1971, US Government Printing Office, Washington DC, 1970, p. 2350.
48 Wilson, p. 229.
49 The *GURC Report*, p. 135.
50 *An Analysis of the Economic Feasibility, Technical Significance and Time Scale for Application of Peaceful Nuclear Explosions in the US, with Special Reference to the GURC Report Thereon (The Long Report)*, Program on Science, Technology and Society, Cornell University, US Arms Control and Disarmament Agency, Washington DC, 1975, p. 25.
51 F. A. Long, *Peaceful Nuclear Explosions*, unpublished ms for ACDA, Washington DC, 1 September 1975, p. 15.
52 Brooks and Myers, pp. 91–92.
53 *The Long Report*, p. 9.
54 *The American Experience with Peaceful Nuclear Explosions*, US Arms Control and Disarmament Agency, US Government Printing Office, Washington DC, 1978, p. 5.
55 *Energy and Technology Review*, p. 17.
56 Ibid.
57 Seaborg and Corliss, p. 177.
58 From an interview with Fred Holzer, Livermore Laboratory, California, 3 December 1986.
59 F. Kreith and C. B. Wrenn, *The Nuclear Impact*, Westview Press, Boulder, Colorado, 1976, p. 14. Tritium can be released during detonation of a fusion device, but can also arise through 'neutron capture' of trace elements of lithium in the surrounding rock. See Borg, *Nuclear Explosions for Peaceful Purposes*, p. 5.
60 *Energy and Technology Review*, p. 17.
61 Kreith and Wrenn, p. 91.
62 S. A. Blumberg and G. Owens, *Energy and Conflict: The Life and Times of Edward Teller*, G. P. Putman's Sons, New York, 1976, p. 440.
63 Borg, *Nuclear Explosions for Peaceful Purposes*, p. 7.
64 *Energy and Technology Review*, p. 17.
65 Seaborg and Corliss, p. 177.
66 From an interview with Fred Holzer, Livermore Laboratory, California, 3 December 1986.
67 Johnson, 'Plowshare at the Crossroads', p. 87.
68 Ibid., p. 90.
69 From an interview with Gerald Johnson, Washington DC, 10 December 1986.
70 Brooks and Myers, p. 92.
71 H. F. Coffer and G. H. Higgins, 'Future Contained Nuclear Explosives Experiments' in Weaver (ed.), p. 50. The largest Soviet nuclear test ever conducted, and the largest by any nuclear weapon state, was a 60 megaton explosion in 1961 (R. Milliken, *No Conceivable Injury*, Penguin, Ringwood, 1986, p. 143).
72 J. T. Arnett, *Industry and Nuclear Explosives: Facts About an Emerging Technology and Its Immense Potential for Benefit*, Atomic Industrial Forum Inc., New York, July 1973, p. 28.

73 Ibid., p. 41.
74 US Congress, JCAE, Hearings on AEC Authorizing Legislation for Fiscal Year 1973, pp. 2376–378.
75 US Congress, Subcommittee on Public Lands of the Senate Committee on Interior and Insular Affairs, Hearings, 11 May 1973 (*The Haskell Hearings*), US Government Printing Office, Washington DC, 1973, p. 44.
76 Kreith and Wrenn, p. 218.
77 *Negotiations on a Comprehensive Test Ban 1963–71*, US Arms Control and Disarmament Agency, Unclassified Document Series, DDS Ref. 587, 22 December 1971, p. xviii.
78 US Congress, JCAE, Hearings on AEC Authorizing Legislation for Fiscal Year 1971, p. 626.
79 Ibid.
80 Ibid., p. 627.
81 Ibid., p. 609.
82 Ibid., p. 625.
83 Ibid., p. 631.
84 Ibid., p. 640.
85 US Congress, JCAE, Hearings on AEC Authorizing Legislation for Fiscal Year 1972, US Government Printing Office, Washington DC, 1971, p. 2306.
86 Ibid.
87 Ibid.
88 Ibid., p. 2309.
89 Ibid., p. 2313.
90 Ibid., pp. 2313–14.
91 Ibid., p. 2327.
92 Ibid., p. 2323.
93 Ibid., p. 2372.
94 Ibid., p. 2375.
95 Ibid., p. 2350.
96 D. Ford, *The Cult of the Atom*, Simon and Schuster, New York, 1982, p. 133.
97 Ibid., p. 134.
98 From an interview with Ed Fleming, Livermore Laboratory, California, 3 December 1986.
99 US Congress, JCAE, Hearings on AEC Authorizing Legislation for Fiscal Year 1973, p. 110.
100 Ibid., p. 81.
101 Ibid., p. 108.
102 From an interview with Ed Fleming, Livermore Laboratory, California, 3 December 1986.
103 US Congress, JCAE, Hearings on AEC Authorizing Legislation for Fiscal Year 1973, p. 106.
104 Metzger, p. 266.
105 Borg, *Nuclear Explosions for Peaceful Purposes*, p. 6; *Energy and Technology Review*, p. 17.
106 From an interview with Ed Fleming, Livermore Laboratory, California, 3 December 1986.
107 Ibid.
108 *The Haskell Hearings*, p. 9.
109 *Nature*, 16 February 1973, p. 421.
110 *The Haskell Hearings*, p. 33.
111 Ibid., p. 40.
112 Ibid., p. 9.
113 From an interview with Gerald Johnson, Washington DC, 10 December 1986.
114 *The Haskell Hearings*, pp. 4–5.
115 Ibid., pp. 44–46.
116 From an interview with Ed Fleming, Livermore Laboratory, California, 3 December 1986.
117 *The Haskell Hearings*, p. 142.
118 Ibid., p. 140.
119 Ibid., p. 148.

120 Ibid., p. 149.
121 Ibid.
122 From an interview with Ed Fleming, Livermore Laboratory, California, 3 December 1986.
123 Ibid.
124 Ibid.
125 *Energy and Technology Review*, p. 18.
126 Borg, *Nuclear Explosions for Peaceful Purposes*, p. 7.
127 *Energy and Technology Review*, p. 18.
128 Borg, *Nuclear Explosions for Peaceful Purposes*, p. 8.
129 Scoville, p. 50.
130 The *GURC Report*, p. 77.
131 *Energy and Technology Review*, p. 18.
132 US Congress, Senate Foreign Relations Committee, Hearings on the TTBT/PNET, 1977, p. 135.
133 The *GURC Report*, p. 22.
134 Kreith and Wrenn, p. 202.
135 Ibid., p. 210.
136 *The Long Report*, p. 10.

# 12

# India Gives PNEs a Worse Name

*India's 'peaceful nuclear explosion experiment' is, first of all, the test of a bomb. Not only is there no real distinction between a military and peaceful explosion, but even the United States, with all its time and technology, has yet to find a single feasible peaceful use for nuclear explosives. For India to call its explosion 'peaceful' and to abjure all military intent is, in a word, rubbish.*[1]

*The Washington Post*, 21 May 1974

On 18 May 1974 India detonated a plutonium implosion device 100 metres underground in the Pokharan range of the Rajasthan desert. The explosion, reported to be between 12 and 15 kilotons, roughly the size of the Hiroshima bomb, caused a mound of sand to rise above the ground, quickly subsiding into a crater 150 metres in diameter.[2] The Bhabha Atomic Research Centre reported that 'the mound was not ruptured by the cavity gases during its growth and fall, which is consistent with the fact that no radioactivity was released to the atmosphere by the experiment'.[3]

This single event, more than any other, brought sharply into focus the arms control implications of nuclear dynamite. According to Pringle and Spigelman, the Indian explosion had 'the greatest impact on the course of world nuclear development since the first Soviet test in August 1949'.[4] In Michael Brenner's view, India's entry into the nuclear club catalysed United States' thinking about the dangers of nuclear proliferation and eventually transformed US policy.[5]

India called the event a 'peaceful nuclear explosion'. In its official announcement the Indian Atomic Energy Commission described the blast as:[6]

> ... a peaceful nuclear explosion experiment using an implosion device. As part of the program of study of peaceful uses of nuclear explosions, the Government of India has undertaken a program to keep itself abreast of developments in this technology, particularly with reference to its use in the field of mining and earth-moving operations.

Suspicion of Indian motives was openly voiced and widespread. Few believed the blast was a PNE. The United Kingdom declared that 'the Indian nuclear explosion has introduced a new factor with world and regional implications; it reinforces the need to ensure wider acceptance of the Non-Proliferation Treaty to which Her Majesty's Government continues to attach the greatest importance'.[7] Sweden described the explosion as 'a severe set-back in the international work for détente and disarmament'.[8] Even Nigeria, one of the non-aligned states most susceptible to Indian pronouncements on the PNE issue, expressed its regret, saying that Nigeria was 'opposed to all nuclear tests and calls for a comprehensive test ban'.[9] Australia said it viewed with concern:[10]

> any action that may encourage or facilitate further proliferation of nuclear weapons. It is a matter of regret that a number of countries, including India have not become parties to the Non-Proliferation Treaty.

New Zealand stated that the Indian action, 'however carefully defined its purpose, runs counter to the Non-Proliferation Treaty and can only encourage other countries with similar capabilities'.[11]

One of the most forthright statements came from the Secretary-General of OPANAL, the organisation charged with implementing the provisions of the Treaty of Tlatelolco. He declared that:[12]

> Notwithstanding the declared intentions with which a nuclear explosive device was developed by the Indian Government, from the point of view of the Treaty of Tlatelolco we are concerned only with one indisputable fact: India has developed a device which according to all information available fits the definition of a nuclear weapon contained in Article 5 of the Treaty, that is, a 'device which is capable of releasing nuclear energy in an uncontrolled manner and which has a group of characteristics that are appropriate for use for warlike purposes'.

OPANAL proceeded to demolish Indian credibility by not only describing India as a 'nuclear power', but by asking it to sign Protocol II to the Treaty of Tlatelolco to guarantee that it would not use nuclear weapons against treaty parties.[13] India refused.

Pakistan's reaction was, naturally, virulent. Appearing before the CCD on 23 May, Foreign Secretary Agha Shahi expressed Pakistan's 'deep concern' over the 'emergence of a sixth nuclear weapon power'. He dubbed any distinction between nuclear weapons and PNEs 'an exercise in casuistry' and accused India of making a 'surreptitious entry into the nuclear club'.[14] He also accused India of violating the PTBT by permitting radioactivity to cross into Pakistani territory.[15]

The strongest reaction apart from Pakistan's came from Canada, which had helped India build the CIRUS research reactor at Trombay which produced the plutonium used in the Indian 'PNE'.[16] Canada's agreement with India for the operation of the reactor was, according to George Quester, 'the laxest known'.[17] It was, in effect, a 'gentlemen's agreement' between two trusting members of the British Commonwealth, requiring no international inspection and only a vague pledge by India to use the materials involved for 'peaceful purposes'.[18] Quester quotes a 'cruel jest' of the period to the effect that Canada would be the first country to give away nuclear weapons without making them itself.[19] When in 1963 India sought a similarly lax agreement to cover a Canadian reactor to be built in Rajasthan, Canada refused, both because of Pakistani protests and also an emerging realisation that tighter guarantees against any explosive uses were needed.[20] Some Canadian officials also wanted to tighten the conditions governing the Trombay reactor, but the 'commercial people' on the Canadian side demurred.[21]

While the Canadians could be forgiven for not forseeing the possibility of PNEs when their original agreement with India was signed, there was no excuse for not tightening the agreement once such a possibility became apparent. More disturbingly, Quester claims that when the Indians openly insisted after 1963 that PNEs were among the 'peaceful purposes' permitted by the India/Canada agreements, the Canadians 'were not at all resolutely opposed'.[22] While both Canada and the United States stated publicly before the Indian explosion that 'peaceful applications', by definition, excluded explosives of any kind, and while the Canadians made several private representations to the Indians on this point, at no time did they insist that the Indians publicly agree with them or sign an agreement to that effect.[23]

On 22 May 1974 the Canadian Secretary of State for External Affairs, Mr Sharp, announced Canada's reaction to the Indian explosion:[24]

> First, we are concerned as to the effect that India's action, whatever its motivation, will have on international efforts, to which Canada has been an active party, to limit and control the proliferation of nuclear explosion technology for which there can be no distinction between peaceful and potential military application. For all intents and purposes, therefore, India now has developed the capability of producing a nuclear weapon. The development of this technology by India is bound to have serious and widespread repercussions throughout Asia and the world. Secondly, we are very distressed and concerned that this latest member of the nuclear club should be a country with which successive Canadian Governments have carried on over the past two decades extensive co-operation in the nuclear energy field . . .

The statement concluded that while Canada fully respected India's sovereignty and independence, it could not be expected 'to assist and subsidise, directly or indirectly, a nuclear program which, in a key respect, undermines the position which Canada has for a long time been firmly convinced is best for world peace and security'. Canada promptly suspended all nuclear co-operation with India.[25] India responded that it was:[26]

> ... in no way bound to accept the Canadian interpretation, based on the discriminatory NPT to which India is not a signatory, that since peaceful nuclear explosion technology may have potential military application, states, other than the privileged Five, should not have access to that.

In contrast with the reaction of the rest of the world, the American reaction was amazingly low-key. In a statement to the CCD on 21 May the US representative said simply that:[27]

> The United States has always been against nuclear proliferation. This position was adopted because of the adverse impact nuclear proliferation will have on world stability. This remains the position of the United States Government.

Even non-official American reaction was mixed. Professor Long of Cornell University, for instance, declared that:[28]

> Sceptics argue that the Indian government is probably more interested in the knowledge of nuclear weapons which a nuclear explosives development program will bring them, but this is sheer speculation.

Secretary of State Kissinger declined to publicly castigate the Indians or pressure them to abandon their nuclear explosive program, partly because he wanted to avoid endowing the acquisition of a crude nuclear device with great intrinsic value.[29] He denied that the United States had contributed any material for the Indian bomb, instead blaming the Canadians.[30] Only after two years did the Nixon Administration publicly admit that it had supplied heavy water to the Indians without restrictions on its use.

According to Michael Brenner, the US Government, despite having received several warning signals, was genuinely surprised by the Indian explosion and confused over how to react. If it reacted too harshly it might inflate India's new status and destroy any chance of securing its co-operation in nonproliferation efforts. If on the other hand it reacted mildly in order not to provoke India, it might give the impression that it did not take seriously the nonproliferation implications of India's explosion. Brenner notes, however, that:[31]

> The stumbling that was witnessed as the United States searched for the appropriate reaction was due as much to senior policy makers' lack of anticipation, and to inadequate mechanisms for using staff, as to the inescapable difficulties of measuring the cost-effectiveness of the strategies being considered.

While the US reaction was low-key, several countries actually supported or expressed sympathy for India's position. In a message of congratulation to Indian scientists, the Chairman of the French Atomic Energy Commission called the Indian experiment 'the crossing of a new and difficult step towards the mastery of nuclear techniques'.[32] Among the developing countries which expressed understanding of India's position were Thailand, Yugoslavia,

Senegal, Nepal, Afghanistan, Bangladesh and Sri Lanka.[33] While the Eastern Europeans were openly supportive, the Soviet Union was more circumspect. The Chinese reaction was 'one of conspicuous mildness'.[34]

India had not always espoused a pro-PNE position, having been one of the first countries to call for a nuclear test ban. This was, however, before the distinction between nuclear weapon tests and PNEs was widely made and before Plowshare had been extensively publicised. By 1963, when signing the PTBT, India made clear that it was at least aware of the possibilities of PNEs, when it declared that:[35]

> India has welcomed and acceded to the Partial Test Ban Treaty as a step towards disarmament. There is however no reason why the benefits of using atomic explosions in civil engineering works should be denied to mankind, so long as such explosions were subject to international supervision.

It is not clear whether India envisaged future amendments to the treaty or interpreted it as already permitting PNEs. New Delhi has never sought an amendment and by conducting its 1974 explosion underground avoided violating the treaty. According to one Indian commentator, India seems to have decided after signing the PTBT that any modification of the treaty would result in the 'whittling down' of its 'discipline' or its 'amendment in the direction of the discriminatory NPT approach'.[36] The Indians were undoubtedly fearful that if the treaty were amended to permit PNEs, such amendments would be hedged with intrusive verification provisions.

The first real indication that India intended to take advantage of the PNE concept came in October 1964, after the first Chinese nuclear test. The founder of India's nuclear program, Homi Bhabha, revealed that India was '18 months away from exploding either a bomb or a device for peaceful purposes...'[37] On 27 November Prime Minister Shastri told Parliament that he opposed nuclear weapons but supported the expansion of nuclear science for peaceful applications, including developing nuclear explosives, provided they were used for peaceful purposes, such as 'tunnelling'.[38] In 1965 he authorised preliminary preparations for a 'peaceful nuclear explosion'.[39] According to one observer this was:[40]

> ... an unmistakable shift of Indian policy... from favouring a ban for nonproliferation purposes on all nuclear explosions to one that insisted on a distinct class of peaceful nuclear explosions as a necessary capability if a country was not to be deprived technologically.

Even so, up to 1967, Indian policy on PNEs remained ambiguous, probably intentionally so. In August 1965 Indian Ambassador Trivedi was still telling the ENDC that 'all nuclear tests are basically evil; they encourage evil, and the sooner the evil is dealt with the better'.[41]

In January 1967 Glenn Seaborg, on a visit to India, explored with Dr Vikram Sarabhai, chairman of the Indian Atomic Energy Commission, the 'disturbing reports that India might undertake the development of nuclear explosives on the grounds that they were to be used only for peaceful purposes'.[42] Seaborg stressed 'the US position that the development of peaceful nuclear explosives was tantamount to the development of nuclear weapons and that it should be left to the nuclear powers to meet the legitimate needs of other nations for the benefits of these highly sophisticated devices'.[43] Sarabhai, according to Seaborg, 'stated categorically that no such program was in progress or contemplated'.[44] However, he maintained that the decision of each nation on whether to develop PNEs should not be inhibited by a nonproliferation treaty.

During 1967 India observed the Brazilian-led debate over PNEs in the Tlatelolco negotiations in Mexico City and became heavily involved in the NPT negotiations in Geneva. Trivedi was now telling the ENDC that:[45]

> The nuclear weapon powers now want comprehensive controls over the peaceful activities of civil nuclear powers... They even want to prohibit the civil nuclear powers from undertaking peaceful explosions purely for their economic development even if such peaceful pursuits take place under international supervision.

By May 1967 there was, according to C. S. Jha, India's Foreign Secretary at the time, a basic consensus in the Emergency Committee of the Indian Cabinet, that India's development of nuclear technology for peaceful purposes would 'necessarily involve nuclear explosions, especially in view of India's inadequate resources of oil'.[46] This was to become one of the reasons for India's rejection of the NPT. In 1968, in officially announcing its decision not to sign the treaty, India alleged that the treaty:[47]

> ... further institutionalised international discrimination by imposing safeguards on non-nuclear weapon states but not on nuclear weapon states and by prohibiting the autonomous use of nuclear explosives for peaceful purposes by the former but not by the latter.

According to George Quester, by this stage even officials known to be opponents of nuclear weapons were mostly not prepared to rule out PNEs.[48] Beginning in 1970, the year the NPT entered into force, Indian ministers and officials repeatedly asserted that India would conduct a PNE as soon as 'investigations' were complete.

In 1970 the Department of Atomic Energy announced that India intended to develop the technology of underground nuclear explosions.[49] In July 1971 the new Prime Minister, Mrs Gandhi, advised Parliament that India would conduct PNEs. In September 1971 Atomic Energy Commission chairman Vikram Sarabhai announced at the Fourth International Conference on the Peaceful Uses of Nuclear Energy in Geneva, that India had given priority to PNEs.[50] On 2 May 1972 Defence Minister Jagjivan Ram told Parliament that the Atomic Energy Commission was 'studying the technology to conduct underground nuclear explosions for peaceful purposes'.[51]

By 1974 then, India had for several years been signalling its intention to develop nuclear dynamite, using this as a reason for opposing international restraints on its nuclear industry. Unlike the Brazilians, however, the Indians did not use the PNE argument consistently or as the sole justification for a nuclear explosive capability. But they did find it difficult to conceal the political and military justifications, particularly after their successful test.

On 1 August 1974 the Director of the Institute of Defence Studies and Analyses in New Delhi, Mr K. Subrahmanyam, asserted that '... after a few more test explosions, this country may reach the level of credibility and consideration accorded to (for example) Britain today'.[52] There was no mention of PNEs. Other Indian observers described the domestic political advantages of the explosion:[53]

> The PNE is a shot in the arm of Mrs Indira Gandhi and her Congress Party. At the same time it raises India's stature in the world and enhances her prestige. Henceforth she is likely to be treated with greater deference and consideration. India now cannot be treated like a non-entity. Her voice is likely to be given more weight in the international arms control and disarmament negotiations. The Pokharan experiment also raises the fighting morale of the armed forces and instils a spirit of self-confidence among the people at large. It provides a powerful incentive to apply the team-spirit, the attitude of dispensing with redtapism and administrative roadblocks, and the self-reliant approach, adopted in the field of nuclear technology, to our economic, social and political problems.

Such explanations sat uncomfortably with UN Ambassador Rikhi Jaipal's disingenuous claim that the Indian test had 'no military or political implications'.[54]

Clearly, political, military and strategic reasons for the Indian explosion *were* paramount. According to Girilal Jain, Editor-in-chief of the *Times of India*, the Chinese attack on India in October-November 1962, and the first Chinese nuclear detonation in 1964 made it difficult for Indian leaders to foreswear the nuclear option. The war with Pakistan over the creation of Bangladesh in 1971 was another incentive. It was during this conflict that President Nixon ordered the US Seventh Fleet into the Bay of Bengal to dissuade India from attacking West Pakistan. This event, according to many observers, steeled India's resolve to develop nuclear weapons. The eventual victory over Pakistan also gave India 'visions of pre-eminence' in South Asia. Jain writes that:[55]

> It is difficult to say which of the two provided a stronger impetus — the fear of a USA–China–Pakistan axis or the dreams of grandeur. In any case the decision to conduct a nuclear test appears to have been taken some time in 1972, that is, soon after the war with Pakistan, and at the beginning of the process of Sino-US reconciliation.

Mrs Gandhi's explanation of the timing of the test ran somewhat differently:[56]

> [The PNE] was simply done when we were ready. We did it to show ourselves we could do it. We couldn't be sure until we had tried it. We couldn't know how to use it for peaceful purposes until doing it. But it was for peaceful

> purposes only. We did it when the scientists were ready... How could it have been political?

A final *post hoc* justification for India's PNE was provided by K. Subrahmanyam. According to him, the blast was for the sake of disarmament and arms control: it was the only event since the signing of the NPT to shake the nuclear weapon states out of their complacency about their own huge nuclear arsenals.[57]

Was there anything at all to the Indian claim that their nuclear test was a PNE? This claim was, after all, crucial to India's legal argument that it had not violated assurances given to Canada and the United States that material from its CIRUS reactor would be used exclusively for 'peaceful purposes'. Certainly the device itself offered no convincing evidence. As Jozef Goldblat has pointed out, 'a nuclear charge of the type exploded by India is, perhaps, not fit for any practical application', whether a weapon or a PNE.[58] While it was not a 'weapon' in the sense that it had not been optimally designed for delivery to a target, neither was it a PNE in the sense that it could have been usefully employed for some 'peaceful' purpose. The best nuclear devices for PNE applications were not the plutonium fission variety, but fusion devices that produced less radioactivity, were relatively cheap and, in the case of underground applications, relatively compact. The French experience, requiring eight years to 'make the jump' from fission to fusion devices, emphasises the difference between the two types of technology.[59] This makes a nonsense of India's claim to have tested a PNE device, despite Mrs Gandhi's protestation that it was an initial 'cheap shot' to see if it would work.

India's attempts to ascribe its explosion to its interest in particular peaceful uses have also been unconvincing. Dr Sethna, Chairman of the Indian Atomic Energy Commission, described the explosion as 'an experiment to study the cratering and cracking effects on rock'.[60] According to Rikhi Jaipal however, 'at the site chosen for the experiment there were no hard rocks which might have fissured; there were only layers of sand, broken clay and shale'.[61] Scientists at the Bhabha Centre later more plausibly described the aims of the explosion as being to study *inter alia* 'ground-motion, containment of radioactivity and the problems involved in post-shot access of the shot-point environment', although even this did not prove that the explosion was a PNE test.[62]

In September 1974 Trivedi, the Secretary of the Indian Ministry of External Affairs, claimed that it was:[63]

> ... somewhat premature to speak of actual applications, as the data of the experiment of May 18 is still being collected. All information regarding this completely contained explosion so far available has been published. Final results will also be shared with the international scientific community.

He nevertheless claimed that India believed PNEs were the 'cheapest, if not the only way of exploiting some oil wells with high viscosity (say in

Gujarat, where there are such wells not now exploited)'. According to Trivedi, the Indian Oil and Natural Gas Commission had been attracted to nuclear stimulation of both oil and gas 'for quite some time'. Although the May 1974 explosion had not been anywhere near oil or gas deposits, Trivedi further claimed that 'although the indications so far have been encouraging, a definite answer will have to await the conclusions of the findings of the May 18 experiment'. Finally he mentioned, no doubt for good measure, that India had deposits of low-grade copper and other minerals which offered additional possibilities for the use of PNEs.

The Indians subsequently tried to pursue such claims by submitting details of the 'immediate physical consequences of the blast' to the IAEA Technical Committee on PNEs in January 1975. Again this proved nothing, as the blast would have had such 'physical consequences' whether it was a weapon test or that of a PNE. The Indians cited oil reservoir stimulation and the mining of non-ferrous metals as potential applications for their PNEs.[64]

Such spurious claims suffered an increasing credibility problem with the passage of time, for India has not subsequently pursued any 'peaceful' applications nor sought advice from the IAEA. While it participated actively in most of the IAEA bodies on PNEs convened in the 1970s, this indicated no particular sincerity of purpose, since other participants, such as Sweden, Thailand and Australia, which did not choose to develop their own nuclear devices, were equally active. India in fact attempted to use its involvement in these bodies, both before and after 1974, as evidence of its *bona fide* interest in PNEs. In 1977 Jaipal even used a 1970 IAEA report, which had commented favourably on the idea of using PNEs to mine low-grade non-ferrous metal ores in India, to imply IAEA support for India's nuclear explosion.[65] This ignored the fact that it was only because India itself had originated and praised the idea that it found its way into the IAEA report.

But even if the Indian claim that their explosion was a PNE were true, it is difficult to see why a developing country would be interested in a PNE program. Data freely available from the United States indicated the paucity of results from the sixteen-year, multi-million dollar Plowshare program. As Franklin Long has noted:[66]

> ... one must record some astonishment that there has been so much expressed interest in peaceful nuclear explosions on the part of the less developed nations, where labour costs are low and where one would have expected more attention to more labor-intensive alternatives... Part of the problem is that many of the protagonists for peaceful nuclear explosions, including some from the United States, have oversold their product.

As Brooks and Myers have argued:[67]

> It appears very unlikely that the economic advantage to be derived from the use of nuclear explosives will be sufficient to justify an explosives development program other than one whose principal objective is to achieve a nuclear weapons capability.

The Indians argue, of course, that their 'PNE' was cheap — according to their estimates only $400 000, including the cost of the plutonium and preparing the test site.[68] This ignores the enormous cost of the plutonium separation plant, the infrastructure needed to produce and detonate a nuclear device, and the Canadian and American aid which enabled India to begin its nuclear program in the first place and sustain it.

It was ironic that the Indians carried out their so-called PNE at the very time that the United States was dismantling Plowshare and interest in PNEs was elsewhere declining. India nonetheless continued for years to justify its 1974 detonation partly by reference to the influence of Plowshare. Presumably, if Plowshare had not existed it would have pointed to the Soviet PNE program as worthy of emulation.[69]

The blasé Indian attitude to the arms control implications of PNEs also cast suspicion on the country's motives. In September 1974 Trivedi claimed that while India had spoken 'since the fifties' about observance and regulation of peaceful nuclear explosions, it maintained that 'first priority should be given to a treaty to prohibit underground nuclear *weapon* tests [emphasis added]'. Only after 'the prohibition of the use of nuclear energy for weapons has come into effect' could the international community 'appropriately and effectively' consider the question of regulating the peaceful uses of nuclear energy, 'including underground nuclear explosions for economic development'.[70] Trivedi was in essence arguing that regulation of PNEs should only be considered after nuclear disarmament had been achieved. This position, which echoed that of Brazil, was of course intended to ensure that there would be no threat to India's development of nuclear explosives. Rikhi Jaipal, later to become the UN Secretary-General's representative to the Conference on Disarmament, confirmed this in an apologia for the Indian explosion published in 1977:[71]

> What is needed is not the consolidation of an unequal treaty like the NPT, but rather a nuclear disarmament treaty which would ban the making, testing, stockpiling and use of nuclear weapons, and would also provide for measures leading to a rapid lowering of levels of nuclear deterrence, speedy reduction of nuclear weapon stocks and their eventual elimination. *Thereafter one might* [emphasis added] evolve a universal system of international observations of PNEs under safeguards *to avoid health and safety hazards* [emphasis added].

Even as late as 1981, some Indians were still attempting to convince the world, and themselves, that the 1974 detonation was a PNE, and that PNEs were, despite the weight of scientific evidence and opinion elsewhere, useful for India's economic development. In a blatantly selective presentation of the facts, two Indian writers, Mirchandani and Namboodiri, made such claims as:[72]

> It has now been proved that beyond a particular yield, nuclear explosions are economically more viable as compared with conventional chemical explosions... Some of the most promising applications for contained nuclear explosions are natural gas and oil stimulation, formation of underground reservoir [sic], *in-situ* ore leaching, geothermal heat recovery, *in-situ* coal gasification and sealing off of uncontrolled gas well fires.

According to the two writers 'advanced gas stimulation experiments involving multiple explosions have been planned in the USA and economic and technical feasibility is being put to the test'.[73] This was eight years after Rio Blanco. They also claimed that the feasibility of a new Panama Canal 'has been under investigation', without revealing the recommendations of the Interoceanic Canal Commission.[74] The book also alluded to the Cape Keraudren and Cape Thompson projects without revealing the reasons for their cancellation, claiming that nuclear excavation techniques for harbour construction 'are being increasingly investigated'.[75]

The effects of India's 'peaceful' nuclear explosion on efforts to stop the spread of nuclear weapons have been profound. The most disastrous was to boost Pakistan's determination to obtain a similar capability to that of India. In a classic demonstration of the phenomenon of regional nuclear proliferation, Pakistan accelerated its own nuclear program after 1974 and is today widely regarded as having the means to detonate a nuclear device, which it might also care to label 'for peaceful purposes only'. Pakistan, having been active in exposing Indian hypocrisy over the PNE issue, is, however, unlikely to portray its bomb as being a PNE. The architect of Pakistan's uranium enrichment program, A. Q. Khan, stated in February 1987 that:[76]

> The word 'peaceful' associated with a nuclear program is humbug. There is no peaceful bomb. After all, there is only a weak, transparent screen between the two. Once you know how to make reactors — all that Pakistan has mastered as well — it becomes rather easy to produce nuclear weapons.

The Indian PNE also helped blur the distinction between a nuclear weapon state and a non-nuclear weapon one, establishing in the process a dangerous precedent. As one writer has commented, 'the PNE represents an ambiguity in nuclear terms that has not been lost on the Third World'.[77]

On the positive side, the Indian explosion led directly to global tightening of controls on the export of nuclear technology and materials. Initially India alone felt the brunt of such restrictions. Canada, for instance, withdrew further nuclear assistance to India after it refused to place *all* its nuclear activities under IAEA safeguards.[78] Even the Soviet Union, while obliged to publicly accept the Indian explanation for its nuclear test, insisted that stringent IAEA safeguards be applied to Indian facilities and material exposed to heavy water imported from the USSR.[79]

In 1975, partly in response to the Indian explosion, a secret nuclear suppliers group, later known as the London Suppliers Club, met to strengthen the 'brakes' on sensitive nuclear exports, especially to 'unstable' countries.[80] The participants, originally the United States, the Soviet Union, France, Canada, the Federal Republic of Germany and Japan, established Guidelines for Nuclear Transfers. These brought a degree of order and consensus among suppliers, ensured relatively uniform minimum standards and interpretation of NPT obligations, and brought France into the 'Club' even though it was not an NPT party.[81] However, in as much as they did not place an embargo on the export of technologies that can be directly used to make nuclear explosives, these guidelines did not go far enough for President Carter or the US Congress. In 1978 Congress passed the Nuclear Non-Proliferation Act, forbidding the supply of key nuclear materials to any state that had not signed the NPT and which refused to accept 'full-scope' safeguards. The 'major spurs' to this legislation, according to Leonard Spector, were the belief that India had violated its 1956 heavy water sales contract, and suspicion that US officials had been slow to reveal this fact to stem criticism of their lenient attitude towards the Indian test.[82] In the view of Pringle and Spigelman, the legislation was 'a dramatic new turning point in nuclear history'.[83]

The Indian explosion also had a marked effect on the fortunes of nuclear dynamite itself. It demonstrated graphically the proliferation dangers of PNEs. In the United States, the Indian test fuelled Congressional and other criticism of PNEs on arms control grounds and vindicated those who had warned of the proliferation dangers inherent in PNEs. More specifically, it increased the difficulty of obtaining Congressional approval for two US/Soviet bilateral treaties relating to nuclear testing: the 1974 Threshold Test Ban Treaty and the 1976 Peaceful Nuclear Explosions Treaty.[84] Internationally, the Indian 'PNE' increased the pressure for establishment by the IAEA of a PNE 'regime' to enable non-nuclear weapon states to obtain the 'benefits' of PNEs without resorting to an Indian-style explosion. Paradoxically, the Indian explosion also lead to greater international support for the view of the new Carter Administration that PNEs should be banned outright under a CTB. This ultimately stultified the IAEA's efforts to establish a PNE regime.

In the meantime the IAEA strengthened its nuclear safeguards agreements to remove any possible PNE 'loophole'. Prior to the mid-1970s, these agreements were drawn up in accordance with an IAEA document known as INFCIRC/66/Rev. 2, which aimed to prevent nuclear materials being used 'to further any military purpose'. According to David Fischer, when the IAEA Statute was drafted in 1956:[85]

> . . . no-one would have suggested that a state could begin to carry out nuclear explosions without thereby 'furthering a military purpose', whether the ostensible object of the explosions was to make holes in the ground or to test the charges for a nuclear weapon.

It seemed superfluous to specify that IAEA-safeguarded nuclear material should not be used in *any* kind of nuclear explosive. While NPT parties could not avail themselves of this loophole since they had agreed, in signing the NPT, not to acquire 'other nuclear explosive devices', non-NPT states with safeguards agreements with the IAEA could have done so. Since 1975, however, in all its safeguards agreements with non-NPT parties, the IAEA has eliminated any possible ambiguity by requiring that safeguarded material must not be used '. . . for the manufacture of any nuclear weapon or to further any other military purpose or for the manufacture of any other nuclear explosive device'. According to Fischer, these changes were made despite 'much resistance from the non-NPT states concerned',[86] especially Brazil, India and Argentina.[87]

Since 1974, Indian policy on PNEs, except during the brief period of the Janata government, has remained steadfast. India has not renounced nuclear dynamite, nor acknowledged the arms control implications of PNEs. Nor has the country signed the NPT. Prime Minister Desai, who succeeded Mrs Gandhi in 1977, announced his displeasure at the 1974 explosion, saying that it had been carried out only for 'political purposes' and had 'created all the trouble', leading the Americans and Soviets to believe that India's objective was not a peaceful one.[88] He said he was completely convinced that nuclear explosions of any kind were unnecessary for India.[89] Documents were being studied, he said, to find out whether the test had added anything to the knowledge of Indian scientists, but as far as he knew it did not benefit them in any way. Indeed they had been unable to convince him that it served any economic or practical purpose. These views stirred great controversy in India, and Desai would not categorically rule out a further Indian 'PNE'. He announced that 'India would not carry out any more nuclear explosions unless absolutely necessary for its peaceful nuclear program; even then, it would not be an explosion conducted in secrecy... We will be open to other people'.[90] This evinces some sensitivity to the nuclear weapon implications of PNEs. He later told Parliament that he did not rule out the use of PNEs for mining and oil and gas exploration.[91]

Mrs Gandhi, on her return to power, abandoned Desai's approach. She indicated to Parliament that a resumption of nuclear tests was conceivable, if warranted by Indian 'security' requirements.[92] In 1981 it was reported that US officials had detected Indian preparations for a second nuclear test.[93] Up until 1983 the site at Pokharan was kept in a state of readiness and the excavations for the detonation were not filled in.[94] Mrs Gandhi's successor, Rajiv Gandhi, has given no indication that India has changed its mind on PNEs. Like Brazil and Argentina, India continues to cause mischief on the subject in international forums and confound those seeking an international consensus on the need for a ban on all nuclear tests.

Since 1981 it has, however, been Pakistan rather than India that has been suspected of preparing to conduct a nuclear test, although it is unlikely to have the audacity to call its test a 'peaceful nuclear explosion'. As the Pakistanis are only too well aware, India's 1974 nuclear explosion could have been used either to dam a river or destroy Islamabad. The atomic bombing of Nagasaki with just such a plutonium device had all too painfully demonstrated that even a crude nuclear explosive, in Nagasaki's case the first plutonium bomb ever built, can have devastating effects when used as a weapon.

# *Notes*

1 *SIPRI Yearbook 1975*, Taylor and Francis, London, 1975, p. 19.
2 Ibid., p. 16.
3 R. Chidambaram and R. Ramanna, *Some Studies on India's Peaceful Nuclear Explosion Experiment*, Bhabha Atomic Research Centre, Trombay, 1974, p. 7.
4 P. Pringle and J. Spigelman, *The Nuclear Barons*, Avon Books, New York, 1981, p. 374.
5 M. J. Brenner, *Nuclear Power and Non-Proliferation: The Remaking of US Policy*, Cambridge University Press, Cambridge, 1981, p. 68.
6 *SIPRI Yearbook 1975*, p. 16.
7 Ibid., p. 519.
8 Ibid.
9 Ibid.
10 Ibid., p. 517.
11 Ibid., p. 519.
12 Ibid., p. 396.
13 M. A. Guhin, *Nuclear Paradox: Security Risks of the Peaceful Atom*, American Enterprise Institute for Public Policy Research, Washington DC, 1976, p. 26.
14 G. Jain, 'India' in J. Goldblat (ed.), *Nonproliferation: The Why and Wherefore*, SIPRI, Taylor and Francis, London, 1985, p. 145.
15 Ibid., p. 146.
16 Ibid., p. 91; Pringle and Spigelman, p. 377.
17 G.H. Quester, *The Politics of Nuclear Proliferation*, Johns Hopkins University Press, Baltimore, Maryland, 1973, p. 157.
18 L. Beaton, *Must the Bomb Spread?*, Penguin, Harmondsworth, 1966, pp. 71–72.
19 Quester, p. 154.
20 Ibid., p. 158.
21 Pringle and Spigelman, p. 377.
22 Quester, p. 157.
23 R. K. Betts, 'India, Pakistan and Iran' in J. A. Yager (ed.), *Nonproliferation and US Foreign Policy*, Brookings Institution Press, Washington DC, 1980, p. 108; see also L. Scheinman, *The World Nuclear Order*, Resources for the Future, Washington DC, 1987, p. 175.
24 *SIPRI Yearbook 1975*, p. 21.
25 Ibid., p. 517.
26 G. Jain, p. 136.
27 *SIPRI Yearbook 1975*, p. 519.
28 F. A. Long, *Peaceful Nuclear Explosions*, unpublished ms for ACDA, Washington DC, September 1975, p. 19.
29 Brenner, p. 69.
30 Pringle and Spigelman, p. 375.
31 Brenner, p. 69.
32 G. G. Mirchandani and P. K. S. Namboodiri, *Nuclear India: A Technological Assessment*, Vision Books, New Delhi, 1981, p. 117.
33 Ibid., pp. 117–18.
34 Guhin, p. 27.
35 J. P. Jain, *Nuclear India*, Radiant Publishers, New Delhi, 1974, p. 105.
36 G. Jain, p. 124.
37 Pringle and Spigelman, p. 378.
38 L. S. Spector, *Nuclear Proliferation Today*, Vintage Books, New York, 1984, p. 28.
39 Betts, p. 111.
40 R. W. Jones, 'India' in J. Goldblat (ed.), p. 116.
41 Ibid., p. 101.
42 G. T. Seaborg and B. S. Loeb, *Stemming the Tide: Arms Control in the Johnson Years*, Lexington Books, Lexington, Mass., 1987, p. 254.
43 Ibid.
44 Ibid., p. 255.
45 G. Jain, p. 93.

46 Jones, p. 121.
47 Ibid., p. 104.
48 Quester, p. 72.
49 Spector, p. 31.
50 Ibid.
51 K. Subrahmanyam, *Nuclear Myths and Realities: India's Dilemma*, ABC Publishing House, New Delhi, 1981, p. 148.
52 *SIPRI Yearbook 1975*, p. 23.
53 G. Jain, p. 137.
54 R. Jaipal, 'The Indian Nuclear Situation', *International Security*, I, 4, Spring 1977, p. 44.
55 G. Jain, p. 93.
56 Jones, p. 114. According to US Senator Henry Jackson, India had tried unsuccessfully to detonate a nuclear device in February 1974, three months before the Pokharan test. The Indian fizzle was apparently detected on US military seismographs (Spector, p. 37).
57 Subrahmanyam, p. 146.
58 J. Goldblat, 'The Indian Nuclear Test and the NPT' in A.W. Marks (ed.), *NPT: Paradoxes and Problems*, Arms Control Association, Washington DC, 1975, p. 33.
59 M. Kalkstein, *International Arrangements for the Peaceful Applications of Nuclear Explosives*, Stockholm International Peace Research Institute, Stockholm, 1970, p. 19.
60 Betts, p. 106.
61 Jaipal, p. 44.
62 Chidambaram and Ramanna, p. 1.
63 V. C. Trivedi, 'India's Approach Towards Nuclear Energy and Non-Proliferation of Nuclear Weapons' in Marks (ed.), p. 44.
64 IAEA Document GC(XXIX)/544, IAEA, Vienna, 1975, p. 55.
65 Jaipal, p. 46.
66 F. A. Long, 'Peaceful Nuclear Explosions', *Bulletin of the Atomic Scientists*, October 1976, p. 27.
67 D. B. Brooks and H. R. Myers, 'Plowshare Evaluation' in B. Boskey and M. Willrich (eds), *Nuclear Proliferation: Prospects for Control*, Dunellen Publishing Co., New York, 1970, pp. 90–91.
68 *SIPRI Yearbook 1975*, p. 19.
69 Trivedi, p. 44.
70 Ibid., pp. 45–46.
71 Jaipal, pp. 48–49.
72 Mirchandani and Namboodiri, pp. 106–07.
73 Ibid., p. 108.
74 Ibid., p. 111.
75 Ibid., pp. 111–12.
76 C. Smith, 'A Policy of Ambiguity: Nuclear Proliferation in South Asia', *ADIU Report*, Armament and Disarmament Information Unit, University of Sussex, Vol. 9, No. 4, July–August 1987, p. 1.
77 F. Carr, 'The View From the Third World' in W. H. Kincade and C. Bertram (eds), *Nuclear Proliferation in the 1980s*, Macmillan, London, 1982, pp. 238–39.
78 Spector, p. 35.
79 Jones, p. 114.
80 D. Fischer and P. Szasz, *Safeguarding the Atom*, Stockholm International Peace Research Institute, Stockholm, 1985, p. 102.
81 Ibid., p. 103.
82 Spector, p. 37.
83 Pringle and Spigelman, p. 377.
84 See Chapter 15.
85 Fischer and Szasz, p. 76.
86 Ibid., p. 77.
87 Scheinman, p. 245.
88 Mirchandani and Namboodiri, pp. 118–9.
89 Desai also revealed that the Pokharan explosion had left behind substantial quantities of plutonium at the test site, whose hazardous effects were being assessed. It is not known whether this plutonium was underground or not (Mirchandani and Namboodiri, p. 119).

90 Jones, p. 122.
91 Betts, p. 113.
92 Jones, p. 115.
93 Spector, p. 43.
94 Ibid., p. 52.

# 13

# The Demise of Plowshare

*The feasibility of using nuclear explosives for peaceful purposes . . . will not be determined only by proven technology, safety and economics. For in a program, such as Plowshare, which involves a new source of enormous energy about which there is often misunderstanding and controversy, factors — political, sociological, and psychological — outside the scope of the AEC's technical program work to exert influence. These factors are complex . . .* [1]

US Atomic Energy Commission, 1966

UNABLE to placate the environmentalists and arms controllers and having alienated industry, Plowshare's political support withered quickly after Rio Blanco. This was not because of any massive outcry or sustained public campaign, but because doubts about the program, in part generated by the public, but also emanating from within government, gradually seeped into the political process and deprived it of support and funding. Even the AEC, under new leadership and beseiged by the anti-nuclear and environmental movements, lost heart, leaving the Livermore Laboratory and the JCAE as Plowshare's sole patrons. Finally the Joint Committee, exasperated with the AEC's unwillingness to support its own gas stimulation program, fell by the wayside. Plowshare ended as it had begun, with only a handful of supporters, mostly scientists involved in the program itself, including Edward Teller.

This decline in Plowshare's fortunes coincided with a sea change in the balance of political forces in nuclear matters. Plowshare, like other US nuclear programs, had been nurtured from birth by a political, scientific and bureaucratic élite centred on the Livermore Laboratory, the AEC and the JCAE and including Teller, Seaborg, Hosmer and Holifield. Successive Presidents were either not interested in nuclear matters or preoccupied elsewhere. Public support for the peaceful uses of nuclear energy was apparently limitless, as was its ignorance of its implications. Congress was content to leave the seemingly arcane matter of nuclear energy up to the experts. Under these circumstances Plowshare was left, so to speak, largely to its own devices.

By the early 1970s, however, the arms controllers, the environmentalists and the anti-nuclear movement began to challenge the nuclear industry, especially in relation to civil nuclear power. At the same time the interest and influence of ACDA, the State Department, the Bureau of the Budget and Congress in nuclear matters increased. ACDA and State became concerned with nuclear proliferation, especially after the 1974 Indian explosion. An aggressive Bureau of the Budget, during a period of fiscal restraint, sought easy targets for cutbacks in areas of previously profligate government spending. Congressional involvement in nuclear matters also increased exponentially after the Indian 'PNE'. It was further heightened by President Nixon's June 1974 trip to the Middle East, where, without consulting Congress, he agreed to provide Egypt and Israel with light water nuclear reactors.[2] Plowshare had the misfortune to be caught up in these trends.

Congressional support for Plowshare funding moved from a large majority favouring *some* level of funding in 1973, to a bare majority in 1974.[3] In November 1974 Congress abolished the AEC and allocated its functions to two separate bodies, the Energy Research and Development Administration (ERDA) and the Nuclear Regulatory Commission (NRC). Plowshare came under the control of ERDA and the name was dropped in favour of the more prosaic 'PNE'. According to Fred Holzer, while the 'patient would have died anyway', the dismantling of the AEC 'put the last nail in the coffin' of Plowshare.[4]

While this may have been true, ACDA was not taking any chances. It commissioned a major study on the PNE question, followed by a further independent evaluation, to seal the fate of Plowshare once and for all. Incredibly, apart from the Canal Commission's report on nuclear excavation, this was the first time such independent analyses of PNE technology had been attempted in the United States. As Professor Long noted, 'up to very recently, almost the only analyses that we had of this program came from the people who were conducting the program, from its enthusiasts and supporters'.[5] Herbert Scoville concurs, noting that such analyses 'largely relied on the experiences of those who had spent their lives designing nuclear weapons rather than solving the practical problems which would arise in carrying out nuclear explosions in inhabited locales'.[6] For Gerald Johnson it was not at all surprising that 'the concepts of the possible use of nuclear explosives for earthmoving, excavation and other engineering projects were developed by individuals with experience in the development and testing of nuclear weapons'.[7] While it may not have been surprising, a clear conflict of interest existed, leading to glowing reports which fed the high hopes originally held for Plowshare.

The major study commissioned by ACDA was carried out by the Gulf Universities Research Consortium (GURC) of Texas in 1974. Known officially as *PNE Activity Projections for Arms Control Planning*, this was the report which became known unofficially as the *GURC Report*. GURC's terms of reference required it to 'attempt to establish — within explicit ranges of uncertainty — when and under what contingencies PNE activity might grow to commercial proportions', both in the United States and overseas through the year 1990.[8] GURC was expected to take into account major uncertainties, including research and development funding, technical breaks, and regulatory decisions.[9] Only underground applications of PNEs were to be examined because, in ACDA's view, excavation applications were ruled out by the PTBT. GURC was expected to adopt an 'independent, unbiased position on the controversial aspects of PNEs' and 'take into account possible biases resulting from desire for Federal subsidy or from uncritical fascination with novel technologies', an obvious jibe at the AEC attitude towards Plowshare.[10]

The official reasons for conducting such a study related to arms control. The first was the fact that the first review conference for the NPT in 1975 would focus on the obligation of the nuclear weapon states to share PNE technology with non-nuclear weapon states. The second was the likelihood that the widespread use of PNEs would complicate the negotiation of a future CTB. In ACDA's view both of these arms control concerns would be strongly influenced by one variable: 'the rate at which the use of PNEs develops'.[11] A further aim outlined by ACDA in its terms of reference was more revealing of the real reason for the report: 'The project prospects and the uncertainties for PNEs should be defined well enough that within the US government and in international arms control forums we will be working from a set of common premises'.[12] The study was clearly designed to end the schizophrenia of US policy, wherein the AEC promoted the use of PNEs, both domestically and internationally, while US arms control negotiators warned of the proliferation hazards. ACDA clearly hoped to use the *GURC Report* to ensure its own view prevailed.

The GURC team decided from the outset that only five PNE applications were worth analysing: oil shale retorting; stimulation of natural gas; underground storage cavities; copper fracturing; and oil stimulation.[13] A 'PNE Industrial Workshop', involving a wide range of US industrial concerns, was organised to discuss the future prospects of these applications. The general conclusions of the workshop were that: PNE technology was generally at an early stage of development; regulatory problems were forbidding; economic evaluations (other than as general ranges) would not be valid until technology and costs were much better defined; and, there would be little or no use of PNEs for industrial purposes prior to 1990. Not surprisingly, however, industry endorsed a continuing PNE field test program, provided that it focused first on resource and reservoir data, that 'non-nuclear studies' in advance of PNE events were given full and careful consideration, and that industry was more involved in test design and evaluation and not required to assume all or most of the costs.[14]

Released in 1975, the *GURC Report* was, as ACDA undoubtedly anticipated, highly sceptical of the prospects for PNEs. While it pointed out that PNE technology had not been proven to be a 'hopeless failure', it concluded nonetheless that, at least for the period to 1990:[15]

> PNE technology is not ready for immediate application on an industrial scale to any of the potential applications analysed.

GURC gave two major reasons for its pessimistic conclusions. First, it seemed quite doubtful whether there would be a 'persuasive demonstration of cost effectiveness' for any of the proposed applications of PNEs, since the 'extensive technical uncertainties' made economic feasibility of PNEs 'a matter of almost pure speculation'.[16] This would appear to contradict Iris Borg's claim that 'the reasons for the termination of the PNE program in the US did not include the conviction that the proposals were not economically sound'.[17] Second, even if cost effectiveness could be demonstrated, there would be 'a number of political and social uncertainties and regulating delays' involved in any use of PNEs.[18] GURC described the regulatory processes as being 'long, complex and uncertain' and likely to add up to five years' delay to any particular project.[19] The resulting costs would be a significant disincentive to investment capital. According to GURC, 'regulatory processes may readily constitute, alone, an element which halts a proposed project completely'.[20]

Among the potential uses of PNEs, GURC looked in particular detail at what many regarded as the most promising application — the stimulation of natural gas. The report identified a number of major problems with this technology, including radiation and seismic hazards; the problem of identifying the quantity and quality of gas available for stimulation; the unpredictability of 'chimney' formation and rock fracturing; the problem of fractures closing up after a detonation; and the possibility of damage to nearby resources such as oil shale and coal.[21] To add insult to injury, GURC identified a conventional alternative, Massive Hydraulic Fracturing, or MHF, as being likely to prove superior to the use of nuclear dynamite.[22] According to GURC, it seemed probable that MHF technology would make 'substantial advances' by the 1980s, while PNE technology would advance 'little, or not at all', thereby increasing the economic advantage in favour of MHF.[23]

Upon receiving the *GURC Report*, ACDA commissioned a review of the document by a team of specialists, led by Professor Franklin Long and including Gerald Johnson and physicist Richard Garwin. The need for this additional review was justified on the grounds that the US nuclear industry was undergoing great change as a result of the break up of the AEC; the First NPT Review Conference was imminent, and the negotiations on the Threshold Test Ban Treaty between the superpowers had resulted in agreement on the need for a separate PNE Treaty.[24] The review, published in April 1975, described the *GURC Report* as a 'sober, documented analysis', and its conclusions as 'well justified'.[25] In addition to the applications considered by GURC, it examined *in situ* leaching of copper, electric power generation and nuclear excavation. As with the applications considered by GURC, the review concluded that there was little likelihood that PNEs would be in commercial use in such applications in the United States by 1990. The report was, however, even more pessimistic than GURC about the overall future of PNEs. Taking greater account of the 'probable social barriers' to PNEs, Long and his colleagues concluded that the possibility of commercial use of PNEs in the United States before the year 2000 was 'quite small'.[26] They added that:[27]

> If the political environment remains as hostile to PNEs as it currently is, there will be no resumption of PNE field testing by 1990, let alone commercial usage.

*The Long Report* also dismissed the JCAE's claim that the NPT obliged the United States to pursue and share its PNE technology, saying that such an obligation would only exist if 'economically viable uses of PNEs are available'.[28] Finally it concluded that the existence of a vigorous Soviet PNE program was no reason for the United States to continue with Plowshare.[29]

In addition to the *GURC* and *Long Reports*, the IAEA had by 1977 also produced its own pessimistic report on PNEs. Professor Long summarised all three studies when he told the Senate Foreign Relations Committee that they revealed:[30]

> ... a score or so of situations where conceivably PNEs might be useful, where it might be applied to problems of either technical or commercial significance... As far as I can see, the available data on all these various proposals do not, however, permit anything more than the most rudimentary and tentative judgement to be made on the economic utility of PNE for these various applications. This says something about the complexity of the problem, of course, and the incompleteness of the studies so far. But it also is a somewhat interestingly negative statement in that it also says that there is no single use that stands out as obviously a great thing, as obviously economic and important.

As well as sorting out US policy on PNEs, the main impact of all these negative assessments was, as ACDA had hoped, to seal the fate of Plowshare. ACDA smugly described the *GURC Report* as 'something of a landmark in US PNE history' and a 'testament' about PNEs which remained relevant through Plowshare's 'remaining phases', namely its complete abandonment.[31] Clearly ACDA used its two reports to demolish any remaining support for Plowshare in Congress, ERDA and the JCAE. No wonder that ACDA Director Fred Ikle reported that, 'Some bureaucrats [presumably those connected with the remnants of Plowshare] wanted to suppress the findings' of the *GURC* and *Long Reports*.[32]

In 1975 ERDA included no explicit line item for PNE activities in its 1976 budget estimates. According to Fred Holzer, ERDA officials, including those from the former Bureau of Mines, had 'little appreciation' of Plowshare.[33] In 1977 the PNE program, while being allocated $1 million,[34] was described by ACDA Director Paul Warnke as 'static' and 'a study program at the present time'.[35] In 1978 funds for PNE research disappeared completely from the US Budget. ERDA testified to the Senate Foreign Relations Committee that this was 'because of uncertainties as to economic feasibility, lack of public acceptance, and nonproliferation concerns'.[36] By comparison, $2.5 million was budgeted for verification of the unratified 1976 Peaceful Nuclear Explosions Treaty.[37] Later in 1977 ERDA was absorbed by the new Department of Energy, effectively ending the US peaceful nuclear explosions program. In an ironic twist, Plowshare's greatest institutional supporter, the JCAE, was itself abolished in 1977.[38] In 1977 the Federation of American Scientists described peaceful nuclear explosives as 'ineffective, unnecessary, and sometimes dangerous despite twenty years of searching for a suitable application'.[39] For Herbert York, in the final analysis, everything about Plowshare,[40] which had cost at least $160 million over two decades, was 'negative'.[41] By 1979, Gerald Johnson, erstwhile Plowshare Director, was quoted as saying that he was 'firmly convinced that PNEs have no practical economic utility in the United States now, or in the foreseeable future'.[42]

The causes of Plowshare's demise were complex and inter-related. By and large, however, technical problems were not the principal cause. Certainly many technical problems arose during the program, particularly in predicting and controlling the effects of nuclear explosions. However, in purely technical terms the value of PNEs had neither been proved nor disproved. In Alan Wilson's view this was:[43]

> . . . as much a reflection of the relatively modest effort which has been directed to PNE research and development, as it is a consequence of the technical difficulties encountered.

Department of Energy spokesman for the Nevada Test Site, Cris West, says of Plowshare in retrospect: 'The funding just went away for it. From a practical standpoint, most of the stuff we looked at was feasible'.[44] As GURC stated, 'it is not correct to dismiss [PNEs]... from a technical viewpoint, even though the technical uncertainties are severe'.[45]

Plowshare did fail politically, socially, environmentally and commercially. In the case of the excavation program, the problem was the lack of political commitment, consequent financial restraints and problems stemming from a strict interpretation of the PTBT. In the case of underground applications, the main reasons were industrial disenchantment and environmental opposition. According to ACDA:[46]

> It was American industry which, having determined the main directions of the US PNE program, finally passed judgement on it. And although the judgement was made essentially on economic grounds, these coincided closely

with considerations of public health and safety. As it happened, they all seemed to point toward the same conclusion: that however appealing the PNE concept had originally appeared, it was not worth pursuing.

## *Industrial Disenchantment*

As Glenn Seaborg has acknowledged, Plowshare 'failed to kindle economic enthusiasm in industrial circles'.[47] For Fred Holzer this was a 'key problem' for Plowshare.[48] There are myriad reasons why. The first was the uncertainty caused by the 1958–61 nuclear test moratorium and the subsequent negotiation of the PTBT.[49] A second reason was that the presumed economic advantages of nuclear dynamite over conventional explosives were not borne out. As Pringle and Spigelman bluntly put it:[50]

> With overheads met by the weapons program, nuclear explosions seemed cheaper than the same amount of TNT . . . enthusiasm for Plowshare projects gradually waned as more complete cost calculations were demanded.

Glenn Seaborg notes that Plowshare was:[51]

> . . . given virtually a 'free ride' insofar as research and development costs were concerned. Such costs — and they were huge — were virtually all charged to the weapons program.

The *GURC Report* noted that in any commercial operation the most important variables are price of product, capital investment and delay times, 'none of which is improved by existing PNE technology'.[52]

Due to their immense energy, unpredictability and potential for damaging side-effects, arrangements for firing PNEs were expensive and difficult to cost in advance. Radioactivity was the most obvious problem, not just in terms of atmospheric venting, but contamination of the product to be exploited. This implied stringent safety measures and either lengthy delays while waiting for the decay of the radioactivity (in some cases at least a year) or decontamination of the product before it left the site. Another problem was high underground temperatures due to slow heat transfer in rock, which could preclude entry into the immediate area for six months to a year, or more. Eleven months after the 1962 Hardhat shot in Nevada, average working temperatures in the re-entry tunnel were 110 to 130 degrees farenheit.[53] Seismic damage to nearby structures or natural resources was another potential side-effect. As *The Long Report* concluded, 'the immediate region of a PNE is a hazardous place in which to do further work'.[54]

Not only could the side-effects from a nuclear detonation hamper further work but they could add prohibitive costs to a PNE project. They also led to time-consuming regulatory and public decision making processes. According to GURC, up to 300 separate agencies, bodies or groups could be involved in any one PNE project.[55] Extremely close co-operation with several government bureaucracies, both Federal and State, was necessary.

The transportation and storage of nuclear dynamite alone required costly special procedures.[56] Many companies could neither run the financial risk nor last the bureaucratic distance.

In a definitive article on the economics of PNEs, Brooks and Myers list an additional large number of 'imponderables' that make cost-benefit analyses of PNE projects difficult. They include:[57]

*Unknowns that cannot be resolved through relatively modest experiment.* In some cases it would be necessary to carry out the project itself to determine whether it was economically advantageous or even technically feasible. (This was a major imponderable in Project Chariot, the Cape Keraudren project and the Panama Canal project.)
*In projects that could be carried out only with nuclear explosives, it is necessary to decide whether the benefits are as great as if the expenditure of resources had been directed at a different type of project.* Since this involves a comparison of projects of very different kinds, conclusions will not always be clear-cut.
*The costs and benefits associated with less tangible aspects, such as the social costs of relocating population, the benefits of achieving an unforeseen technological or economic breakthrough or the costs resulting from interference with the ecology or destruction of aesthetic values.* There are no completely satisfactory means of assigning values to these, even though they may be very important.
*The costs and benefits associated with political constraints.* These may be very large and non-quantifiable.
*Possible costs resulting from a catastrophic accident.* While such an accident may be unlikely, the costs associated with one would be very high and difficult to estimate.

Such uncertainties, according to Brooks and Krutilla, produced 'a certain looseness in stating cost figures' for PNE projects and 'many problems in project evaluation'.[58]

Quite apart from these 'imponderables' the AEC could not give accurate cost estimates for its proposed PNE 'services'. As far as the cost of the nuclear devices themselves was concerned this was in part due to 'security' considerations. However, according to Fred Holzer, there were also long debates within the AEC about the charges that should be levied for 'commercial' PNE services as a whole. In May 1964, as a direct result of Australian criticism, the AEC finally announced some projected charges, but with the following telling caveat: '... it should be recognised that the costs to be assumed by the AEC as finally negotiated might be significantly different from the projected charges'.[59] Thirteen years later, in 1977, the IAEA *Ad Hoc* Advisory Group on PNEs was still complaining that: 'it is difficult to get reliable estimates of prices of nuclear explosives'.[60]

For a true costing of PNE projects, not only would the cost of the explosive have to be taken into account, but also the hidden research, development and production costs of nuclear explosives. In the case of Plowshare, all research and development costs were borne by government, either directly through the Plowshare budget or indirectly through the nuclear weapons program and the US Army Corps of Engineers. Obviously the AEC could not recoup all these costs without pricing its PNE service out of the commercial market. But neither could it subsidise the commercial use of PNEs indefinitely. During a period of financial stringency the natural temptation of the AEC, prodded by the Bureau of the Budget, was to recoup an increasingly large proportion of its costs from its industrial collaborators. As Brooks and Myers have noted:[61]

> In the United States (and probably the Soviet Union) a large part of the development and production costs [of PNEs] are charged to the weapons program. Where the line is drawn on this will affect significantly the economic calculations.

This exacerbated doubts on the part of potential American and foreign customers about the economics of PNEs. It is the major reason why Sentinel Mining withdrew from the Cape Keraudren project.

Potential industrial partners were in any case troubled by the chequered history of budgetary support for Plowshare, which was not generously funded compared with other AEC activities and had never been given the facilities and resources required. Despite the publicity given to Plowshare, the AEC, the Livermore Laboratory and NEVOO saw their primary role as the development and testing of nuclear weapons. Plowshare therefore remained a 'stepchild' of the weapons program. While it certainly benefited from nuclear weapons testing, the priority given to the weapons program affected both the level of funding for Plowshare and the number of dedicated Plowshare experiments that could be conducted. This made commercial collaborators uncertain about the political and financial commitment that successive US Administrations were prepared to make to the project.

Another difficulty was that private companies could not simply buy nuclear explosives 'off the shelf', as they could conventional explosives, nor could they have access to classified information about nuclear devices. This was despite Edward Teller's bland assurances that: 'Clean nuclear weapons would be the same as conveniently packaged high explosives. They would be nothing more'.[62] Custody of nuclear devices would always remain strictly with the AEC. Since PNEs were, in effect, nuclear weapons, the AEC also adopted, quite properly, a very restrictive 'need to know' policy to prevent unauthorised disclosure of sensitive information. Teller railed against this restriction, as he did against the entire policy of secrecy in nuclear matters, charging that it was holding back the peaceful exploitation of the atom for the benefit of mankind.[63]

Another serious problem facing the widespread commercial use of PNEs was the absence of large-scale production facilities for nuclear dynamite.[64] The US government would naturally be reluctant to construct such facilities unless assured that the product would be purchased. For its part industry would be reluctant to place firm orders without government guarantees as to delivery schedules, prices and the future regulatory climate. GURC described this as a 'chicken-or-the-egg' puzzle, an apt description of the entire government/industry relationship over nuclear dynamite.[65]

Large-scale production and use of PNEs would also increase the possibility of accidents, misfires, theft and sabotage. As the *GURC Report* commented: 'With this number of shots a major accident becomes a very real possibility; one major incident would no doubt drastically modify public opinion and therefore the whole future of the PNE program'.[66] According to Wolfgang Panofsky of the Stanford Linear Accelerator, a large-scale PNE industry would create:[67]

> ... an industrial technological base and traffic in nuclear explosives of a volume which involves dangers of diversion, highjacking, and so forth, which would dwarf the risks which are now under so much discussion both before the Congress and the people in connection with the plutonium economy inherent to civilian nuclear power...

Finally, nuclear dynamite had to compete with non-nuclear technology. As the *GURC Report* stated, 'PNE technology faces serious competition from non-nuclear alternatives'.[68] As Ulf Ericsson put it:[69]

> In purely commercial projects... the cost of nuclear explosives, of their emplacement and of the special measures against health hazards will have to meet not only the condition of absolute commercial feasibility but that of advantage over other techniques. The nuclear advantage will have to be clear, because in marginal cases the non-economic complications of nuclear operations will favour other techniques.

As Ericsson noted of nuclear dynamite's chief competitor:[70]

> The products of chemical explosions are so much more congenial to biological life than radioactive substances that they present almost no problems at all.

According to Herbert York, Plowshare technology 'never made the difference between the possible and the impossible — there were always other ways, always alternatives'.[71]

## *Environmental Concerns*

Along with industrial scepticism, environmental concerns also helped sound the death knell of Plowshare. As the *GURC Report* noted, the US public was 'deeply concerned ... with environmental protection in general; and a considerable amount of political hostility has been focused on the hazards of nuclear devices in particular'.[72] For Iris Borg, mounting public concerns about environmental consequences and waning industrial interest were linked, since public objections to the use of PNEs added 'a new and unknown dimension to cost estimates'.[73]

Considerable progress was made by Plowshare towards attentuating one significant environmental hazard of PNEs: the release of radioactivity. So successful was this effort that in 1971 Alan Wilson told the Fourth International Conference on the Peaceful Uses of Nuclear Energy that it was 'approaching a practical limit'.[74] Yet this was still not enough to satisfy a sceptical public. In the words of the Livermore Laboratory:[75]

> Although studies show that civil nuclear explosions would produce only a few additional millirems of radiation per person per year (a dose equivalent to one round-trip cross country airplane flight or a week's vacation in the high Sierra Nevada), an influential segment of the public found this unacceptable despite the program's considerable economic promise.

Although opposed to Plowshare, Herbert York describes public fears of radioactivity from Plowshare projects as 'exaggerated'.[76] In some quarters even a *demonstrably* 'clean' device would not have been enough. The issue was emotional and no amount of scientific argumentation would have sufficed. In 1968 Ulf Ericsson, noting that '... political and emotional problems are involved in the peaceful use of nuclear explosives', had warned that:[77]

> There will remain the need to prevent non-rational beliefs of nuclear explosions as evil, fears of radioactivity as a menace, of non-rational accusations of nuclear weapons spread or of nuclear weapons tests despite sufficient control.

However the AEC and the nuclear industry only had themselves to blame for such a situation. Public suspicion of Plowshare was only natural in the light of attempts by the AEC throughout the 1950s and 1960s to cover up the real effects of radioactivity; the many irregularities involved in the establishment and maintenance of civil nuclear power stations; the 1953 Bravo test in the Pacific in which Japanese fishermen and Marshall Islanders were severely affected by fallout; and the generally high-handed behaviour of the nuclear industry in its dealings with the public.

In another sense too, public concern was justified. The AEC was clearly unwilling to face up to the problem of the environmental effects of tens of thousands of PNEs being detonated for commercial purposes in the United States each year. The magnitude of the problem can be seen from the AEC's estimate that economic exploitation of the gas fields in New Mexico, Colorado, Utah, Arizona and Wyoming, would require 10–15,000 nuclear shots.[78] While the Livermore Laboratory spoke confidently of a 'few additional millirems' being added to the environment, it failed to mention how many explosions this involved or in what locations. The Commission simply did not know, for instance, whether radioactivity from such a large-scale program would contaminate ground water.[79] As Peter Metzger noted in 1972:[80]

> In addition to inconveniencing the public by causing a great deal of structural damage from the earth shocks produced, there will be left behind, deep underground, thousands of hardened puddles of radioactive glass, the 'fallout' from the atomic explosives themselves. Since the criterion for where these

> puddles end up is the presence of natural gas and not the ability of the rock formation to contain the waste for very long periods, the chances are that sooner or later the wastes will contact an underground water supply and, through fractures, be brought to the surface and enter the biosphere.

In the *GURC Report's* view the public concerns expressed about PNEs were identical to those expressed about nuclear power plants: the possibility of a major accident; lack of knowledge about the effects of low level radiation; and the adequate disposal of radioactive wastes.[81]

Irrational and exaggerated as they may have been, public fears of radioactivity were clearly a major difficulty for Plowshare. The *GURC Report* concluded in 1975 that 'opposition to PNE experiments (away from the Nevada Test Site) has increased in effect in the past three or four years'.[82] It also noted that: 'Public concern apparently will be far greater than the real hazard; but the cost of dealing with the problem cannot be escaped'.[83] As the general guidance on radiation exposure of man issued by the International Commission on Radiological Protection in 1977 stated:[84]

> Undertakings involving exposure of man to radiation must have overall advantages when compared with other methods of achieving the same objective. Any undertaking which results in the radiation exposure of man must be justified in the sense that the total benefit from the undertaking outweighs the total detriment. The total benefit can only be assessed after economic, social and other advantages have been evaluated. It follows that radiological protection cannot be considered in isolation from economic, social and other issues.

## *The Longevity of Plowshare*

With all these problems the wonder is not that Plowshare died when it did but that it survived so long. Who was responsible for Plowshare's longevity? Much of the responsibility must go to Edward Teller. As an internationally renowned and vocal scientist Teller promoted Plowshare, with varying degrees of success, with US Presidents and other high officials. William Broad relates that Teller:[85]

> ... consorted with generals, financiers, industrialists and politicians. The military looked on him as a prophet. Richard M. Nixon, then Vice-President and second only to McCarthy as a fervent anti-Communist, sought his advice. Nelson Rockefeller became a fast friend. In the 1960s, at Teller's invitation Ronald Reagan became the first governor of California to visit the Livermore weapons lab.

Herbert York has noted that:[86]

> ... ever since he led the winning side in the fight over the H-bomb, [Teller's] main influence over the course of American nuclear affairs has been primarily of a political nature ... exercised not so much through formal advisory channels

> but rather more through direct and personal relationships with a host of high-level decision makers in the American government and business establishment. And not only has Teller been a key figure in the promotion of the uses of nuclear technology, but also, beginning with his fight against the nuclear test moratorium of 1958–61, he has been the leading personality in opposing the various attempts to contain the nuclear arms race.

Also influential were Teller's writings, through which he promoted Plowshare and other uses of nuclear energy. These include *Our Nuclear Future* (1958), *The Reluctant Revolutionary* (1964), *Energy From Heaven and Earth* (1979) and *The Legacy of Hiroshima* (1982). A single quotation from a chapter in the latter, entitled 'How To Be An Optimist In the Nuclear Age', demonstrates Teller's single-minded devotion to Plowshare:[87]

> Quite a few of these Plowshare projects would be not only economical but highly profitable. Used for peaceful purposes, nuclear explosives can bring the dreams of truly massive engineering feats into the realm of economic reality. Used to increase, conserve, and distribute the world's supply of water, nuclear explosives can make deserts bloom and help stave off the threat of large-scale hunger posed by the world's exploding population.

These exaggerated claims for Plowshare are reminiscent of Teller's over-optimistic predictions in the 1980s about the x-ray laser, a key component of President Reagan's Strategic Defense Initiative (or 'Star Wars').[88] Teller's technical work on PNEs, *The Constructive Uses of Nuclear Explosives* (1968), was particularly influential and widely quoted. The Indians, for example, used it to justify their interest in PNEs.

Teller played a key role at pivotal times in Plowshare's history, such as rescuing it from the threatened test ban in 1958, initiating the Cape Keraudren project, and testifying on the program's behalf before the JCAE. While Teller was helped in this by other nuclear scientists and Plowshare officials, in Herbert York's opinion, 'Teller pushed harder and longer than anyone else, continuing long after most of the rest of us had had second thoughts'.[89]

But one man, however influential, cannot sustain an entire program such as Plowshare. Glenn Seaborg, who presided over the AEC during Plowshare's heyday, also bears a major responsibility. Seaborg attempted to 'save' Plowshare from the PTBT, from the strict legal interpretation of that treaty, and from the possibility of a comprehensive test ban treaty. He also promoted Plowshare internationally in an attempt to ensure its survival at home; defended it before the JCAE and the 269 Committee; argued for an international PNE service in the NPT context (again to help save Plowshare domestically); and encouraged such countries as Egypt, Australia and Brazil in their attempts to secure the benefits of nuclear dynamite. Seaborg also pursued the heavy element program and other Plowshare applications and failed to curb the excesses of the AEC's publicity machine in selling Plowshare to the nation and the world. Of the excavation program in particular, Seaborg now acknowledges in retrospect that: 'The lure of the program perhaps made us less than completely objective in weighing the arguments against'.[90] Despite Seaborg's retrospective claims to have been pro-arms control, particularly in relation to a comprehensive test ban, he was in part responsible for helping make Plowshare an obstacle to disarmament. In his testimony to the JCAE in 1965, for example, he seemed to see no implications for arms control in the Plowshare program.

Apart from Teller and Seaborg, the JCAE itself was also responsible for Plowshare's longevity. This was partly the result of the Committee's blanket endorsement of all matters atomic, and partly a by-product of its determination to minimise the hazards of radioactivity. As Teller told the Committee in 1960:[91]

> The Joint Congressional Committee has done most admirable work in dispelling unfounded fears about radioactivity.

The Committee also supported Plowshare more directly by securing funding and stimulating industrial interest. Plowshare director Kelly told the JCAE that the success of the gas stimulation projects and industrial participation in them was 'due in no small part to the foresight and initiative of this committee'.[92] When the Bureau of the Budget removed the Gasbuggy experiment from the 1967 Budget, the JCAE attempted to have it reinstated.[93] JCAE members developed close informal contacts with Plowshare scientists at Livermore.[94]

In the final analysis however, the longevity of Plowshare must also be attributed to several general factors pertaining to the project's historical and institutional setting. The first is the tendency of programs and organisations to assume a life of their own. For Michael Brenner, Plowshare was 'inspired by the general desire to promote nuclear power, and by the [Livermore] laboratory's particular interest in winning support for a program in which it had an organisational stake'.[95] It is usually difficult for governments and bureaucracies to abolish existing programs, especially those that are not disastrous failures but are simply limping along. There is a tendency in such cases, as JCAE hearings into Plowshare often revealed, to believe that success is just around the corner if only more funds and effort are expended. There is also a reluctance to 'waste' funds already spent by closing down such programs.

A second general reason for Plowshare's survival was its relative obscurity. Despite the publicity campaigns, it was still a very minor part of the AEC's overall activities. It was dwarfed not only by the nuclear weapons program but by the controversial civilian nuclear power program. Plowshare was able to hum along in obscurity until financial largesse for things nuclear turned into miserliness.

Finally, Plowshare's survival must be attributable to the general awe with which nuclear science was viewed up until the 1970s. The magic of the atom was regarded as beyond the ken of the ordinary citizen. Decisions about nuclear power were seen as the province of nuclear scientists alone. Plowshare, emanating as it did from such luminaries as Edward Teller, obviously benefited greatly from this phenomenon. The result, in the words of Herbert Scoville, was that:[96]

> ... many projects were evaluated on unrealistic bases, thereby leading to over-optimism about their feasibility. Research costs could be written off as part of the weapons program and tended to be largely ignored. Non-nuclear technology was often assumed to stand still while nuclear technology made major advances.

It was only when Plowshare began running up against engineering and other 'lay' assessments, beginning with the Panama Canal Commission, that the program was called into question. Only when 'non-scientific values began to intrude into the preserves that had once been governed almost exclusively by men of science', did Plowshare collapse.[97] As Alan Wilson wisely warned university and industry participants in Plowshare in 1969:[98]

> ... the judicious approach... would be to avoid assuming too readily that, because of their experience and specialised competence, [scientists] are qualified to dismiss the concerns of those who are not so experienced and trained. It should never be overlooked that a mite of humility can upon occasion avert a veritable mount of humiliation and that all wisdom need not derive from a single intellectual domain.

As Professor Smyth had cautioned the JCAE way back in 1960:[99]

> The technical application of scientific knowledge is not necessarily beneficial. Technical knowledge gives men greater power to control their environment. Unfortunately, it does not automatically enhance their wisdom nor improve their morals.

# *Notes*

1 G. T. Seaborg *et al.*, *Plowshare*, Division of Technical Information, US Atomic Energy Commission, Oak Ridge, Tennessee, 1966, p. 37.

2 D. L. Clarke, *Politics of Arms Control: The Role and Effectiveness of the US Arms Control and Disarmament Agency*, The Free Press, New York, 1979, p. 174.

3 *PNE Activity Projections For Arms Control Planning* (the *GURC Report*), Gulf Universities Research Consortium, Galveston, Texas, US Arms Control and Disarmament Agency Document, ACDA/PAB-253, Washington DC, 1975, p. 19.

4 From an interview with Fred Holzer, Livermore Laboratory, California, 3 December 1986.

5 US Congress, Senate Foreign Relations Committee, Hearings on the TTBT/PNET, 28 July, 3 August, 8 and 15 September, 1977, US Government Printing Office, Washington DC, 1977, p. 127.

6 H. Scoville Jr., 'Peaceful Nuclear Explosions — An Invitation to Proliferation' in A. W. Marks (ed.), *NPT: Paradoxes and Problems*, Arms Control Association, Washington DC, 1975, p. 47.

7 G. W. Johnson, 'Plowshare At The Crossroads', *Bulletin of the Atomic Scientists*, June 1970, p. 83.
8 The *GURC Report*, Vol. II, p. 212.
9 Ibid., p. 216.
10 Ibid., p. 213.
11 Ibid., p. 212.
12 Ibid., p. 213.
13 Ibid., Vol. I, p. 42.
14 Ibid., pp. 37–38.
15 Ibid., Vol. I, p. ii.
16 US Congress, Senate Foreign Relations Committee, Hearings on the TTBT/PNET, 1977, p. 136 and the *GURC Report*, Vol. I, p. v.
17 I. Y. Borg, *Nuclear Explosions for Peaceful Purposes*, Lawrence Livermore National Laboratory Paper No. UCRL-95026, Rev. II, November 1986, p. 17.
18 US Congress, Senate Foreign Relations Committee, Hearings on the TTBT/PNET, 1977, p. 136.
19 The *GURC Report*, Vol. I, p. ii.
20 Ibid.
21 Ibid., pp. 77–78.
22 The *GURC Report*, Vol. I, p. 102. MHF involves pumping fluid at very high pressures into a gas well until the formation is fractured. Carefully selected sand or other material is added to the liquid as a 'proppant' to prop open cracks created by the pressure. While MHF, like PNEs, also has its technical uncertainties, it was being used successfully in the mid-1970s (see the *GURC Report*, pp. 79 and 102 and *An Analysis of the Economic Feasibility, Technical Significance and Time Scale for Application of Peaceful Nuclear Explosions in the US, with Special Reference to the GURC Report Thereon (The Long Report)*, Program on Science, Technology and Society, Cornell University, US Arms Control and Disarmament Agency, Washington DC, 1975, p. 16).
23 The *GURC Report*, Vol. I, p. 100. Alan Wilson disputed this in 1979, saying that 'the range of uncertainty is such that the apparent cost advantage of MHF may be illusory' and that 'for the present, neither method can be said to offer much prospect of producing gas at today's prices' (see A. R. W. Wilson, 'Technical Aspects of Peaceful Nuclear Explosions Relevant to their Possible Role in the Future Proliferation of Weapon — Usable Nuclear Materials' in *Nuclear Energy and Nuclear Weapon Proliferation*, SIPRI, Taylor and Francis, London, 1979, p. 313).
24 See Chapter 15.
25 *The Long Report*, p. ii.
26 Ibid., p. iii.
27 Ibid., p. 32.
28 Ibid., p. 36.
29 Ibid., p. 33.
30 US Congress, Senate Foreign Relations Committee, Hearings on the TTBT/PNET, 1977, p. 127.
31 *The American Experience with Peaceful Nuclear Explosions*, US Arms Control and Disarmament Agency, US Government Printing Office, Washington DC, 1978, p. 4.
32 Clarke, p. 115.
33 From an interview with Fred Holzer, Livermore Laboratory, California, 3 December 1986.
34 M. I. Shaker, *The Nuclear Non-Proliferation Treaty: Origin and Implementation 1959–1979*, Vol. I, Oceana Publications, London, 1980, p. 391.
35 US Congress, Senate Foreign Relations Committee, Hearings on the TTBT/PNET, 1977, p. 12.
36 Ibid., p. 55.
37 Ibid., p. 82.
38 Clarke, p. 175.
39 US Congress, Senate Foreign Relations Committee, Hearings on the TTBT/PNET, 1977, p. 106.
40 Ibid., p. 127.
41 From an interview with Herbert York, La Jolla, California, 1 December 1986.
42 Quoted in G. T. Seaborg and B. S. Loeb, *Stemming the Tide: Arms Control in the*

*Johnson Years*, Lexington Books, Lexington, Mass., 1987, p. 351.
43 Wilson, 'Technical Aspects of Peaceful Nuclear Explosions', p. 312.
44 *Washington Times*, 2 September 1987, p. 5.
45 The *GURC Report*, Vol. I, p. 186.
46 *The American Experience*, p. 7.
47 G. T. Seaborg and B. S. Loeb, *Kennedy, Khrushchev and the Test Ban*, University of California Press, Berkeley, 1981, p. 248.
48 From an interview with Fred Holzer, Livermore Laboratory, California, 3 December 1986.
49 R. Sanders, *Project Plowshare*, Public Affairs Press, Washington DC, 1962, p. 166.
50 P. Pringle and J. Spigelman, *The Nuclear Barons*, Avon Books, New York, 1981, p. 220.
51 Seaborg and Loeb, *Stemming the Tide*, p. 347.
52 The *GURC Report*, Vol. I, p. 73.
53 D. B. Brooks and J. V. Krutilla, *Peaceful Use of Nuclear Explosives: Some Economic Aspects*, Resources for the Future, Inc., Washington DC, 1969, p. 16.
54 *The Long Report*, p. 5.
55 The *GURC Report*, Vol. I, p. 12.
56 *The Long Report*, p. 7.
57 D. B. Brooks and H. R. Myers, 'Plowshare Evaluation' in B. Boskey and M. Willrich (eds), *Nuclear Proliferation: Prospects for Control*, Dunellen Publishing Co., New York, 1970, pp. 88–90.
58 Brooks and Krutilla, p. 17.
59 Australian Atomic Energy Commission, *Peaceful Uses of Nuclear Explosions — An Evaluation, for Australian Purposes, of Proposed Civil Engineering and Mining Applications*, AAEC, Sydney, 1964, p. 219.
60 *Report of the Ad Hoc Advisory Group on Nuclear Explosions for Peaceful Purposes*, IAEA Document GOV/1854, 26 August 1977, IAEA, Vienna, 1977, p. 35.
61 Brooks and Myers, p. 89.
62 E. Teller and A. L. Latter, *Our Nuclear Future*, Criterion Books, New York, 1958, p. 172.
63 E. Teller, 'The University's Role in Nuclear Explosives Engineering Research' in L. E. Weaver (ed.), *Education for Peaceful Uses of Nuclear Explosives*, University of Arizona Press, Tucson, Arizona, 1970, p. 298.
64 The *GURC Report*, Vol. I, p. 140.
65 Ibid., p. 181.
66 Ibid., p. 23.
67 US Congress, Senate Foreign Relations Committee, Hearings on the TTBT/PNET, 1977, p. 142.
68 The *GURC Report*, Vol. I, p. ii.
69 U. Ericsson, *The Question of Nuclear Explosions for Peaceful Purposes by Non-Nuclear Weapon States and the Possibility of Misuse of Such Technology for the Production of Nuclear Weapons*, Document A/CONF. 35/DOC. 3, Conference of Non-Nuclear Weapon States, Geneva, 1968, p. 6.
70 Ibid., p. 3.
71 From an interview with Herbert York, La Jolla, California, 1 December 1986.
72 The *GURC Report*, Vol. I, p. 184.
73 Borg, *Nuclear Explosions for Peaceful Purposes*, p. 10.
74 A. R. W. Wilson, 'Current Status of Civil Engineering and Mineral Resources Development Applications of Peaceful Nuclear Explosions' in Proceedings of the Fourth International Conference on the Peaceful Uses of Nuclear Energy, 6–16 September 1971, IAEA, New York and Vienna, 1972, p. 217.
75 *Energy and Technology Review*, Lawrence Livermore National Laboratory, September 1982, p. 20.
76 From an interview with Herbert York, La Jolla, California, 1 December 1986.
77 Ericsson, pp. 3 and 8.
78 F. Kreith and C. B. Wrenn, *The Nuclear Impact*, Westview Press, Boulder, Colorado, 1976, p. 176.
79 US Congress, Subcommittee on Public Lands of the Senate Committee on Interior and Insular Affairs Hearings, 11 May 1973 *(The Haskell Hearings)*, US Government Printing Office, Washington DC, 1973, p. 148.
80 H. P. Metzger, *The Atomic Establishment*, Simon and Schuster, New York, 1972, p. 236.

81 The *GURC Report*, Vol. I, p. 162.
82 Ibid., Vol. I, p. 2.
83 Ibid., pp. 77–78.
84 *Report of the Ad Hoc Advisory Group on Nuclear Explosions for Peaceful Purposes*, p. 28.
85 W. J. Broad, *The Star Warriors*, Simon and Schuster, New York, 1985, p. 39.
86 H. F. York, *The Advisors: Oppenheimer, Teller and the Superbomb*, W. H. Freeman and Co., San Francisco, 1976, p. 145.
87 E. Teller, *The Legacy of Hiroshima*, Doubleday and Co., New York, 1982, p. 91.
88 'Red Flag At Weapons Lab', *Time*, 18 January 1988, p. 38.
89 From correspondence with Herbert York, La Jolla, California, 25 September 1986.
90 Seaborg and Loeb, *Stemming the Tide*, p. 347.
91 US Congress, JCAE, Hearings on AEC Authorizing Legislation for Fiscal Year 1961, US Government Printing Office, Washington DC, 1960, p. 50.
92 US Congress, JCAE, Hearings on AEC Authorizing Legislation for Fiscal Year 1971, US Government Printing Office, Washington DC, 1970, p. 619.
93 US Congress, JCAE, Hearings on AEC Authorizing Legislation for Fiscal Year 1967, US Government Printing Office, Washington DC, 1966, pp. 1360 and 1394.
94 From correspondence with Gerald Johnson, Palos Verdes, California, 17 January 1987.
95 M. J. Brenner, *Nuclear Power and Non-Proliferation: The Remaking of US Policy*, Cambridge University Press, Cambridge, 1981, p. 307.
96 Scoville, p. 50.
97 D. S. Greenberg, *The Politics of American Science*, Penguin, Harmondsworth, 1969, p. 262.
98 A. R. W. Wilson, 'Educational Needs of Industry' in Weaver (ed.), p. 326.
99 US Congress, JCAE, Hearings on AEC Authorizing Legislation for Fiscal Year 1961, p. 12.

# 14

# A Fading Regime: The IAEA Dumps PNEs

*Even though it may be reasonably argued that it is premature to set up any international service for peaceful nuclear explosions until their practical value is established, it can also be argued that if the international regime is created and is on the books awaiting proof of the practicality of such explosions, there will be less incentive for non-nuclear states to attempt to develop their own peaceful nuclear explosions.*[1]

William Epstein, 1976

WITH its newly-won mandate from the United Nations General Assembly, the IAEA began its work on the peaceful uses of nuclear explosions in 1970. The Agency moved slowly. Its secretariat, according to former IAEA Director of External Affairs David Fischer, regarded the PNE business as at best a distraction and at worst a positive hindrance to its primary functions.[2] Few resources were devoted to the work. Staff were not employed full-time on PNE-related assignments but 'doubled up' from the Department of Technical Operations. Far from preparing for the inauguration of an international PNE regime, the IAEA ran its PNE program on a shoestring.

At this stage Plowshare was still very much alive, and the Americans, at least at the technical level and within the AEC, were still keen on promoting nuclear dynamite internationally. Fischer recalls many arguments with United States representatives over the value of pursuing the PNE idea. He notes that, until Plowshare collapsed after 1973, the United States was the prime mover in sustaining interest in PNEs within the IAEA. In his view the Americans have only themselves to blame for the problems caused by the PNE issue for arms control and nonproliferation efforts. Alan Wilson agrees that the Americans, as well as the Soviets, created the PNE 'problem' by their over-zealous promotion of nuclear dynamite.[3]

The first task of the IAEA in its work on PNEs was the report on 'international observation' requested by the UN General Assembly in 1969. The IAEA Board of Governors approved a draft of the report in June 1972. It defined the aim of observation as being to verify that PNE projects in non-nuclear weapon states did not violate the 'intent and letter' of the NPT's ban on the acquisition of nuclear weapons or other nuclear explosive devices by such states. The document also detailed the type of agreement, between the IAEA and the nuclear weapon and non-nuclear weapon states involved, which would be required for such observation.[4] David Fischer recalls drafting the document with an American looking over one shoulder and a Soviet representative looking over the other. It was clear, Fischer says, that they were not at all concerned about observing PNE detonations to prevent the conducting state from gaining weapons-related benefits. Rather, they were concerned with ensuring that no weapons-related benefits leaked out to non-nuclear weapon states. The Soviets declared that the purpose of international observation was 'to exclude the possibility of using peaceful nuclear explosions as a means for non-nuclear weapon States to obtain special information necessary for the production of nuclear weapons'.[5] Hence observation was to be outwards from the project site rather than inwards.[6] As Fischer put it, more bluntly: 'The international observers keep an eye on the non-nuclear weapon state rather than on the PNE'.[7]

In 1969 the IAEA also began to carry out the brief assigned to it by the 1968 Conference of Non-Nuclear Weapon States, to compile and disseminate information on PNEs. It circulated to members information on both the current status and potential applications of PNE technology. In 1970 it published a bibliography on PNEs containing 1759 references to literature published until June 1969. This was updated in 1980. Information on the geological, geophysical and seismological aspects and potential uses of PNEs was also included in the Agency's International Nuclear Information System (INIS), a world-wide computer data base for nuclear science and technology.[8]

Between 1970 and 1975 the Agency established four technical panels to keep itself and IAEA members informed of developments in the PNE field. These examined such subjects as the phenomenology of contained and cratering explosions, the application of PNEs for industrial purposes, and the health, safety and economic aspects of PNEs.[9] They were comprised of experts from countries that had played a major part in developing PNEs or related technology. Interestingly, Australia and India were the only two non-nuclear weapon states represented on all four panels. Australia's Alan Wilson chaired the January 1975 panel on the health, safety and economic aspects of PNE projects.

In classic UN fashion, the work of the IAEA on PNEs began to feed on itself. For instance, the Technical Panel on Practical Applications of Contained Peaceful Nuclear Explosions for Industrial Purposes, which counted among its members Argentina, Australia, India and Mexico, recommended that the Agency develop detailed procedures for responding to members' requests for assistance in obtaining PNE services. In April 1974 a group of experts, improbably called the Working Group to Provide Detailed Procedures for the Agency to Use in Responding to Requests for PNE-Related Services, was convened to advise on the elaboration of such procedures, with representatives both from potential PNE supplier states and potential user states. Australia was a member, as were Argentina, Egypt, France, Sweden, Thailand, the United Kingdom, the United States and the Soviet Union.[10] The group concentrated on procedures for the early stages of a PNE project, including guidelines for requesting states and procedures for the Agency in concluding observation agreements with requesting and supplier states.[11] After adoption of these guidelines by the Board of Governors, IAEA members were invited to indicate their willingness to offer explosive or other PNE-related services to other states. This enabled the IAEA to establish by the end of 1974 a register of so-called PNE Supplier States and PNE Consultant States.

In September 1974, due to 'increased interest' in PNEs, the Board of Governors established a special unit in the Department of Technical Operations of the IAEA secretariat to deal with the technological, economic, safety and legal aspects of PNEs.[12] The Board also reiterated the Agency's 'readiness and willingness' to provide 'ancillary' PNE services not directly related to the detonation of a nuclear explosive.[13]

The alleged increased interest in PNEs cited by the Board of Governors was questionable. Between 1970 and 1974 the IAEA only received three serious enquiries about PNE services, none of which resulted in the use of a PNE. Czechoslovakia, for instance, sought information in relation to gas stimulation, while Romania asked about stimulation of oil deposits. In 1971 Madagascar requested information on the possible use of a PNE for removing a rock obstacle in a harbour construction project. The IAEA approached France, the Soviet Union, the United Kingdom and the United States, all of which expressed a willingness to supply experts to appraise the feasibility of the idea. This was conveyed to the Malagasy authorities, but nothing further was heard.[14]

Despite this lack of actual demand for PNE services, the influential developing countries were by 1974 becoming increasingly impatient with the IAEA and the nuclear weapon states for not moving fast enough to establish an international PNE regime. While for most developing countries the establishment of such a regime would have had more symbolic than practical importance, the fulfilment of Article V of the NPT was nonetheless regarded as an essential *quid pro quo* for their renunciation of nuclear weapons and an important test of the good faith of the nuclear weapon states. This view was supported by SIPRI, which urged that an international PNE regime should be established in its entirety to help ensure that the NPT had a 'reasonable chance of survival'.[15] Ignoring the guidelines and procedures for PNE projects already worked out by the IAEA,[16] SIPRI commented, exaggeratedly:[17]

> In practice, no steps at all have been taken to set up the mechanisms for the 'appropriate international observation' and 'appropriate international procedures' for PNEs.

The Swedish government supported this view, proposing to the CCD that it begin negotiations on the international PNE agreement envisaged in Article V of the NPT.[18] Sweden suggested that the agreement specifically state that the benefits of PNEs should be available on a non-discriminatory basis 'to those countries that forego production of nuclear devices'. The IAEA, Sweden suggested, should play a critical role in evaluating the technical feasibility of all PNE projects, including economic, health and safety aspects, and in 'arranging for and controlling the actual explosion'. The 'overall advisability' of the project should however, in Sweden's view, be determined by an unspecified 'political international body'.

It is not difficult to imagine why this proposal was not greeted with universal acclaim in the CCD. On the one hand India, Argentina and Brazil, which refused to renounce the acquisition of nuclear weapons, could hardly be expected to participate in negotiating a treaty designed to exclude them from being beneficiaries. On the other hand, neither the Americans nor the Soviets were keen to devolve as much authority over PNEs to the IAEA as the Swedes envisaged, or to countenance establishment of yet another body dealing with PNEs.

At the 1974 session of the UN General Assembly the developing countries pushed through a resolution inviting the United States and the Soviet Union to give the First NPT Review Conference in 1975 an account of the steps they had taken to establish a PNE regime since the entry into force of the NPT in 1970.[19] India voted against, while Argentina and Brazil abstained. These so-called threshold[20] states, while accusing the nuclear weapon states of delaying the establishment of a PNE regime in order to avoid sharing PNE technology with the rest of the world, in fact wanted no progress towards a regime that might ultimately constrain their own freedom to develop and use PNEs themselves. For their part the Western states were increasingly inclined to soft-pedal the PNE issue, having become convinced that the international promotion of PNEs, even under IAEA auspices, was legitimising the claims of the 'threshold' states that PNEs were useful and necessary for their economic development. After the 1974 Indian explosion had confirmed their worst fears, Western sentiment moved sharply towards forswearing the promotion, development and use of PNEs altogether. The United States, with Plowshare in decline, joined these ranks.

The Soviets, while continuing to laud the possibilities of nuclear dynamite for the Soviet Union's own economic development, remained acutely conscious of the arms control implications of PNEs. They continued to be quietly but determinedly opposed to an international PNE regime, both because of fear that interest in PNEs might be further stimulated and because Soviet PNE services might be demanded. They denied being in breach of Article V, by arguing that PNE technology was not yet fully developed. As they told the CCD in late 1974:[21]

> ... at present, when peaceful nuclear explosions still have no practical application because of inadequate technology, there are no grounds for speaking of failure to implement that part of the Treaty [Article V]. It would be more accurate to say that the Treaty's provisions with regard to such explosions have not so far been implemented, since there has been no practical need for such explosions.

This was of course in spectacular contrast to their claim that several Soviet PNE applications were 'proven technology'.

Having generated interest in PNEs among the developing countries in the first place, the United States (and, to the extent that it accepted responsibility, the Soviet Union) now faced the dilemma of how best to quench such interest. As a study for the US Council on Foreign Relations put it:[22]

> Either nuclear states can maintain a low-profile scepticism toward the utility of PNEs without foreclosing the possibility of offering assistance if asked or they can establish a formal mechanism to provide PNEs and thereby probably stimulate widespread interest. Of these choices, the former seems preferable. Since the goal is to plug both the real and symbolic aspects of the PNE loophole, subtle suppression that leaves open the option of formal renunciation is better than enthusiastic or even modest marketing.

In September 1974 a non-governmental conference on the NPT was held in Divonne, France.[23] Herbert Scoville Jr., of the US Arms Control Association, urged that the euphoria of 1968 about PNEs should not continue unevaluated at the First NPT Review Conference. Alleging that a 'rosy vision' of PNEs had been promoted in some non-nuclear weapon states to 'serve nationalistic political objectives', he described India's 'PNE' as a graphic reminder of how nuclear dynamite could subvert the non-proliferation regime.[24] The conference, in apparent agreement with Scoville, concluded that the benefits of PNEs were 'questionable' and called for a moratorium on such explosions while the UN Secretary-General studied the PNE problem.[25]

## *The First NPT Review Conference*

The First NPT Review Conference convened in Geneva in May 1975. As expected, the PNE issue was contentious. The Western countries expressed great scepticism about the usefulness and feasibility of PNEs. The United States favoured giving them up altogether, having virtually already done so itself.[26] Sweden also suggested that, if studies showed that on balance no considerable benefits could be derived from PNEs, states should forego them.[27] The Conference President, Mrs Inga Thorssen of Sweden, urged the establishment of an international PNE regime so that individual states, including the nuclear powers, could be dissuaded from embarking on their own PNE programs.[28] Several non-nuclear weapon states, mostly developing and non-aligned (Ghana, Mexico, Nigeria, Peru, the Philippines, Romania, Syria and Yugoslavia), proposed immediate consultations to set the place and date for a meeting of NPT parties to conclude an international PNE agreement. This was even less acceptable to the nuclear weapon states than it had been in 1968. The nuclear powers, including the Soviet Union, continued to prefer to deal with PNEs, if at all, within the IAEA. The structure and procedures of the Agency demonstrably allowed them to wield influence much more readily than at a special conference, which could establish its own rules of procedure and where they were likely to be outvoted.[29]

The compromise eventually reached on the issue signalled a victory for the nuclear weapon states and their anti-PNE allies. Australia and Canada, meeting in secret, skilfully drafted the PNE section of the Conference's Final Document to reflect their own preferences and then sold it to the rest of the meeting, including the non-aligned countries. The text reaffirmed the paramount role of the IAEA in the PNE issue, declaring the Agency to be 'the appropriate international body, referred to in Article V of the Treaty, through which potential benefits from peaceful applications of nuclear explosions could be made available to any non-nuclear weapon State'.[30] The declaration then lamely urged the IAEA:[31]

> ... to expedite work on identifying and examining the important legal issues involved in, and to commence consideration of, the structure and content of the special international agreement or agreements contemplated in Article V of the Treaty.

Finally, the document referred the contentious issue of the arms control implications of PNEs to the CCD.

The document subtly cleared the way for continued procrastination in the IAEA over the establishment of an international PNE regime and for the eventual disappearance of the idea altogether. In addition, in offering PNE services to *all* non-nuclear weapon states, not just NPT parties, the document was an attempt to lure Argentina, Brazil and India into the nonproliferation fold. Taking at face value the threshold states' oft-expressed enthusiasm for the alleged economic benefits of PNEs, the Conference offered them such benefits without NPT membership, but subject to the safeguards of a future PNE regime. Most of the Conference participants probably did not harbour any illusion that the threshold states would accept this offer and indeed they never have. Yet the fact that the offer was made effectively neutralised the threshold states' claim that they were forced to develop their own PNEs because they could not obtain the benefits of PNE technology any other way.

The CCD lost no time in considering the issue of the arms control implications of PNEs, as referred to it by the Conference. From 4–18 July 1975 it held a series of meetings on the subject, with the participation of experts. According to SIPRI, the following were among the conclusions that could be drawn from the deliberations:[32]

> 1. It is impossible to develop nuclear devices which would be capable only of peaceful application. All such devices could also be used as weapons; they are transportable and could be carried to targets by various delivery vehicles. Thus, any state conducting peaceful nuclear explosives acquires a nuclear weapon capability. And since, from the point of view of arms control, it is the capability rather than the declared intention of states that matters, the proliferation of peaceful nuclear explosives would be equivalent to the proliferation of nuclear weapons.
> 2. As distinct from contained underground explosions, the use of nuclear explosions for excavation purposes would inevitably release radioactive products into the atmosphere. Should the debris cross the territorial limits of the state under whose jurisdiction or control the explosion was conducted, the state in question could be brought to task for violating the 1963 Partial Test Ban Treaty.
> 3. Any agreed restraints on underground nuclear weapon testing by the nuclear powers, placing a limit on the number of explosions and/or their yield, must be accompanied by corresponding limitations on peaceful nuclear explosions, because peaceful programs could be used to obtain weapon-related information.

As if this scathing report was not enough, the IAEA was also planning its own examination of the PNE issue.

## *The IAEA* Ad Hoc *Advisory Group on PNEs (1975–77)*

Prior to the Review Conference, Australia, on the initiative of Alan Wilson, had begun pressing the IAEA to conduct a study of the PNE issue at the 'political' rather than the technical level. The aim was to examine internationally for the first time the economic and technical feasibility of PNEs and the possibility of fulfilling Article V of the NPT. At first sight it is not clear why Australia should have taken such an initiative, particularly when the strong inclination of the United States was now to play down the PNE issue and at least delay the establishment of a regime. Australia was in fact hoping that by thoroughly airing the very substantial problems facing the use of nuclear dynamite, the PNE issue could be excised from the international agenda. In Canberra's view, it was insufficient for the superpowers simply to contend, however justly, that PNEs offered no genuine 'benefits' and that Article V should remain unfulfilled. This only promoted, as Ian Smart has pointed out, 'unfounded sympathy for Indian and Brazilian insistence' that the benefits of PNEs existed but the nuclear weapon states were cynically intent on withholding them from others.[33] Rather, the international community had to be educated about PNEs. As SIPRI noted:[34]

> It seems... that only a wide dissemination of the data on which judgements must be based will be able to alter the insistence of some states on the possible 'benefits' of PNEs.

Australia was especially concerned that while the technical experts involved in the IAEA panels were becoming educated about the PNE problem, the message had not filtered back to their political masters.

At first the Australian idea was ridiculed, particularly by the United States and the Soviet Union, which felt that such a study would create more interest in PNEs than it would quash. The Australians persisted, and in February 1975 submitted a draft resolution to the Board of Governors seeking the establishment of an *Ad Hoc* Advisory Group on Nuclear Explosives for Peaceful Purposes to consider 'certain matters' relating to PNEs. The draft was co-sponsored by the Federal Republic of Germany, Japan, the United Kingdom and Thailand.[35] The United States and the Soviet Union were notably missing from the list of co-sponsors. The Board did not however take an immediate decision on the resolution because of the impending NPT Review Conference.

At the Review Conference the Australians provided impetus for their proposal by including in the Final Document a call for the IAEA to 'broaden its consideration of this subject [PNEs] to encompass, within its area of competence, all aspects and implications of the practical applications of nuclear explosions for peaceful purposes'.[36] The IAEA was further urged to establish 'appropriate machinery' in which 'intergovernmental discussion can take place' on the PNE issue.

In June 1975, after the Conference, Australia resubmitted its resolution, which was adopted by the Board of Governors on 11 June 1975 in virtually the same terms as originally proposed. The *Ad Hoc* Advisory Group on PNEs was to have the following terms of reference:[37]

> (a) To examine the aspects of PNEs coming within the Agency's sphere of competence with particular reference to:
> (i) procedural aspects relating to possible requests for PNE-related services
> (ii) legal aspects and treaty obligations
> (iii) health and safety matters
> (iv) economic aspects, including comparisons with non-nuclear alternatives.
> (b) To advise the Board on the factors involved in the establishment and operation of an international service for PNEs.
> (c) To advise the Board, within the Agency's sphere of competence, on the structure and content of agreements necessary under Article V of the NPT.

The *Ad Hoc* Advisory Group on Nuclear Explosions for Peaceful Purposes, chaired by Alan Wilson, held 42 meetings from September 1975 to August 1977. Its long report is a strange animal, an obvious product of political as well as technical compromises. Despite the group's terms of reference, the report claims not to represent a 'comprehensive treatment of the subject'.[38] It has no summary of the group's views and makes no recommendations about the future of PNEs or the political and legal framework for their conduct. It even fails to rank the possible uses of PNEs in order of technical feasibility, 'since this would involve value judgements which may become obsolete as the technology develops'.[39] While purportedly adopted by consensus, the report is hedged by interpretive statements and, in one case, footnoted reservations.

One problem faced by the group was disagreement between the United States and the Soviet Union over the alleged benefits of PNEs. The United States, with Plowshare wilting, was keen to play down such benefits as much as possible. While originally sceptical about the work of the group, the Soviets soon discovered it was an opportunity to document their PNE prowess. At the 1976 session of the CCD they delivered a major statement in praise of PNEs, obviously designed to influence the group's work. The Soviets claimed that PNEs opened up new opportunities for complex engineering tasks which it would be inconceivable to carry out by traditional methods. The statement then enumerated virtually all the ideas which Plowshare had tackled and found to be economically and technically impracticable.[40]

Despite this Soviet bullishness, the group's report was a masterful swamping of Soviet views. Skilfully drafted to secure consensus, it dutifully recorded the Soviet belief that PNEs were 'proven technology', while making clear that this did not mean they would work outside the Soviet Union. The report noted, for example, that the use of PNEs for sealing runaway oil and gas well fires 'must be considered proved and economic for the particular circumstances of the Bukhara incidents' in the Soviet Union. The report concluded that 'this method might be used once or twice a decade throughout the world'.[41] Again, in regard to oil stimulation, the report noted that the Soviet Union considered the PNE technique 'proven technology' but that 'the mechanism for recovery enhancement is not well understood'. This was tantamount to saying that it worked well for the Soviets but no-one knew why.[42]

Reading between the lines, the report overall was highly pessimistic about the future of PNE technology. Its implied message was that there were no conclusive economic or technical benefits to be derived from PNEs. This pessimism pervaded the group's examination of every single potential use of PNEs, even when, in an effort to accommodate Soviet views, the claimed 'successes' of the Soviet Union were perforce mentioned. Clearly the report was calculated to discourage developing countries from even contemplating such problem-infested technology. It catalogued the classic problems of PNE technology as:[43]

- possible failure to predict accurately the effects of explosions
- health and safety hazards from radioactive contamination of air, ground, water or products resulting from PNEs, such as gas or copper
- seismic damage to structures

- damage to the natural environment
- nuclear accidents or theft
- cost over-runs, and
- failure of demand for project services to develop.

Among the most telling comments were those on the economics of PNEs. The report stated that:[44]

> ... generalisation about PNEs should be suspect; analysis should be subject specific. Only when details of a specific, intended project are analysed can an evaluation be made with any confidence. Even then the remaining problems may be formidable.

On the environmental implications of PNEs, the report intoned:[45]

> Projects employing explosions for peaceful purposes may affect large expanses of land occupied by many people. Consequently the technology of PNEs must be reliable and safe and their environmental effects should be well understood.

The report was particularly sceptical about nuclear excavation. In relation to harbour construction it stated that 'although harbours have been constructed using conventional explosives the PNE approach is technically and economically unproved'.[46] In relation to other excavation applications it said: 'The available information neither establishes nor refutes the possibility that nuclear excavation may allow significant cost savings in specific situations'.[47]

The report's analysis of future international legal structures for PNEs was similarly discouraging. No single system was recommended. Rather, four complicated alternatives, plus variations, were provided, undoubtedly reflecting differing views within the group. Obviously to accommodate group members Argentina and India, all four alternatives were drafted to avoid stating clearly that non-nuclear weapon states should only obtain the benefits of PNEs from nuclear weapon states, and that they should not be assisted to produce and use their own 'home-made' nuclear dynamite. The report also totally ignored the controversy over interpretation of the Treaty of Tlatelolco's PNE provisions.

Even so, Argentina and India disassociated themselves from a key principle proposed for the conduct of PNEs, namely that 'all aspects of the international arrangements must fully take into account that all nuclear explosive devices designed for peaceful purposes could also be employed in some fashion as weapons'.[48] A footnote recorded that Argentina and India had 'reservations' about this paragraph because 'according to them, the essential content of this principle, embodied in the second part of the sentence, is technical and they know of no technical assessment which could form the basis of this statement'.[49] This quasi-technical mumbo-jumbo conveniently overlooked the CCD's study of the arms control implications of PNEs in July 1975, which concluded that there was no difference between PNEs and nuclear weapons.[50]

The four alternative international legal instruments described in the report were:[51]

I. An international 'umbrella' agreement establishing binding principles and procedures, to which would be annexed a Model Project Agreement. This agreement could be drafted by the IAEA and refined by a diplomatic conference convened by the Agency.
II. General principles established by the Agency, but not embodied in a legally-binding international agreement. States involved in a PNE project would be bound only by those principles required by their individual treaty obligations (an obvious attraction to non-NPT parties such as India, Brazil and Argentina).
III. Bilateral master agreements between the Agency and individual PNE supplier or consultant states and separate agreements for individual PNE projects, taking account of Agency guidelines.
IV. Memoranda of Understanding. These would be on a good faith basis and involve the creation, through individual project agreements, of rights and obligations for PNE projects which would be binding under international law.

The *Ad Hoc* Advisory Group's report was accepted by the IAEA Board of Governors on 23 September 1977 and promptly consigned to history with a decision to distribute the report widely and to 'keep the subject matter of the report under review'.[52] According to David Fischer, by the time the report was completed, most IAEA members had come to regard it as a waste of time and were merely going through the formalities.[53] The IAEA secretariat, for its part, was glad to see the end of a study into a technology which no-one, with the exception of the Soviet Union, appeared genuinely interested in pursuing.

At the 1977 session of the UN General Assembly there was some debate on the report, but only three states, Bangladesh, Nepal and India, seem to have found the report encouraging. Bangladesh and Nepal simply called for full implementation of Article V. India, bemoaning the fact that the potential benefits of PNEs were sometimes ignored, disingenuously claimed that the report had identified 'major practical applications, both existing and potential' for PNEs. It added that the participation of a large number of IAEA members in the Advisory Group was itself an indication of 'growing' interest in PNEs.[54] Czechoslovakia, no doubt reflecting Soviet views, claimed that the report was a good starting point for what it modestly described as a 'successful continuation of work in the PNE field'.[55] This, in Czechoslovakia's view, should begin with elaboration of the 'structure and content' of international agreements to make the benefits of PNEs accessible to non-nuclear weapon states. The General Assembly resolution on the study was an anticlimax, again the product of compromise between conflicting views on the future of PNEs, and again favouring the Western approach. Adopted by consensus, the resolution simply noted with appreciation the IAEA Board of Governor's decision to keep the PNE issue under review.

Strangely, the IAEA was to attempt one more piece of report writing on PNEs before the issue faded away altogether. At the request of the IAEA Director-General, Dr Ulf Ericsson of Sweden prepared a report on Extended Procedures for the Agency to Use in Responding to Requests for Services Related to PNEs.[56] Dr Ericsson's Working Paper was prepared for a meeting of the *Ad Hoc* Advisory Group on PNEs scheduled to be held in Vienna from 19–22 December 1977. But this meeting was never held and 'extended procedures' for PNE requests never developed.

If the Australian intention in proposing the comprehensive study in 1975 had been to discourage international interest in PNEs, then it worked beyond Australia's wildest dreams. According to SIPRI, the report resulted in considerable growth in scepticism about the 'technical feasibility and economic viability' of PNEs.[57] Only the Soviets appeared encouraged. The release of the report in 1977 was followed by a marked increase in the annual number of Soviet PNEs.[58] And for Argentina, Brazil and India, the PNE issue continued to have an invaluable political utility. For the rest of the world, however, the promise of PNEs dissipated quite rapidly after 1977.

Even Egypt, the only country to have come anywhere near using the IAEA PNE 'service' lost heart. In 1976, having been snubbed by the Americans, the Egyptian Ministry of Electricity and Energy asked the IAEA for assistance in studying the Qattara Depression proposal. An IAEA fact-finding mission was sent to Egypt in April-May 1976, but this failed to result in a request for specific PNE services.[59] There were, apparently, many unanswered questions about the Qattara project, including possible seismic effects on Alexandria and other communities, and the environmental effects of flooding a large area of Egypt.[60] A West German engineering firm which studied the proposal recommended that, if the project were to proceed, conventional excavation methods should be used.

Egypt was the last country ever to seek assistance from the IAEA in regard to a PNE project. Nor is the Agency aware of any requests made direct to the nuclear weapon states themselves. In 1978 all mention of PNEs disappeared from the IAEA *Annual Report*. The *Ad Hoc* Advisory Group languished and no further technical bodies on PNEs were established.

In February 1979 the IAEA Board of Governors approved new technical assistance guidelines which stated the principle that the peaceful uses of nuclear energy excludes 'research on, or development, testing or manufacturing of a nuclear explosive device'. Reflecting language incorporated into the more recent safeguards agreements concluded between the Agency and such non-NPT states as Argentina, this wording, in Lawrence Scheinman's view, removed PNEs 'from the ambit of acceptable nuclear activity'.[61]

In stark contrast to the First NPT Review Conference, there was no pressure at the Second Review Conference, held in 1980, for establishing a PNE regime. The conference, preoccupied with disarmament and the non-explosive peaceful uses of nuclear energy, ignored PNEs. While it failed to agree on a Final Document, this had nothing to do with the PNE issue. The IAEA Director-General announced inscrutably that Article V of the NPT had been 'overtaken by events' and that PNEs might entail greater risks than the benefits they would bring.[62] One of these mysterious 'events' was obviously the gloomy 1977 IAEA report. Another was the atrophy of Plowshare after 1973, which effectively removed one of the main potential suppliers of PNE services. A third was the Indian explosion of 1974, which increased exponentially the awareness of the arms control implications of PNEs. Finally, in 1977 there was the advent of the so-called trilateral negotiations between the United Kingdom, the United States and the Soviet Union on a comprehensive nuclear test ban treaty. As Lamm, a Hungarian writer, quaintly puts it:[63]

> The regulation of peaceful nuclear explosions became somewhat eclipsed within the IAEA, and the related talks were also suspended as a consequence of the emergence of the same problem at the disarmament negotiations about the comprehensive ban of [sic] nuclear weapon tests.

At the suggestion of several Western countries, Australia, Austria, Canada, Denmark, Finland, Norway and Sweden, the Second Review Conference expressed its satisfaction that the trilateral negotiators would be considering the PNE issue as an integral part of their CTB negotiations.[64] Produced by Main Committee II chaired by Australia, the text also contained the unchallenged assertion that:[65]

> Peaceful uses of nuclear explosions are at an early stage of development and no application has reached the stage at which projects can be subjected to the economic assessment judged appropriate by the *Ad Hoc* Advisory Group.

The conference noted that potential benefits from peaceful applications of nuclear explosions had not been demonstrated. It assigned a continuing watching brief on the issue to the IAEA. At the Third NPT Review Conference in 1985, the PNE issue was non-existent and the IAEA's watching brief again renewed.

A large part of the credit for the decline in international interest in PNEs after 1975 must go to Australia. From its beginnings as one of the most enthusiastic proponents of nuclear dynamite in the 1960s, Australia emerged in the 1970s as a wily and determined foe of nuclear proliferation through the PNE route. Its role at the First and Second NPT Review Conferences and its stubborn advocacy and masterful chairmanship of the *Ad Hoc* Advisory Group from 1975 to 1977 could be regarded as Australia's finest, if least known, contribution to the cause of arms control. As Alan Wilson puts it, Australia had played a crucial role in heading off a proliferation danger born of 'ignorance and deviousness'.[66]

# Notes

1 W. Epstein, *The Last Chance*, The Free Press, New York, 1976, p. 177.
2 From an interview with David Fischer, Canberra, 4 September 1986.
3 From an interview with Alan Wilson, Sydney, 7 August 1987.
4 International Atomic Energy Agency (IAEA) Document GOV/COM. 23/2.
5 M.I. Shaker, *The Nuclear Non-Proliferation Treaty: Origin and Implementation 1959–1979*, Vol. I, Oceana Publications, London, 1980, p. 430.
6 From an interview with David Fischer, Canberra, 4 September 1986.
7 D. Fischer and P. Szasz, *Safeguarding the Atom*, Stockholm International Peace Research Institute, Stockholm, 1985, p. 98.
8 IAEA Document GOV/COM. 23/2.
9 Ibid.
10 NPT Review Conference Document NPT/CONF./12.
11 IAEA Document GOV/COM. 23/2.
12 Ibid.; IAEA, *Annual Report 1975*, GC(XX)/565, IAEA, Vienna, 1976, p. 32.
13 *Report of the Ad Hoc Advisory Group on Nuclear Explosions for Peaceful Purposes*, IAEA Document GOV/1854, 26 August 1977, IAEA, Vienna, 1977, p. 6.
14 IAEA Document GOV/COM. 23/2.
15 *SIPRI Yearbook 1976*, Almqvist and Wiksell, Stockholm, 1976, p. 9.
16 These were *Guidelines for International Observation by the Agency of PNEs* and *Procedures for the Agency to Use in Responding to Requests for Services Related to PNEs* (IAEA Document GOV/COM. 23/2).
17 *SIPRI Yearbook 1975*, Taylor and Francis, London, 1975, p. 35.
18 Conference of the Committee on Disarmament, Geneva, Documents CCD/PV. 647.
19 UN General Assembly, Resolution 3261D (XXIX), 9 December 1974.
20 Threshold states are those considered to be capable of and inclined towards developing nuclear weapons. The countries most commonly perceived as being in this category are Argentina, Brazil, Israel, India, Pakistan and South Africa, although India has demonstrated its capability and Israel is widely believed already to have an untested nuclear capability.
21 Quoted in Epstein, p. 177.
22 T. Greenwood, H.A. Feiveson and T.B. Taylor, *Nuclear Proliferation: Motivations, Capabilities and Strategies for Control*, McGraw-Hill, New York, 1977, p. 68.
23 Sponsored by the US Arms Control Association and the Carnegie Endowment for International Peace.
24 H. Scoville Jr., 'Peaceful Nuclear Explosions — An Invitation to Proliferation' in A.W. Marks (ed.), *NPT: Paradoxes and Problems*, Arms Control Association, Washington DC, 1975 p. 47.
25 O. Marwah, 'Epilogue: The NPT Review Conference, Geneva, 1975' in O. Marwah and A. Shulz (eds), *Nuclear Proliferation and the Near-Nuclear Countries*, Ballinger Publishing Co., Cambridge, Mass., 1975, p. 304.
26 *SIPRI Yearbook 1976*, p. 378.
27 Ibid., p. 377.
28 Marwah, p. 307.
29 *SIPRI Yearbook 1976*, p. 377.
30 Ibid., p. 409.
31 Ibid.
32 *SIPRI Yearbook 1976*, p. 378.
33 I. Smart, 'Nonproliferation Treaty: Status and Prospects' in Marks (ed.), p. 25.
34 *SIPRI Yearbook 1978*, Taylor and Francis, London, 1978, p. 353.
35 IAEA Document GOV/COM. 1756, 5 June 1975.
36 *SIPRI Yearbook 1976*, p. 409.
37 IAEA Document GC(XIX)/544.
38 *Report of the Ad Hoc Advisory Group on Nuclear Explosions for Peaceful Purposes*, p. 4.
39 Ibid., p. 5.
40 *UN Disarmament Yearbook 1976*, United Nations, New York, 1977, pp. 109–10.
41 *Report of the Ad Hoc Advisory Group on Nuclear Explosions for Peaceful Purposes*, p. 54.
42 Ibid., p. 51.

43 Ibid., p. 37.
44 Ibid., p. 36.
45 Ibid., pp. 21–22.
46 Ibid., pp. 86–87.
47 Ibid., p. 97.
48 Ibid., p. 106
49 Ibid.
50 Eighteen-Nation Disarmament Committee Document, ENDC/PV. 293.
51 *Report of the Ad Hoc Advisory Group on Nuclear Explosions for Peaceful Purposes*, p. 115.
52 NPT Review Conference Document NPT/CONF.III/II, 28 June 1985.
53 From an interview with David Fischer, Canberra, 4 September 1986.
54 *UN Disarmament Yearbook 1977*, United Nations, New York, 1978, p. 128.
55 Ibid.
56 IAEA Document GOV/COM. 23/15.
57 *SIPRI Yearbook 1981*, Taylor and Francis, London, 1981, p. 323.
58 *SIPRI Yearbook 1980*, Taylor and Francis, London, 1980, p. xxxii.
59 *Report of the Ad Hoc Advisory Group on Nuclear Explosions for Peaceful Purposes*, p. 81; NPT Review Conference Document NPT/CONF. III/II, 28 June 1985.
60 H.S. Rowen and R. Brody, 'Nuclear Potential and Possible Contingencies' in J.A. Yager (ed.), *Nonproliferation and US Foreign Policy*, Brookings Institution Press, Washington DC, 1980, p. 212.
61 L. Scheinman, *The IAEA & World Nuclear Order*, Resources for the Future, Washington DC, 1987, p. 244.
62 *SIPRI Yearbook 1981*, p. 324.
63 V. Lamm, *The Utilisation of Nuclear Energy and International Law*, Budapest, 1984, p. 147.
64 NPT Review Conference Document NPT/CONF. II/CII/36, quoted in Lamm, p. 146.
65 *SIPRI Yearbook 1981*, p. 354.
66 From an interview with Alan Wilson, Sydney, 7 August 1987.

# 15

# Worse Than No Solution: The Peaceful Nuclear Explosions Treaty

> *... it is clear that peaceful nuclear explosions, which are indistinguishable from nuclear weapons tests, can be used by other countries as an excuse to justify nuclear weapons development... By completing a treaty allowing such explosions, the United States and [the] Soviet Union give new and unwarranted respectability to India and other nations which undoubtedly will use the new treaty to argue that their conduct of PNE programs has been vindicated.*[1]
>
> US Arms Control Association, 1976.

IN early 1974 the United States and the Soviet Union agreed to investigate the possibility of further restrictions on nuclear testing. According to the American Federation of Scientists, 'the dying Nixon Administration felt intensely interested in some kind of agreement at the Moscow Summit', scheduled for mid-1974.[2] When it became evident that agreement could not be reached on strategic arms limitations, the Administration turned its attention to nuclear testing. A team of US experts was dispatched to Moscow for technical talks. Agreement on a Threshold Test Ban Treaty (TTBT), banning nuclear weapon tests above 150 kilotons, was reached during the Moscow summit meeting and signed by President Nixon and General Secretary Brezhnev on 3 July 1974. Ratification of the treaty was delayed, however, because the two parties had agreed to negotiate a separate agreement covering peaceful nuclear explosions.

The United States' preference from the outset of the TTBT talks had been to impose the same threshold limitation on PNEs as on nuclear weapon tests. As former US Secretary of State Henry Kissinger recalls, while neither the United States nor the Soviet Union wanted to 'close off' the possibility of using nuclear dynamite, it was the US which was most anxious to avoid a loophole whereby weapons could be developed under the guise of PNEs.[3] With Plowshare winding down, the United States had in any case little to lose by imposing restrictions on nuclear dynamite.

In a complete reversal of their positions in the 1960s, it was now the Soviet Union, not the United States, which, in Glenn Seaborg's words, 'was chiefly interested in having a PNE program'.[4] While agreeing that PNEs should be regulated, the Soviets insisted that some way be found to permit PNEs over the 150 kiloton limit agreed for nuclear weapon tests. This limit had been fixed somewhat arbitrarily and with no consideration of the implications for nuclear dynamite.[5] If applied to PNEs it would have severely hampered Soviet plans for the Kama-Pechora Canal Project, which envisaged the use of excavation shots well above 150 kilotons. (The Soviets either overlooked the venting provisions of the PTBT, which rendered nuclear excavation impracticable, or they still harboured hopes that the treaty could some day be amended.)

Some 'in principle' understandings between the two sides were reached during the TTBT negotiations regarding measures to ensure that no weapons-related benefits could be obtained from PNEs after the TTBT entered into force. These included prior notification of the time and place of each detonation and, most importantly, on-site inspection.[6] According to US delegation member Warren Heckrotte, the Soviets said explicitly that they would 'allow US observers at PNE events'.[7] For Heckrotte, 'it appeared that the long-held Soviet opposition to on-site inspection was about to change'.

However negotiations on detailed PNE provisions proved too difficult to conclude in time for the summit meeting at which the TTBT was to be signed. According to Heckrotte:[8]

> It was recognised by both sides that it would take more time than was then available to develop verification procedures ensuring that PNEs did not provide opportunities for weapons testing that were denied under the TTBT's 150 kiloton yield limit.

'At Soviet insistence' according to another US negotiator, Fred C. Ikle, it was agreed that the TTBT should leave the PNE question open for 'subsequent negotiations'.[9]

The TTBT therefore provided in Article III that 'underground nuclear explosions for peaceful purposes shall be governed by an agreement which is to be negotiated and concluded by the Parties at the earliest possible time'.[10] The treaty effectively skirted around the problem of distinguishing between PNEs and weapon tests by defining the latter as those carried out at designated nuclear weapon test sites. All other nuclear tests would be considered PNEs.

In a statement to the CCD on 16 July 1974, US Ambassador Martin stated that a PNE agreement was 'integrally related to the purpose of the (TTBT), namely to prevent military testing at yields greater than 150 kilotons'.[11] In an obvious allusion to the Indian explosion conducted just two months earlier, Martin emphasised that the PNE treaty 'would not be applicable to the problem posed by the development of nuclear explosive capability by a non-nuclear weapon state'. It was, he said:[12]

> ... clearly impossible for a non-nuclear weapon state to develop a capability to conduct nuclear explosions for peaceful purposes without, in the process, acquiring a device which could be used as a nuclear weapon.

This is not how the Indians saw it. In a paper presented at the Divonne Conference on the NPT in September 1974, the Secretary of the Indian Ministry of External Affairs, Trivedi, helped justify the Indian explosion on the grounds that the TTBT specifically excluded PNEs. He failed to mention that a separate treaty regulating PNEs was envisaged.[13]

Sweden was also critical of the TTBT but for different reasons. Its ambassador, Mrs Thorsson, noted in a statement to the CCD on 30 July 1974, that the new treaty gave PNEs a 'new and distinct place':[14]

> Together with the attention already given to them under the NPT, and the recent carrying out of a peaceful nuclear explosion by India — an event which has adversely affected the efforts to stop the spread of nuclear-explosive capability — PNEs have obtained a political importance...

She suggested that the proposed PNE treaty provide for 'international observation' of all PNEs conducted by the two parties, even those carried out for their own purposes outside the framework of Article V of the NPT. According to Mrs Thorsson this would be 'a good example and would considerably ease the task of arranging international observation of PNEs made by countries still outside the NPT'.[15]

Opposition to the proposed PNE treaty also arose in the United States. Wolfgang Panofsky told the Senate Foreign Relations Committee in September 1974 that:[16]

> ... a substantial disservice to the future expectation of mankind to prevent non-proliferation of nuclear weapons has been done by attempting to provide for a separate PNE agreement in the TTBT and thereby endorsing the exaggerated significance which some nations appear to be giving to the economic and prestige importance of peaceful nuclear explosions.

Panofsky recommended strongly that efforts to negotiate a separate PNE treaty be abandoned.

When negotiations on the Peaceful Nuclear Explosions Treaty began in Moscow on 7 October 1974, it was not clear what the final outcome would be. In particular no-one knew whether a higher threshold for PNEs than for nuclear weapon tests would be possible without allowing weapons-related benefits to be derived from such explosions.[17] The negotiators laboured for over a year in a vain attempt to solve this problem, which lay at the heart of the PNE issue. According to Warren Heckrotte, US attempts to 'ferret out' the intent of the Soviet statement during the TTBT negotiations to the effect that observers would be permitted at PNE projects, 'did not meet with prompt success'.[18] Both sides eventually concluded that the problem was, in Glenn Seaborg's words, 'too difficult'.[19] As ACDA Director Warnke explained:[20]

> The negotiations failed to disclose any means for carrying out individual nuclear explosions for peaceful purposes above 150 kilotons without raising the risk of making available military benefits otherwise precluded by the threshold ban on nuclear weapon tests.

The solution eventually reached was to limit individual PNEs to 150 kilotons, the same as for nuclear weapon tests under the TTBT, since the size of an explosion was easier to verify than whether it was a PNE or a weapon test.[21] Under-Secretary of State Habib testified to the Senate Foreign Relations Committee that:[22]

> The central problem of not allowing weapon-related benefits otherwise precluded by the TTBT was solved by ensuring that no individual explosion would have a yield exceeding 150 kilotons. In other words the thresholds of the two treaties are identical.

As there was no restriction under the TTBT on the number of weapon tests conducted (although the treaty did commit the parties to keeping the number of tests to a minimum) there could now be no advantage (and indeed a lot of trouble involved) in either side attempting to use PNEs to obtain nuclear weapon testing experience.

The possibility of permitting individual PNEs over 150 kilotons at some future stage, when verification might be feasible, was set aside in Article III.3 of the treaty, 'to be considered by the Parties at an appropriate time to be agreed'. (See Annex 4 for full treaty text.) The Soviet Union later said, rather cheekily, that it considered this provision to be of 'fundamental' importance, since it did not 'make the limit set for the yield of nuclear explosions for peaceful purposes dependent on that of nuclear weapon test explosions'.[23] As Jozef Goldblat notes, the United States was unlikely ever to agree to a higher threshold for PNEs than for weapon tests.[24]

More importantly, the Soviets insisted that, in order to allow their proposed large-scale excavation projects to proceed, multiple detonations, involving individual PNEs with yields up to 150 kilotons each, should be permitted. Multiple detonations were later defined by the treaty as those 'for which the time interval between successive individual explosions does not exceed five seconds' and the devices for which were emplanted not more than 40 kilometres apart.[25] The United States was opposed, arguing that there was no practical way to determine the yields of the individual detonations comprising such explosions. The usual method of detecting nuclear tests, long-distance seismic (teleseismic) monitoring, would clearly be unable to accurately separate out the individual seismic signals.

Heckrotte reports 'a sharp collision' over this matter that lasted some weeks.[26] Finally the head of the Soviet delegation asked whether the US would agree 'in principle' to group explosions if a satisfactory verification technique was available. The US responded positively. Heckrotte notes:[27]

> With that response in hand, and not until then, the Soviets proceeded to describe a technique by which, they claimed, the individual yields would be measured. This involved placing instruments to measure ground motion at appropriate locations in the vicinity of the explosions. We questioned them on how we could have any confidence in such measurements. They responded that US personnel would be there to make the measurements. The Soviets had finally agreed 'in principle' that US observers or inspectors could be present at some PNE events.

Heckrotte observes however that much work remained to translate this agreement in principle into explicit provisions. Agreement, according to Henry Kissinger, only came after 'prolonged wrangling'.[28] Heckrotte describes sharp disagreements between the two delegations over at least two dozen concrete issues, ranging from terminology (the Soviets thought the term 'inspectors' carried invidious implications) to technology. Heckrotte recounts a particularly 'deep and serious' stand-off over the question of inspection of monitoring equipment.[29] According to him, haggling continued on the last issue to be settled (which he does not name) right up until a few hours before the final departure from Moscow of the US delegation.[30]

The yield measurement technique eventually agreed upon was a direct, hydrodynamic technique known by the Americans as SLIFER and by the Soviets as MIS. It involves placing an electrical cable directly into the PNE detonation hole or into a nearby drill hole. When the explosive is fired, the lower end of the cable is systematically crushed by the expanding shockwave. This can be measured by electronic equipment at the surface as a function of time, enabling yield to be deduced.[31]

Henry Kissinger claims that the reason why the United States insisted on vigorous on-site inspection for the PNET was that 'the locale of PNEs would be impossible to specify by international agreement in advance'.[32] This explanation is unconvincing. For one thing there was nothing to stop the treaty requiring advance notification of the 'locale' of PNEs prior to each detonation, and in fact this was a requirement the treaty eventually included.[33] Second, it would not have mattered where single PNE detonations up to 150 kilotons were conducted, or whether they were weapon tests. No military benefits additional to those gained from nuclear weapon tests up to 150 kilotons conducted at the usual nuclear weapon test sites would be derived from such PNEs. Verification would have needed to be no more exacting than for the TTBT, where the chief problems were to tell whether detonations at designated weapon test sites were under the 150 kiloton limit and whether suspicious 'events' over that limit anywhere else were nuclear tests or earthquakes.

The real reason for the exacting verification requirements of the PNET was not the locale of PNE detonations, but the possibility that multiple 150 kiloton devices, when detonated simultaneously, might appear to be one large detonation. The Energy Research and Development Administration confirmed in 1977 that 'in the case of the PNET, on-site monitoring was only required to establish that the yield of individual explosions did not exceed 150kt'.[34]

Further evidence lies in the fact that the treaty provides for compulsory on-site inspection only for multiple explosions whose aggregate yield is likely to exceed 150 kilotons.[35] For multiple explosions totalling between 100 and 150 kilotons, on-site inspection is optional, to be determined by consultations between the parties.

A more subtle motive for US insistence on rigorous on-site inspection provisions, and indeed on such detailed verification provisions all round, was possibly a desire to make the use of PNEs so complicated that the Soviets would no longer be bothered with them. Since the United States was itself no longer interested in PNEs, all the restrictions envisaged in the PNET would in practice fall solely on the Soviets. Paul Warnke hinted at such a motive when he said that the Soviets would 'either have to permit these on-site inspection procedures or forego the kinds of PNE uses which would bring these on-site procedures into effect'. He added, 'we would prefer the latter because we would like to see an end to all nuclear explosions'.[36]

A less conspiratorial view is that the US insisted on such exacting verification because, as Warren Heckrotte notes, without a PNE program of its own, the United States was 'concerned only about the effectiveness of PNE verification and not about its effects on our security'.[37] He points to an irony that had always characterised negotiations on the PNE issue between the superpowers:[38]

> We seek verification to have some assurance that the Soviets don't cheat on the treaty. The Soviets, for their part, seek verification to have some assurance that we don't cheat with respect to the purpose of the tools of verification.

The PNET's verification provisions, set out in a protocol to the treaty, are indeed exacting. As well as authorising the use of so-called National Technical Means (NTMs), such as long-distance seismic and satellite monitoring, the protocol requires each party to provide detailed information on *all* intended PNE detonations.[39] The detail of the data required increases with the size of the explosion and includes: the purpose, location, date and aggregate yield of the planned explosion; the number of explosives; the planned yield of each explosive and its location relative to other explosives; and the emplacement depth and time interval between each explosive in a group. Also required is a detailed description of the project and a report on the geological and geophysical characteristics of the explosion site which could influence yield determination.

The party conducting the explosion is required to inform the other party, not later than two days beforehand, of the planned time of each explosion with a precision of one second. Not later than ninety days after each explosion, the conducting party must provide confirmation of the actual time and yield of the explosion and any other data provided prior to the detonation. Any variations in the predicted and actual data are to be explained.

For all multiple explosions over 150 kilotons the treaty provides for rigorous on-site inspection.[40] Observers ('designated personnel') are permitted to: verify information provided by the conducting party, including satisfying themselves that an actual peaceful use is involved, observe the emplacement of the explosives, thereafter observe the emplacement hole until detonation and finally, observe the detonation itself. For any group explosion with a planned aggregate yield exceeding 500 kilotons, observers could, in addition, install a local seismic network to ascertain that no undeclared explosions were being conducted along with the announced ones.[41] Multiple explosions exceeding 1.5 megatons were banned altogether.[42]

On-site inspection procedures are spelled out in detail, including the number of observers permitted, the geographical extent of their access, and the provision of certain information, such as maps of the area.[43] Certain privileges and immunities for observers (and their equipment) and standards of housing and work facilities are specified, to give the inspectors freedom to assert their rights and carry out their functions effectively. The treaty, however, ignores the Swedish proposal for 'international observation'. The observers mentioned in the treaty are Americans and Russians. Finally the treaty provides for a Joint Consultative Commission to be established to handle data exchange and compliance problems.[44]

An Agreed Statement of 13 May 1976 which accompanies the PNET, specifies that a 'peaceful application' of an underground nuclear explosion does not include the *developmental* testing of any nuclear explosive, including PNEs. Such testing must be carried out at a nuclear weapon test site specified by the TTBT. Hence the PNET only permits and regulates the actual use of PNEs in the field, not the development and testing of PNE devices. Not only does the PNET thereby attempt to preclude weapons testing during PNE projects, but it also precludes PNE developmental testing in the field. (As Goldblat points out, such a distinction may be difficult to draw in practice.[45] For example, an attempted application which fails could be construed as a developmental test rather than a peaceful use.)

As well as PNEs carried out by the Soviets and the Americans on their own territory, the treaty also covers any PNE conducted by the two parties elsewhere. It therefore adds a new set of restrictions to the regulation of PNEs envisaged in Article V of the NPT. In particular, by limiting the size of individual PNEs to 150 kilotons, it constrains the way in which large-scale PNE projects anywhere can be tackled. The Cape Keraudren project, for instance, which envisaged multiple blasts of 200 kilotons each, would be illegal under the PNET.

The PNET was the first treaty to deal specifically with the regulation of PNEs. Until the conclusion of the Intermediate-Range Nuclear Forces (INF) Agreement in 1987, it was also the most complicated arms control agreement ever negotiated. Writing in 1983, Warren Heckrotte noted that 'no other arms control treaty approaches it in substance and technical provisions'.[46] It was also historic because of Soviet agreement to on-site inspection on its territory, thereby establishing a valuable precedent for future arms control agreements.[47] As former ACDA Director Fred Ikle noted, 'for the first time ever, we have obtained Soviet agreement to on-site inspection, thus the objection has been broken in the precedent'.[48]

Not all observers were impressed by the new treaty. Herbert Scoville, former ACDA Assistant Director for Science and Technology, argued that by endorsing multiple explosions for excavation projects the PNET gave an open invitation to states to violate the Partial Test Ban Treaty. Scoville claimed that Article I of the PNET protocol, by permitting the emplacement of multiple explosives at a much shallower depth than was necessary to preclude venting, implied acceptance of nuclear excavation projects.[49]

Scoville also argued that the on-site inspection provisions were no breakthrough:[50]

> The complex and highly specialized procedures for inviting designated observers to a predetermined location to witness a preplanned explosion bears little relationship to the on-site inspections sought in the late 1950s and 1960s in connection with test ban negotiations. The principle that US negotiators then sought to establish involved the right to dispatch US or Soviet inspection teams, upon acquisition of suspicious information suggestive of nuclear testing, to any location, anywhere in the USSR or the United States, at any time.

Scoville sarcastically declared that he looked upon the PNET's on-site inspection provisions 'with great amazement, wonder, and admiration for those who must have had to negotiate those kind of arrangements with the Soviet Union'. He added: 'My only regret is that this great talent was wasted on such an unimportant project'.[51]

The PNET indeed has no precedential value for dealing with PNEs under a CTB, since it provides no mechanism for determining whether a 150 kiloton PNE is a nuclear weapon test, or indeed a developmental test (as opposed to an actual use) of a PNE device. There is, for instance, no provision in the treaty for device inspection or provision of blueprints. Article V.5 of the protocol specifically rules this out:

> Designated personnel [the observers] shall not have or seek access by physical, visual or technical means to the interior of the canister containing the explosive, to documentary or other information descriptive of the design of an explosive nor to equipment for control and firing of explosives.

During Senate hearings into the PNET in 1977 Senator John Glenn (Democrat, Ohio) asked, 'even with on-site inspection... would it be possible, say, to test a device for a delta class ballistic missile (warhead) and call it a PNE and explode it which would be in effect a test by another name?' ACDA Director Warnke replied:[52]

> I think there is no question, Senator Glenn, that that could be done, but, of course, there would be no point in doing it under these treaties because the limit is the same for both. So it does not matter what they call it. Now, when you have a comprehensive test ban that would be a problem if you had a PNE exception.

Despite this, the PNET's Agreed Statement half-heartedly attempted to prevent weapons-related information being obtained from any PNE, by declaring that 'associating test facilities, instrumentation or procedures related only to testing of nuclear weapons or their effects with any explosion carried out in accordance with the Treaty does not constitute a "peaceful application"'. The treaty made no specific provision for verifying that such instrumentation or procedures were not used. Presumably such illegal activities were intended to be detected as a by-product of the detailed verification measures for determining the yield of each PNE detonation. In any case, as Warnke pointed out, violations of this sort would be of little consequence, since each party can obtain all the weapons-related information it needs from its nuclear weapon tests. Providing PNEs are detonated below 150 kilotons, there is no advantage in monitoring them with diagnostic equipment to obtain military-related benefits.

The PNET would only have provided a model for permitting PNEs under a CTB regime if individual PNEs over the 150 kiloton limit had been permitted and an appropriate verification procedure devised. It is unlikely that such a method could have been found. As the CIA testified in 1977:[53]

> If PNEs are permitted under a CTB, we have not yet identified a way, in terms of assured verification provisions, of assuring that weapon-related benefits are not obtained from such explosions regardless of their yield.

Apart from its verification provisions, the PNET was also unprecedented in committing the United States and the Soviet Union to co-operate in the PNE field and to keep the IAEA informed of this activity. According to Warren Heckrotte, this resulted from Soviet attempts to water down stringent on-site inspection provisions with so-called 'co-operative arrangements'.[54] Heckrotte records that 'considerable discussion and time were devoted to developing a text that acknowledged the role of cooperation without compromising the integrity of the verification procedures'.[55] Ironically this came at a time when Plowshare, to all intents and purposes, no longer existed.

The PNET parties also undertook to 'promote the development of the international agreement or agreements and procedures' provided for in Article V of the NPT and to assist the IAEA in this regard.[56] This was a cynical gesture since both superpowers had long since concluded that interest in nuclear dynamite should not be excited by fulfilling the provisions of Article V. As Paul Warnke testified to the Senate Foreign Relations Committee in 1977, 'the logical extension' of a comprehensive test ban treaty, 'a regime where the threshold for nuclear weapon tests would be zero', was not an international PNE service, but 'a ban on PNEs as well'.[57]

From the perspective of the non-nuclear weapon states the two parties added insult to injury by also agreeing not to carry out PNEs in any country unwilling to accept the international observation and procedures contemplated in Article V of the NPT and the PNET protocol.[58] The two potential PNE supplier states thereby locked any possible future PNE regime — whose establishment both now opposed — into a web of regulations that they had devised without consultation with their potential customers. This Soviet/American duopoly would, as a result, be even better placed than before to call the tune on the conditions under which Article V of the NPT would, if ever, be implemented.

The PNET took eighteen months of unexpectedly difficult negotiations, over six separate sessions, to conclude.[59] Yet much of the impact of the treaty remains theoretical. Although it was signed by President Ford and General Secretary Brezhnev on 28 May 1976,[60] the Ford Administration initially declined to send either the TTBT or the PNET to the Senate for advice and consent, owing to anticipated opposition by conservatives.[61] The two treaties were finally submitted in July 1976. During hearings in the Senate Foreign Relations Committee, both were decried as a sham by those favouring a CTB. According to these critics the Nixon Administration should have set its sights higher and strived for a treaty banning all nuclear tests, including PNEs — even if Soviet interest in PNEs would have made the negotiation of such a treaty impossible.[62] As the Arms Control Association scathingly put it:[63]

> Bilateral actions by the two superpowers affect the world's future security and well-being, and cosmetic 'arms control' agreements drawn up solely for their mutual convenience, to keep all possible options open, are not good enough.

As for nuclear dynamite itself, the Association had this to say:[64]

> The United States has quite properly, but belatedly, all but abandoned efforts to develop nuclear explosives for peaceful purposes. Years of experimentation and millions of dollars in research into ways of using nuclear explosives for excavation, underground engineering, and electric power generation have all led to the conclusion that PNEs cannot compete with conventional means of accomplishing the same objectives, when all economic, environmental, and political considerations are taken into account.

In view of such criticisms and the imminent Presidential election, the Committee delayed action on the treaties. The Democratic presidential candidate, Jimmy Carter, had proposed a five-year US/Soviet agreement prohibiting all nuclear explosions, including PNEs, which would obviate the need for the PNET.[65]

Once elected, Carter decided not to press for the ratification of the Threshold and PNE Treaties. The President made scathing remarks about them, reiterating his proposal for a CTB. As ACDA Director Warnke stated: 'Our objective is a comprehensive test ban that will end all nuclear explosions, whether for peaceful purposes or for weapons test purposes'.[66] While Carter conceded that the Soviets might be permitted some explosions for projects such as river diversions, he hoped to persuade them that 'whatever the benefits [of PNEs] may be, they are far offset by the risks and by the proliferation consequences of continuing with the nuclear program'.[67]

After Soviet rejection of his attempt to alter the Vladivostok agreement on strategic arms, President Carter decided in 1977 to send the TTBT and PNET back to the Senate for advice on ratification. According to Herbert Scoville the reasons were 'purely political' and not based on any arms control criteria.[68] Nonetheless, criticism of the treaties was just as fierce as during the first round of hearings. Among the influential critics of the PNET to appear before the Senate Foreign Relations Committee were George Rathjens, Chairman of the Federation of American Scientists; Herbert Scoville Jr., Vice-President of the Arms Control Association; Professor Franklin Long of Cornell University; and Wolfgang Panofsky.

A major focus of criticism was the possibility that near-nuclear states like Brazil and India could see the PNET as 'legitimising' PNEs. Herbert Scoville testified that Indian representatives were already arguing that the existence of the PNET meant that a future CTB should not ban PNEs.[69] Indian defence commentator, K. Subrahmanyam, for instance, was claiming that:[70]

> Neither the United States nor the Soviet Union are in a position to contest India's contention on peaceful nuclear explosion[s] as long as they subscribe to the concept of PNE and are parties to the Peaceful Nuclear Explosions Treaty.

For Herbert Scoville, the only solution to the PNE problem was to include in a comprehensive test ban treaty a five-year moratorium on PNEs, after which the position could be reconsidered. In his words, 'this moratorium would provide an opportunity to study [the problem] further, would not foreclose peaceful nuclear explosions forever, and nevertheless would permit going ahead with the urgently needed CTBT'.[71] Professor Long agreed, but suggested a ten-year moratorium.[72]

Senator Frank Church (Democrat, Idaho) quizzed ACDA's Paul Warnke on whether or not the PNET was a bad precedent:[73]

> *Senator Church:* By agreeing that peaceful tests may be permissible, even though they are underground, and even though they may not exceed 150kt, is that a precedent for other nations who want to develop nuclear weapons to say 'We are conducting peaceful tests, they are being conducted underground and since the United States and the Soviet Union have reserved this right we are still conforming with your established practice formalised by your treaty?'
>
> *Mr Warnke:* In our opinion, it would be far better to have a comprehensive test ban which precluded any nuclear explosions. We believe, however, that this treaty would be an improvement on the existing situation under which there is no limit on the kilotonnage, or megatonnage for that matter, of PNEs. This does settle it. As a result, I think it interferes significantly with some potential uses of PNEs, and as a consequence I believe it furthers the

> non-proliferation cause. I would point out that under a regime under which there was no restraint on PNEs the Indian Government went ahead and developed what they termed a peaceful nuclear explosion.

This was extremely misleading. For a start the Indian explosion was all of 12 kilotons, way below the limits set by the TTBT and the PNET. Second, the PNET was a bilateral treaty not open to accession by India even if it had wanted to. As Warnke himself later said, 'As far as the non-nuclear weapon states are concerned, we wouldn't want them, of course, to be parties to these treaties because it would appear to give them permission to go ahead both with nuclear weapons up to 150kt and with PNEs'.[74] Senator Church said he saw the PNET as an unfortunate precedent for a full-scale CTB in its treatment of PNEs as a separate, legitimate testing activity to be separately catered for.[75] Under Secretary of State Habib interjected, saying that the PNET was helpful to the negotiation of a CTB, because it demonstrated the principle that weapons benefits could be obtained from PNEs.[76]

In the closing hour of the hearings, Senator George McGovern (Democrat, South Dakota) asked Wolfgang Panofsky whether there was merit in the United States publicly foreswearing PNEs and inviting others like the Soviet Union to do likewise. Panofsky replied:[77]

> ... it is a matter of a political question whether it would be productive to challenge the Soviets in public or in private on this particular point since it is the Soviets who are at this time the chief advocates of peaceful nuclear explosions.

In fact, from the beginning, the PNET was protected by something of a conspiracy of silence between the United States and the Soviet Union. While in such forums as the CCD and the IAEA the United States expressed scepticism about the potential of PNEs and highlighted their arms control and proliferation drawbacks, in the bilateral context it conceded to Soviet insistence on the right to conduct PNEs as the price to be paid for the TTBT, itself a sop to worldwide demands for limitations on nuclear testing.

Despite the harsh criticism of both the TTBT and the PNET, some critics recognised however that the Carter Administration had been unfairly lumbered with the treaties. They conceded that the treaties should perhaps be ratified solely to avoid giving the Soviets an impression of American 'bad faith' in negotiating them in the first place, and to give a fillip to US/Soviet relations under a new Administration. Such critics proposed, however, that ratification proceed only on the clear understanding with the Soviets that the treaties represented simply a first step towards a CTB. The possibility of ratification by President Carter ended, however, when he withdrew the two treaties from Senate consideration after a threat by some senators to require additional verification provisions for the TTBT.[78]

The TTBT and the PNET remain unratified to this day, although both the Soviet Union and the United States have undertaken to abide by the

150 kiloton limit for both weapon tests and PNEs.[79] While the verification provisions of the treaties have not been implemented, the United States has trained and prepared on-site investigation teams for verifying the PNET.[80] In the absence of ratification, the Soviet Union has never invited them to observe any of its numerous PNEs. Even if the treaty had been ratified, however, none of the Soviet PNEs conducted since 1976 have been large enough to require the on-site inspection provisions of the treaty to be invoked.

In contrast to the Carter Administration, the main concern of the Reagan Administration was not the worthlessness of the TTBT and PNET overall, but rather with the verification provisions of the TTBT. The Administration accused the Soviets of violating the 150 kiloton threshold and said it would not ratify the treaty until improvements were made. President Reagan stated in a message to the US Senate on 13 January 1987 that, 'as written, the TTBT relies solely on teleseismic detection and yield measurement systems and on inadequate, and unverifiable data exchange'.[81] He made no such accusations about the verifiability of the PNET. Nor did he accuse the Soviets of violating the PNET.

Nevertheless, the PNET's close link with the TTBT has meant that its ratification has been held hostage to the resolution of the dispute over its companion treaty. It has also meant that the PNET has been increasingly lumped in with the TTBT as requiring (unspecified) improvements to its verification provisions.

Ironically, the main improvement envisaged for the TTBT is the use of an improved version of the direct, hydrodynamic yield measurement technique provided for in the PNET. In July 1982 President Reagan invited the Soviets to visit Nevada to make yield measurements using this method, now known as CORRTEX.[82] The Soviets declined, arguing that improved verification procedures would only help the United States improve its nuclear weapon testing procedures.

In July 1986 the two sides began talks on nuclear testing in Geneva, during which the United States proposed improvements to TTBT verification, while the Soviet Union advocated a comprehensive test ban treaty.[83] On 10 October 1986 at the Reykjavik summit with General Secretary Gorbachev, President Reagan announced that he would ask the US Senate to recommence ratification hearings on the TTBT and PNET, but reserved the right to withhold ratification until the verification provisions for both treaties were improved.[84]

At the hearings, conducted by the Senate Foreign Relations Committee on 13 and 15 January 1987, opinion was divided on whether improvements were necessary to the TTBT. By contrast there was little criticism of the adequacy of the PNET. According to Roger Batzel, director of the Livermore Laboratory, the verifiability of the PNET had not been an issue 'in practice' since the estimated yield of Soviet PNEs since 1976 had been significantly less than 150 kilotons.[85] In Batzel's view the data exchange envisaged in the PNET would provide valuable information about each Soviet PNE, which would 'significantly improve our understanding of the geophysical properties of the Soviet Union' and enable the United States 'to focus various National Technical Means (NTMs) on these events to study such aspects as detection thresholds and seismic signal propagation in the USSR'.[86] In addition, the PNET's Joint Consultative Commission, Batzel testified, could be a valuable channel for exchange of technical data and compliance questions for the TTBT as well as the PNET. In other words, ratification of the PNET as it stands would assist materially in verifying compliance with its companion treaty.

Batzel stated that the primary improvements that could be suggested for the PNET consisted of implementing the existing provisions, but at lower thresholds than envisaged in the treaty.[87] He did not say what thresholds he had in mind. The dearth of suggestions for the improvement of the PNET suggests that its original negotiators did their job well.

Among the supporters of ratification of the two treaties as they stood were Paul Warnke, former CIA Director William Colby, and Wolfgang Panofsky. With the world-wide decline in interest in PNEs it seems that the arms control community no longer sees the PNET as a threat to its endeavours. Rather, they have apparently concluded that the treaty, while of limited practical value, is a useful step towards the goal of a comprehensive ban on nuclear testing. On 19 February 1987 the Senate Foreign Relations Committee recommended ratification of the TTBT and PNET as negotiated.

In September 1987 President Reagan and Soviet Foreign Minister Shevadnardze indicated that the United States and the Soviet Union had agreed to allow each other to monitor a nuclear test on each other's territory to improve verification techniques for the TTBT and PNET. The Soviets agreed to the CORRTEX method being used by the Americans, while the Soviets would use their own techniques.[88] These tests, known as the Joint Verification Experiment (JVE), were successfully conducted on 17 August 1988 at the Nevada test site and on 14 September 1988 at the Soviet test site at East Kazakhstan. Earlier in 1988, on 31 March, the United States presented to the Soviets a draft verification protocol for the PNET which would permit the use of CORRTEX and similar methods to verify the treaty.[89] By June 1988, even before the JVE, the two sides were reportedly 'close to agreement' on the new protocol.[90]

Still the PNET remains unratified more than a decade after its signature. In Jozef Goldblat's view this is regrettable, since it undermines confidence in arms control.[91] It remains to be seen whether current developments will end the hiatus.

# Notes

1 US Congress, Senate Foreign Relations Committee, Hearings on the TTBT/PNET, 28 July, 3 August, 8 and 15 September 1977, US Government Printing Office, Washington DC, 1977, pp. 112–113.
2 Ibid., p. 105.
3 H. Kissinger, *Years of Upheaval*, Little, Brown and Co., Boston, 1982, p. 1167.
4 G. T. Seaborg and B. S. Loeb, *Kennedy, Khrushchev and the Test Ban*, University of California Press, Berkeley, 1981, p. 295.
5 Raymond Garthoff records that the figure of 150 came about when Kissinger, during the Moscow talks, asked for his brief for the TTBT negotiations and was told there was none. Someone recalled that 100 kilotons would be a good round number and it was so agreed. The next day Nixon proposed this to the Soviets. Before they could respond, Kissinger corrected him to a 150 kiloton limit, having called Washington overnight to check and learned that Secretary of Defense Schlesinger insisted on not going below 150 kilotons. The Soviets agreed to this limit (R. L. Garthoff, *Detente and Confrontation*, The Brookings Institution, Washington DC, 1985, p. 427). ACDA was not consulted and had no choice but to accept the 150 kiloton limit (D. L. Clarke, *Politics of Arms Control: The Role and Effectiveness of the US Arms Control and Disarmament Agency*, The Free Press, New York, 1979, p. 96).
6 Conference of the Committee on Disarmament, Geneva, Document CCD/PV. 643.
7 W. Heckrotte, 'Negotiating with the Soviets', *Energy and Technology Review*, Lawrence Livermore National Laboratory, May, 1983, p. 12.
8 Ibid.
9 M. A. Guhin, *Nuclear Paradox: Security Risks of the Peaceful Atom*, American Enterprise Institute for Public Policy Research, Washington DC, 1976, p. 66
10 *Arms Control and Disarmament Agreements*, US Arms Control and Disarmament Agency, Washington DC, 1982, p. 168.
11 CCD/PV. 643.
12 Ibid.
13 V. C. Trivedi, 'India's Approach Towards Nuclear Energy and Non-Proliferation of Nuclear Weapons' in A. W. Marks (ed.), *NPT: Paradoxes and Problems*, Arms Control Association, Washington DC, 1975, p. 44.
14 CCD/PV. 647.
15 Ibid.
16 Statement by Wolfgang Panofsky to US Senate Foreign Relations Committee, 12 September 1974, reprinted in US Congress, Senate Foreign Relations Committee, Hearings on the TTBT/PNET, 13 and 15 January 1987, US Government Printing Office, Washington DC, 1987, pp. 271–2.
17 According to a statement by Wolfgang Panofsky to the US Senate Foreign Relations Committee, 15 September 1977, reprinted in US Congress, Senate Foreign Relations Committee, Hearings on the TTBT/PNET, 1987, p. 255.
18 Heckrotte, p. 13.
19 Seaborg and Loeb, *Kennedy, Khrushchev and the Test Ban*, p. 296.
20 US Congress, Senate Foreign Relations Committee, Hearings on the TTBT/PNET, 1977, p. 8.
21 According to Dr Siegfried S. Hecker of the Los Alamos Laboratory, it was Los Alamos that 'originated' this proposal and which provided the technical input that guided the phase of the negotiations that dealt with on-site measurement of PNE yeidls (US Congress, Senate Foreign Relations Committee, Hearings on the TTBT/PNET, 1987, p. 373).
22 US Congress, Senate Foreign Relations Committee, Hearings on the TTBT/PNET, 1977, p. 5.
23 Quoted in J. Goldblat, 'The Nuclear Explosion Limitation Treaties' in J. Goldblat and D. Cox (eds), *Nuclear Weapon Tests: Prohibition or Limitation?*, SIPRI, Oxford University Press, Oxford, 1988, p. 134.
24 Ibid.
25 Article II, Peaceful Nuclear Explosions Treaty.
26 Heckrotte, p. 13.
27 Ibid., pp. 13–14.
28 Kissinger, p. 1167.

29 Heckrotte, p. 15.
30 Ibid., p. 16.
31 Heckrotte, p. 15; US Congress, Senate Foreign Relations Committee, Hearings on the TTBT/PNET, 1987, p. 16.
32 Kissinger, p. 1167.
33 Article II, Protocol to the Peaceful Nuclear Explosions Treaty.
34 US Congress, Senate Foreign Relations Committee, Hearings on the TTBT/PNET, 1977, pp. 60–61.
35 Article III, Protocol to the Peaceful Nuclear Explosions Treaty.
36 US Congress, Senate Foreign Relations Committee, Hearings on the TTBT/PNET, 1977, p. 15.
37 Heckrotte, p. 19.
38 Ibid., p. 15.
39 Article II, Protocol to the Peaceful Nuclear Explosions Treaty.
40 Ibid.
41 Article II.4, Protocol to the Peaceful Nuclear Explosions Treaty.
42 Article III, Peaceful Nuclear Explosions Treaty.
43 Articles III–VII, Peaceful Nuclear Explosions Treaty.
44 Article V, Peaceful Nuclear Explosions Treaty.
45 Goldblat, p. 135.
46 Heckrotte, p. 12.
47 The Antarctic Treaty provides for on-site inspection of Soviet Antarctic bases. Only a few months after signing the PNET, the Soviet Union expressed its willingness to seek an agreement prohibiting all nuclear weapon tests, where on-site 'clarification of relevant circumstances' could be envisaged on a 'voluntary basis' (*SIPRI Yearbook 1977*, Almqvist and Wiskell, Stockholm, 1977, p. 355). Subsequently, the Soviet draft CTB treaty of 22 November 1976 contained, for the first time, a provision for on-site inspection of suspicious events, even though this was not compulsory. Today the concept of on-site inspection is accepted by the Soviets as a natural and essential part of disarmament and arms control agreements.
48 US Congress, Senate Foreign Relations Committee, Hearings on the TTBT/PNET, 1977, p. 124.
49 Ibid., p. 107.
50 Ibid., p. 111.
51 Ibid., p. 122.
52 Ibid., p. 17.
53 Ibid., p. 30.
54 Heckrotte, p. 14.
55 Ibid.
56 Article VII.1, Peaceful Nuclear Explosions Treaty.
57 US Congress, Senate Foreign Relations Committee, Hearings on the TTBT/PNET, 1977, p. 8.
58 Article VII.2, Peaceful Nuclear Explosions Treaty.
59 *Arms Control & Disarmament Agreements*, p. 171.
60 The White House postponed the original signing ceremony, scheduled for 12 May, in a 'panic reaction' to Ford's loss of the Nebraska primary to Ronald Reagan (R. L. Garthoff, *Detente and Confrontation*, The Brookings Institution, Washington DC, 1985, p. 550).
61 Garthoff, p. 550.
62 US Congress, Senate Foreign Relations Committee, Hearings on the TTBT/PNET, 1977, p. 103.
63 Ibid., p. 113.
64 Ibid.
65 *SIPRI Yearbook 1977*, p. 412.
66 US Congress, Senate Foreign Relations Committee, Hearings on the TTBT/PNET, 1977, p. 15.
67 Ibid.
68 Ibid., p. 119.
69 Ibid., p. 107.
70 K. Subrahmanyam, *Nuclear Myths and Realities: India's Dilemma*, ABC Publishing House, New Delhi, 1981, p. 27.

71 US Congress, Senate Foreign Relations Committee, Hearings on the TTBT/PNET, 1977, p. 109.
72 Ibid., p. 137.
73 Ibid., pp. 12–13.
74 Ibid., p. 19.
75 Ibid., p. 13.
76 Ibid.
77 Ibid., p. 156.
78 The *Arms Control Reporter*, Institute for Defense & Disarmament Studies, Brookline, Mass., 1987, p. 605.A.1.
79 Ibid.
80 US Congress, Senate Foreign Relations Committee, Hearings on the TTBT/PNET, 1977, p. 61.
81 The *Arms Control Reporter*, p. 605.D.8.
82 Ibid., p. 605.A.1.
83 *UN Disarmament Yearbook 1986*, United Nations, New York, 1987, p. 140.
84 Ibid., p. 605.A.2; *Arms Control Chronicle*, Canadian Centre for Arms Control and Disarmament, Ottawa, 19, January-February 1987, p. 15; US Congress, Senate Foreign Relations Committee, Hearings on the TTBT/PNET, 1987, p. 6.
85 US Congress, Senate Foreign Relations Committee, Hearings on the TTBT/PNET, 1987, p. 223.
86 Ibid., p. 224.
87 Ibid.
88 *Arms Control Chronicle*, 21, August 1987, p. 16.
89 *Arms Control Update*, No. 5, April 1988, p. 3.
90 *Arms Control Chronicle*, 27, August 1988, p. 13.
91 Goldblat, p. 139.

# 16

# PNEs and a Comprehensive Test Ban Treaty

*There is no valid reason to extend the hope that the negotiating effort which failed to achieve its goal of distinguishing peaceful from military explosions in the context of the TTBT will succeed in the context of a treaty in which military explosions are banned entirely.*[1]

Wolfgang Panofsky, 1977

A COMPREHENSIVE nuclear test ban treaty or CTB has long been regarded as one of the most effective means of halting the nuclear arms race. It has also been recognised that peaceful nuclear explosions constitute one of the major stumbling blocks to the achievement of such a treaty. The problem is: how to permit the development and use of PNEs, while at the same time banning *all* nuclear weapon tests by *all* states for *all* time. This is a much more difficult challenge than negotiators faced in dealing with PNEs under the NPT or the PNET.

Under the NPT the problem was one of preventing non-nuclear weapon states from acquiring and using their own PNEs, just as they were to be prevented from acquiring any other type of nuclear device. The solution was to ban the acquisition of *all* types of nuclear devices by such states, while assuring them (fallaciously as it turned out) that the benefits of PNEs could be obtained through an international PNE service provided by the nuclear weapon powers. Under the PNET, once the 150 kiloton limit was agreed to, the problem was simply to ensure that PNEs conducted by the superpowers stayed under this limit. Under a comprehensive test ban, the trick is to prevent the nuclear weapon states themselves from deriving weapons benefits from PNE detonations. It is the most fundamental of the many challenges posed to arms control by nuclear dynamite.

Following the successful conclusion of the PTBT in 1963, the issue of a comprehensive test ban slipped from the international agenda. In the desultory discussions on the issue that did occur, the PNE problem was largely ignored. This was not only because the verification issue dominated the debate, but because it seemed that the PNE problem was being resolved in the NPT negotiations. Only Sweden during this period highlighted the link between the PNE issue and a CTB. At its most extreme, the Swedish position envisaged that the NPT should ignore the PNE problem altogether, leaving it to be entirely handled during negotiations on a CTB.

## *Consideration by the Eighteen-Nation Disarmament Committee (ENDC)*

The PNE controversy in relation to a CTB was resurrected in 1968 when the ENDC for the first time included the cessation of nuclear testing among the measures which could be discussed under its agenda item 'Further measures relating to the cessation of the nuclear arms race and nuclear disarmament'.[2] Most ENDC members appeared at this stage to envisage that PNEs could continue to be conducted under an otherwise total ban on nuclear testing. Canada and Sweden, for instance, both maintained that a test ban should permit PNEs, provided that specific permission be sought for each PNE conducted.[3] India agreed that a CTB and the PNE problem should be considered together.[4] In its view a test ban should ban all nuclear *weapon* tests by all states, nuclear and non-nuclear. Peaceful nuclear explosions would then be considered under a separate international instrument. India did not spell out how or by whom PNEs would be supplied under a CTB regime. Judging by its reaction to Article V of the NPT, it was unlikely to be sympathetic to the idea that the nuclear weapon states should be the sole suppliers.

The United Kingdom opined that PNEs could be treated as 'permitted exclusions' under a CTB provided they conformed to internationally-agreed arrangements.[5] The UK envisaged that IAEA experts, possibly along with experts from the nuclear weapon states, would check 'associated instrumentation' for each PNE to satisfy the international community 'as far as possible' that the explosion in question was indeed for peaceful purposes.

Italy, in a working paper tabled on 23 August 1968, proposed separating the PNE issue from that of nuclear weapon tests in order to facilitate agreement on the former. Under the Italian proposal, nuclear weapon states would be required to notify the UN of all proposed PNE detonations, along with such information as the approximate date, location, depth, purpose and yield. All explosions not notified would be considered weapon tests. The state carrying out the detonation would be obliged to permit experts from non-nuclear weapon states to observe the shot, to familiarise themselves with PNE techniques. This proposal, like India's, unfortunately did not resolve the fundamental problem of nuclear weapon states deriving military benefits from PNEs under a CTB regime.

Italy also proposed for the first time a moratorium on PNEs under the seabed, pending the establishment of 'international regulations safeguarding the interests of all nations'.[6] Luckily for the United States and Australia this idea never took on, for at that time they were beginning consultations on the Cape Keraudren project. Italy later amended its PNE proposals to take into account the newly-agreed provisions of the NPT. Instead of states notifying the UN of PNE detonations, as Italy originally suggested, notification would be made to the IAEA. Similarly, the non-nuclear weapon states would submit their proposed lists of expert observers to the IAEA rather than to the nuclear weapon state carrying out the project.[7]

On 28 August 1968, in a joint memorandum, the eight non-aligned members of the ENDC (Brazil, Burma, Egypt, Ethiopia, India, Mexico, Nigeria and Sweden) declared that the close link between PNEs on the one hand and both the NPT and a CTB on the other, increased the urgency of a 'universal and comprehensive solution' to the PNE issue within the context of a CTB.[8]

On 1 April 1969 Sweden submitted to the ENDC the draft provisions of a CTB treaty.[9] Sweden envisaged a three-tier approach to a CTB, the PTBT being the first tier, an underground nuclear weapon test ban the second, and a separately negotiated PNE agreement the third.[10] The preamble of its draft treaty affirmed the principle that 'any technological by-products' derived from PNEs by the nuclear weapon states should be made available to non-nuclear weapon and nuclear weapon states alike. While Article I, in language identical to that of the PTBT, banned all underground nuclear weapon test explosions and all *other* underground nuclear explosions, this was subject to one exception: explosions 'for construction or other peaceful purposes' which take place in conformity with 'an international agreement to be negotiated separately'.[11]

Egypt and the United Kingdom asked whether the international agreement envisaged by Sweden was the same as presaged by Article V of the NPT. The NPT, the UK pointed out, permitted bilateral agreements in addition to a 'special international agreement or agreements'.[12] Sweden responded that its 'international agreement' was 'one and the same' as that envisaged in the NPT.[13] The Swedes explained that the NPT only dealt with PNEs detonated by nuclear weapon states on the territory of non-nuclear weapon states, whereas under a CTB all PNEs, including those detonated in the territory of the nuclear weapon states themselves, would require international supervision. Sweden's proposed 'special international agreement' would provide the over-riding regulations for all PNEs, including those conducted under bilateral agreements.[14]

India commended Sweden's approach, as did Brazil, in the following terms:[15]

> The working paper presented by the Swedish delegation deals with the question of peaceful nuclear explosions in a logical and well-balanced manner: it sets forth the general and universal rule of prohibition that is applicable to all

countries, whether nuclear or non-nuclear, without any loopholes, and leaves the question of the regulation of nuclear explosions for peaceful purposes — which must be negotiated separately in another context — to be the subject of a special international agreement.

Just why India and Brazil should be so pleased with the Swedish draft is not difficult to fathom. Unlike Italy's proposal, it envisaged leaving the PNE question until last. India and Brazil were interpreting this to mean, probably incorrectly, that nuclear weapon testing by the superpowers would stop, while states like themselves could legally continue to work towards a PNE capability even to the point of conducting explosions. The Soviet Union on the other hand, in what was probably a correct interpretation of the Swedish draft, criticised it for appearing to suggest that no PNEs could be conducted by anyone, including the nuclear weapon states — either for their own purposes or for the benefit of other states under 'bilateral agreements' — until the proposed PNE agreement was concluded. This, the Soviets argued, was contrary to the NPT.[16]

## *Consideration by the Conference of the Committee on Disarmament (CCD)*

At the end of 1969 the newly-enlarged ENDC changed its name to the Conference of the Committee on Disarmament (CCD). Not until 1971 did the CCD resume the debate on PNEs. It did so in the context of renewed discussions on a CTB.

On 29 April 1971 the Netherlands proposed, no doubt to the chagrin of the United States and the Soviet Union, that the procedures being worked out by the IAEA for PNE services to non-nuclear weapon states should be applied to *all* PNEs carried out under a CTB regime, including those conducted on the territory of nuclear weapon states. The Netherlands suggested that after entry into force of a CTB only previously tested nuclear devices should be used for PNE projects.[17]

In September 1971 Sweden tabled amendments to its 1969 draft treaty elements, including, for the first time, a separate article on PNEs. As in the Swedes' previous Article I, Article II exempted from the test ban 'nuclear explosions which are carried out for construction or other peaceful purposes'. This time, however, instead of PNEs being regulated by 'an international agreement to be negotiated separately', Sweden now envisaged a Protocol II to be annexed to a CTB.[18] While no draft protocol was attached at this stage, the new text was a considerable advance on previous Swedish advocacy of a separate PNE treaty.[19]

With an obvious eye to India's rumoured development of a 'peaceful' nuclear capability, Pakistan also tabled a working paper, proposing that a CTB, like the NPT, divide states into nuclear weapon and non-nuclear weapon states, with different provisions for each.[20] While nuclear weapon states would be permitted to conduct PNEs (presumably under international control), non-nuclear weapon states would not be permitted to conduct *any* nuclear explosions, but could obtain the benefits of PNEs under a separate PNE agreement. Although Pakistan was far from being a supporter of the NPT and, like India, had refused to sign it, these proposals were basically the arrangements envisaged under the NPT's Article V. They still did not, however, address the critical problem of how to prevent nuclear weapon states from obtaining military benefits from PNEs detonated either for their own purposes or for other states. This was not, of course, Pakistan's main concern.

In a joint memorandum of 30 September 1971, several non-aligned members of the CCD — Burma, Egypt, Ethiopia, Mexico, Morocco, Nigeria, Sweden and Yugoslavia — issued a new call for a CTB. On the subject of PNEs, they asserted that these would be 'regulated' and that the IAEA 'should play an important role in this context'.[21] Significantly, neither Argentina, Brazil or India associated themselves with this memorandum. Surprisingly, neither did Pakistan.

In 1972 Pakistan nervously noted that some of the countries with the capacity to produce nuclear weapons had not signed the NPT and had declared their intention to explode 'so-called peaceful nuclear devices'.[22] Pakistan declared that, as there was no difference between nuclear weapons and 'so-called' PNEs, production of such devices by a non-nuclear weapon state (read India) would inevitably mean that it had become a nuclear power. In direct opposition to the Indian view, Pakistan said that if the question of so-called PNEs was left unresolved in the context of a CTB, the nonproliferation regime 'will prove to be a mere illusion'. Pakistan's statement was remarkable both for its warning of the Indian 'peaceful nuclear explosion' detonated less than two years later, and for the laudatory references to the NPT, a document which Pakistan still refused to sign.

When in May 1974 India realised Pakistan's worst nightmare, the PNE issue erupted again in the CCD.[23] Interest in the PNE issue was further aroused by the signature of the TTBT by the United States and the Soviet Union in July 1974. In July 1975 a group of experts on the PNE issue convened by the CCD concluded that:[24]

> With a treaty banning all nuclear weapon tests, the incentive for seeking military benefits from peaceful nuclear explosions would be even greater. An effective verification system would, therefore, be needed to ensure that weapon tests were not carried out under the guise of peaceful explosions conducted by nuclear weapon states on their territory, or elsewhere under Article V of the NPT. Failing such a system, a choice would have to be made between peaceful nuclear explosions and a comprehensive test ban.

In September 1975 the Soviet Union submitted to the UN General Assembly a draft CTB treaty.[25] This was the first time since the PTBT negotiations that a nuclear weapon state had proposed a full draft treaty covering underground nuclear weapon tests. The draft contained a much more detailed PNE article than the Swedish draft treaty elements. Article III proposed that a ban on nuclear weapon tests not apply to underground PNEs conducted by the nuclear weapon states. Such PNEs would be regulated as follows:[26]

> (a) if detonated on the territory of non-nuclear weapon states, they would have to conform with the provisions of Article V of the NPT
> (b) if detonated on the territory of the nuclear weapon states themselves, they would have to conform with a special agreement negotiated by such states, 'with due regard for the recommendations of the IAEA on the subject and which will be concluded as speedily as possible'.

The latter proposal, granting only 'due regard' to the IAEA and presumably no regard to the views of the non-nuclear weapon states, was unlikely to be popular with the nuclear 'have-nots'.

There was, moreover, nothing in the Soviet draft to indicate how the envisaged 'special agreement' would ensure that military benefits were not obtained from ostensibly peaceful nuclear explosions or how parties were to distinguish between a PNE and a nuclear weapon test prior to the 'special agreement' entering into force. There was no ban on PNEs in the interim since Article I specifically banned only nuclear weapon tests. The provision that the special agreement be concluded 'as speedily as possible' did not guarantee that it would be concluded before the entry into force of a CTB. This fitted well with Soviet policy that a solution to the PNE problem should follow the conclusion of a CTB. Hence the Soviet draft CTB treaty did little to advance realistic debate on the PNE issue. It again put PNEs into the 'too hard' basket. It was in part because of these considerations that the United States and the United Kingdom rejected the Soviet draft treaty.[27]

At the 1976 session of the CCD, the Netherlands suggested that in view of the difficulty of verifying that PNEs were not weapon tests, PNEs might have to be abandoned altogether under a CTB regime. This was the first time a non-nuclear weapon state had explicitly supported this view of the United States and the United Kingdom.[28] The Soviet Union and India, on the other hand, continued to maintain that a solution to the PNE problem could be found after a nuclear weapon test ban had been agreed upon. The Indians even had the temerity to claim that the PNE issue was 'outside the scope' of disarmament problems.[29] The Soviet Union and other committee members, including Italy (again out of step with its Western partners), stressed that a CTB should not create obstacles to the sharing of the benefits of PNEs with non-nuclear weapon states.[30]

Debate at the 1976 session of the UN General Assembly, which immediately followed the CCD session, revealed a growing belief that the PNE question should be resolved in the context of a CTB, not in a separate agreement. This view was expressed most strongly by the United States and the United Kingdom. Several other countries, including Denmark, Iran,

Nepal (surprisingly in view of its close links with India), the Netherlands and Norway, while noting that a satisfactory solution to the problem had to be found, questioned, with varying degrees of emphasis, the value of PNEs given the evident risks involved. Ireland and New Zealand asserted that the PNE question was so difficult that, in the light of the doubtful value of nuclear dynamite, a PNE moratorium was essential.[31] Sweden now suggested that PNEs be abandoned altogether to facilitate the negotiation of a CTB.

The PNE controversy also arose in the voting on two nuclear test ban resolutions before the Assembly. The United States, the FRG and the Netherlands abstained on the joint Western/non-aligned resolution partly because it gave no attention to the PNE question, apart from a misleading reference to 'control and supervision of PNEs in the PNET'.[32] The United States noted that the draft failed to call for a verification system which would provide adequate assurance against military benefits being obtained from PNEs.[33] The United States and United Kingdom abstained on the Soviet draft test ban resolution also partly because of its failure to address the PNE problem. The PNE issue had clearly emerged, along with the question of verification and whether the participation of all nuclear weapon states in a test ban was essential, as one of the three major issues confronting the negotiation of a CTB.

At the 1977 session of the CCD, two draft CTB treaties were tabled containing PNE provisions. In February the Soviet Union tabled the draft treaty it had introduced at the 1975 United Nations General Assembly. The PNE provisions remained unchanged.[34] In July Sweden tabled a complete draft CTB treaty, but with only slightly revised wording for its Article II on PNEs, compared to its 1971 draft treaty 'elements'. PNEs were now to be carried out 'under international supervision and control', as well as in conformity with the still missing Protocol II.[35] In tabling the treaty the Swedes made clear that the safeguards in the protocol would apply to all parties, whether nuclear or non-nuclear.[36] Sweden said, however, that it envisaged that PNEs would only be conducted under a CTB regime if they were of 'over-riding national or international importance'. It joined the Netherlands in urging that PNEs be temporarily banned if they threatened the achievement of a test ban.[37]

## *The Trilateral Negotiations on a Comprehensive Test Ban Treaty (1977–80)*

In June 1977 the Soviet Union and the United States held bilateral discussions on the subject of a comprehensive test ban treaty. They jointly informed the CCD that the talks had been useful. They also announced that they would be joined in early July by the United Kingdom to begin negotiations on a draft CTB. Such a draft might be elaborated into a multilateral treaty by the CCD.

However the three powers revealed little in the course of 1977 about how the talks were progressing. SIPRI commented that:[38]

> The matters discussed trilaterally are surrounded by secrecy. What is known about them is that the main subjects of discussion are the questions of verification, participation, peaceful nuclear explosions and entry into force.

As far as the PNE issue was concerned, the problem, as in the TTBT negotiations, was that the Soviets were insisting on the right to conduct PNEs under a test ban treaty. According to ACDA official Nick Carrera, the United States, on the other hand, was making a 'very strong and well-supported' case that PNEs should be banned outright because it was impossible to prevent military benefits being obtained from such explosions.[39] Demarches were also made to this effect in London and Moscow. Despite continuing support for PNEs at the Livermore Laboratory and in some parts of ERDA, the Americans also gave the Soviets voluminous material on why they considered PNEs were not economic.[40] At the 1977 session of the UN General Assembly, President Carter described the aim of the United States during the trilateral negotiations as being a fair and verifiable agreement to end all explosions of nuclear devices, whether for military or peaceful purposes.[41] The United Kingdom, Canada, the FRG, the Netherlands and New Zealand concurred with this aim.

New Zealand, a non-CCD member, sent a detailed exposition of its views to the CCD in July 1977, asserting that 'no existing or foreseeable technology or verification procedure can distinguish between peaceful and weapon test explosions'.[42] New Zealand proposed a moratorium on PNEs until satisfactory arrangements to safeguard arms control were agreed. In a nod to Soviet preoccupations it also said that consideration might need to be given to an exception for 'emergency PNEs', such as for sealing oil and gas leaks. Another suggestion, made almost in passing, was that the nuclear weapon states should agree to pool information on PNE devices, 'including internal design and all other descriptive information', to ensure that no military advantage was derived from PNEs.[43] The perennial problem with this idea was that PNE designs were similar to those of advanced nuclear weapons.

The socialist countries on the other hand, led by the Soviet Union and supported by some of the neutral and non-aligned states such as Austria, Finland and Nigeria, continued to argue that the PNE question should be solved separately from a CTB and in accordance with the provisions of the NPT.[44] Another group of countries, including Afghanistan, Argentina, India and Nepal (now obviously falling into line with Indian policy) took the position that a test ban should not include any measures that discriminated against non-nuclear weapon states with regard to the peaceful uses of nuclear energy. This undoubtedly referred to PNEs.

By this time the United States was preparing a detailed critique of PNEs entitled 'US Perspectives on Peaceful Nuclear Explosions', in an effort to convince the Soviets to change their position.[45] The paper declared that the opposition of the United States to permitting PNEs under a CTB was based on 'concerns regarding the military benefits accruing to a country (even a nuclear weapon state) conducting PNEs and the implications for future proliferation of nuclear explosives technology'.[46] Moreover, the paper continued, 'on the basis of extensive studies over a period of twenty years, we have concluded that the potential economic benefits (if any) of PNEs are not sufficient to over-ride these very serious concerns'. No application had been identified by the United States, the paper said, where PNEs offer important, clear-cut advantages over alternative approaches. With regard to nuclear excavation projects, the paper asserted that, even in the unlikely event that they could be conducted without violation of the PTBT, 'release of radioactivity, with associated environmental consequences, cannot be avoided'.[47] Finally, the paper pointed out the political dangers of a nuclear weapon state continuing with PNEs:[48]

> First, there would be an implied obligation to offer similar services to non-nuclear weapon states desiring to exploit their indigenous resources. Shipping hundreds of nuclear explosive devices to remote locations in foreign countries would pose security risks which would be entirely unacceptable — particularly in an era when terrorist activities have become almost commonplace. Second, large-scale use of PNEs by the nuclear weapon states would provide a strong incentive and justification for the non-nuclear weapon states to proceed to acquire their own PNEs.

By the time the paper was handed to the Soviets, however, Moscow had caved in on the PNE issue, persuaded finally that there would be no CTB if it continued to insist on the right to conduct PNEs.[49]

On 2 November 1977 Soviet General Secretary Brezhnev announced in Moscow that the Soviet Union was prepared to accept a suspension of all underground tests, including a moratorium on PNEs, for a definite period of time, 'as long as a CTB lasts'. This was later clarified to mean a three-year moratorium but with prospects for a permanent ban on PNEs.[50] The position of the Soviet Union had therefore come full circle from favouring a ban on all explosions in the early 1960s, to favouring a ban only on nuclear weapon tests in the late 1960s and early 1970s, and then back again to apparently favouring a truly comprehensive test ban treaty. According to Herbert York, this decision was only reached after great debate within the Soviet hierarchy. The chief Soviet delegate, in announcing the decision in Geneva, was visibly distressed at the turnaround. Warren Heckrotte described the Soviet delegation as 'sad looking'.[51] Herbert York reports that a Soviet delegate told the Americans: 'You won!' York says it was the first time he had ever seen dissension within the Soviet camp on a major arms control issue.[52]

The Soviet decision was a major shift in position and a landmark event in the history of PNEs. It allowed the trilateral negotiators to settle the matter for the time being and move on to other stumbling blocks. The

Americans welcomed the Soviet decision in low-key fashion, stating that it went 'in the direction' of the US position favouring the inclusion of a ban on PNEs in a CTB.[53] The new Soviet position also clearly undercut the position of supporters such as India. It remained to be seen how India would have reacted to the trilateral draft treaty had it been sent to the CCD for consideration. Possibly India would have insisted that it was simply an arrangement between the three nuclear weapon states which had no implications for the type of CTB that the CCD should negotiate.

On 16 March 1978 the trilateral negotiators reported to the CCD that they were continuing negotiations on the provisions of a treaty 'prohibiting nuclear weapon tests and a protocol covering nuclear explosions for peaceful purposes which would be an integral part of the treaty'.[54] This was the first time this formula had been used and indicated that the Soviets had now also abandoned their insistence that negotiations on a PNE agreement be held subsequent to agreement on a CTB. According to Heckrotte and Carrera, the idea of a PNE protocol was a face-saving device for the Soviets, designed to lessen the humiliation of having so abruptly changed their PNE policy.[55] The CTB now envisaged would effectively ban PNEs for all time.

At the First Special Session of the UN General Assembly Devoted to Disarmament in mid-1978, the international community appeared to endorse the solution to the PNE question agreed to by the trilaterals. In paragraph 51 of its Final Document, adopted by consensus, the session called for the urgent conclusion of the trilateral negotiations on a 'treaty prohibiting nuclear-weapon tests, and a protocol covering nuclear explosions for peaceful purposes, which would be an integral part of the treaty'.[56] It was still not at all clear, however, that a treaty in this form would survive passage through the CCD. The Final Document hinted at this by asserting that the results of the trilaterals should be 'submitted for full consideration by the multilateral negotiating body [the Committee on Disarmament, successor to the CCD] with a view to the submission of a draft treaty to the General Assembly at the earliest possible date'. The Final Document also failed to use the phrase 'comprehensive test ban treaty' (which many states now took to mean a complete ban on all nuclear test explosions for all time, including PNEs) but rather referred to the 'cessation of nuclear weapon testing'.[57]

At the resumed session of the CCD following the Special Session, the United Kingdom, on 8 August, reviewed on behalf of the three trilateral partners the progress made to date in the trilateral negotiations. The United Kingdom confirmed that it had been agreed that the treaty being negotiated would ban all nuclear weapon tests in all environments, and that a protocol, which would be an integral part of the treaty, would apply to PNEs. The three parties had also agreed that after a certain period they would review the operation of the treaty, the implication being that the PNE provisions in particular would be examined.[58] In commenting on the UK's presentation, Japan proposed that the treaty ban PNEs until agreement was reached on appropriate international supervision and procedures and that the treaty parties agree to continue negotiations on such procedures in good faith.[59] This proposal was clearly a misreading of the intentions of the trilateral negotiators and was politely ignored. The United States reiterated its call for the outright prohibition of *all* nuclear explosions, including PNEs.

In 1979 the CCD was replaced by the Committee on Disarmament (CD) created by the First Special Session. It contained eight new non-nuclear weapon state members: Algeria, Australia, Belgium, Cuba, Indonesia, Kenya, Sri Lanka and Venezuela. It also included, for the first time, China and France. The CD had as its first agenda item the issue of a 'nuclear test ban' treaty. Discussion of this item in 1979 was marked by disappointment at the failure of the trilaterals to produce a draft treaty or at least a comprehensive progress report.[60]

Towards the end of the 1979 session the United Kingdom, on behalf of its trilateral partners, did provide another brief report, in which it stated that negotiations had continued intensively throughout the year and had concentrated on the question of verification. The CD remained disappointed with the lack of candour shown by the trilaterals.[61]

At its 1979 session the UN General Assembly, in a resolution initiated by Australia and New Zealand, for the first time specifically requested the multilateral disarmament negotiating body, the CD, to initiate negotiations on a CTB 'as a matter of the highest priority'.[62] No specific mention was made of PNEs, but the resolution reaffirmed the conviction of the General Assembly that 'a treaty to achieve the prohibition of all nuclear test explosions by all states for all time is a matter of the highest priority'.[63] While the Assembly was meeting, the trilateral negotiations were in fact slowing down as a result of difficulties over the ratification of the second Strategic Arms Limitation Agreement (SALT II).[64]

In July 1980 the trilateral negotiators finally succumbed to pressure from the CD to table a full report on their deliberations. On the PNE question the report confirmed that the draft CTB treaty envisaged would be accompanied by a protocol covering PNEs, which would be 'an integral part of the treaty'. The protocol would 'take into account', in ways unspecified, the provisions of Article V of the NPT. In the protocol the parties would establish a moratorium on PNEs whereby they would:[65]

> . . . refrain from causing, encouraging, permitting or in any way participating in, the carrying out of such explosions until arrangements for conducting them are worked out which would be consistent with the treaty being negotiated, the Treaty Banning Nuclear Weapon Tests in the Atmosphere, in Outer Space and Under Water and the Treaty on the Non-Proliferation of Nuclear Weapons.

The trilateral parties had further agreed that:

> Without delay after entry into force of the treaty, the parties will keep under consideration the subject of arrangements for conducting nuclear explosions for peaceful purposes, including the aspect of precluding military benefits.

> Such arrangements, which could take the form of a special agreement or agreements, would be made effective by appropriate amendment to the protocol.

Amendments would require the approval of a majority of the parties, including permanent members of the UN Security Council — that is, the nuclear weapon states. This was very close to treaty language, indicating that the PNE problem had effectively been 'solved' in the trilateral negotiations. The reaction of the CD to the 1980 Trilateral Report, as it came to be known, was critical, although there was no criticism of the PNE provisions.[66]

The trilaterals subsequently failed to reach agreement on a CTB under the Carter Administration. The new Reagan Administration suspended the talks pending a review of US CTB policy. In July 1982, having concluded that a CTB was now only a long-term objective, the Administration decided not to resume the negotiations.[67]

## *Post-Trilateral Consideration of the PNE Issue*

Three years after the trilaterals ended, even though no CTB negotiations had resumed, the debate on the PNE issue flared in the CD as a result of the tabling of draft treaty language for a CTB by the Soviet Union, Sweden and Australia respectively, and the establishment of a subsidiary body to consider the CTB issue.

The Soviet Union tabled new draft test ban treaty 'elements' in the Committee on 16 February 1983.[68] While still providing for a moratorium on PNEs, the text gave somewhat stronger emphasis than the trilateral 'elements' to the need to establish procedures for carrying out PNEs in the future.[69] It also failed to mention, unlike the trilaterals, that such procedures should be consistent with the PTBT and the NPT.[70]

Such changes were undoubtedly intended to regain the support of the pro-PNE non-aligned states such as India. The Soviet position that year in the more 'political' atmosphere of the UN General Assembly was even more mischievous. Its annual resolution in the First Committee failed to mention the PNE problem or the solution agreed to in the trilateral negotiations, referring only to a ban on nuclear *weapon* tests. Again, this was intended to maximise non-aligned support. It was also one reason for consistent Western opposition to the Soviet resolution, even by those states which were otherwise enthusiastic about a nuclear test ban.

In 1983 Sweden also tabled a revised version of its draft CTB in the CD.[71] It did not provide for a simple ban on all nuclear explosions, which was Sweden's preferred position, but emulated the trilateral agreement by providing for a moratorium on PNEs pending negotiation of a PNE regime. The Swedish draft was, however, an advance on the trilateral draft in that it moved the PNE provisions into the body of the treaty, whereas the trilaterals had envisaged a protocol. This would make the PNE provisions more difficult to amend than if they were in a separate protocol — even if such a protocol was 'an integral part of the treaty' as the trilateral negotiators insisted *ad nauseam*. The Swedish draft was also an advance on the trilaterals in not calling for the parties to consider the establishment of a PNE regime 'without delay after the entry into force' of a CTB. The Swedes were being even vaguer about the possibility of future arrangements for PNEs than the trilaterals had been. The Swedish draft read as follows:[72]

> *Article II*
> 1. Each Party to this Treaty undertakes not to carry out any nuclear explosion for peaceful purposes and accordingly to refrain from causing, encouraging, assisting, permitting or in any other way participating in the carrying out of any such explosion until international arrangements for conducting them are worked out which would be consistent with this treaty and the obligations of each Party under other relevant international treaties.
> 2. The Parties undertake to keep under consideration the question of arrangements for conducting nuclear explosions for peaceful purposes on a non-discriminatory basis, including the aspect of precluding military benefits. Such arrangements may take the form of a special agreement or agreements constituting an integral part of this Treaty.

Australia also tabled in the CD in 1983 the following draft first article for a CTB treaty:[73]

> *Article I*
> 1. Each Party to this Treaty undertakes not to carry out any nuclear weapon test explosion, or any other nuclear explosion.
> 2. Each Party to this Treaty undertakes, furthermore, to refrain from causing, encouraging, assisting, permitting or in any other way participating in the carrying out of any nuclear weapon test explosion or any other nuclear explosion.
> 3. Each Party to this Treaty undertakes to take all necessary measures to prohibit and prevent any activity in violation of the provisions of the Treaty anywhere under its jurisdiction or control.

This was the first time that any country had tabled draft language that provided explicitly for a complete ban on PNEs as part of a CTB.

These three draft texts were dissected in the CD's subsidiary body on the CTB issue, which had been established, after prolonged wrangling, in April 1982. While this *Ad Hoc* Working Group on a Nuclear Test Ban lacked a mandate to negotiate a treaty, it did provide a forum for lively debate on the key issues of verification and compliance. In 1983 the group's mandate was broadened slightly to include the issue of the scope of a future CTB. This permitted the first full-blown discussion of the PNE issue since the trilateral negotiations had ended.

Far from indicating that the trilaterals had solved the PNE problem, the debate revealed that the issue was as divisive as ever. The Western countries, led by Australia, the Netherlands and the United Kingdom (which tabled a detailed working paper),[74] argued for a complete ban on PNEs for all time, on the grounds that it was unlikely that arrangements could ever be devised to totally preclude the possibility of military benefits being obtained from PNE detonations. In a unique move for a nuclear weapon state, the United Kingdom stated it would be prepared to renounce permanently the right to conduct PNEs as part of an agreement on a CTB.[75] Brazil, on behalf of the 'threshold' states, argued that this approach was discriminatory, inconsistent with the obligations entered into under the NPT (which Brazil still had not signed), the progress achieved in the trilateral negotiations, and the existence of the (unratified) PNET.

The Soviet Union charged that since the suspension of the trilateral talks, the Western participants had changed their position on PNEs to one of outright opposition. This conveniently ignored the fact that the trilateral report envisaged in effect a permanent, total ban on PNEs (albeit one dressed up to save Soviet face), and that the Soviet Union had itself modified the agreed near-treaty language of the trilateral report in tabling its latest draft treaty 'elements'. The Soviets said they continued to attach great importance to PNEs — albeit secondary to the goal of a complete prohibition of nuclear *weapon* tests — and were convinced that an appropriate PNE regime could be worked out after the conclusion of a CTB.[76] The Soviets were obviously backtracking on the trilateral agreement.

Despite the optimism that had prevailed during the trilateral negotiations, the PNE issue, in relation to a CTB, remains unresolved. If and when negotiations on a CTB resume, the PNE issue could still derail the negotiations or lead to a treaty that is not universally accepted. As Jozef Goldblat has noted:[77]

> The difference of opinion over peaceful nuclear explosions is very important. It would have to be settled before a multilateral test ban treaty could be seriously contemplated and concluded. Leaving peaceful nuclear explosions unconstrained would render the treaty ineffective.

## Notes

1 US Congress, Senate Foreign Relations Committee, Hearings on the TTBT/PNET, 28 July, 3 August, 8 and 15 September 1977, US Government Printing Office, Washington DC, 1977, p. 137.

2 *The United Nations and Disarmament 1945–1970*, Department of Political and Security Affairs, United Nations, New York, 1970, p. 244.

3 Ibid., pp. 244–45.

4 Ibid., p. 245.

5 Eighteen-Nation Disarmament Committee Document, ENDC/PV. 387.
6 ENDC/PV. 234.
7 ENDC/PV. 414.
8 *The United Nations and Disarmament 1945–1970*, p. 246.
9 ENDC/PV. 242, reproduced in *The United Nations and Disarmament 1945–1970*, pp. 490–92.
10 ENDC/PV. 404.
11 Ibid.
12 Ibid.
13 ENDC/PV. 415.
14 Ibid.
15 ENDC/PV. 409.
16 ENDC/PV. 415.
17 Conference of the Committee on Disarmament, Geneva, Document CCD/PV. 512.
18 CCD/348.
19 *Negotiations on a Comprehensive Test Ban 1963–71*, US Arms Control and Disarmament Agency, Unclassified Document Series, DDS Ref. 587, Washington DC, 22 December 1971, p. 94.
20 CCD/PV. 576.
21 *Negotiations on a Comprehensive Test Ban 1963–71*, p. 98.
22 CCD/PV. 576.
23 See Chapter 12.
24 *SIPRI Yearbook 1976*, Almqvist and Wiksell, Stockholm, 1976, p. 378.
25 Ibid., p. 306.
26 Ibid., p. 354.
27 Ibid., p. 309.
28 *UN Disarmament Yearbook 1976*, United Nations, New York, 1977, p. 91.
29 *UN Disarmament Yearbook 1977*, United Nations, New York, 1978, p. 96.
30 *UN Disarmament Yearbook 1976*, p. 91.
31 Ibid., p. 93.
32 Ibid., pp. 94–95.
33 Ibid., p. 95.
34 CCD/523.
35 CCD/562/Rev.1.
36 *UN Disarmament Yearbook 1977*, p. 99.
37 Ibid., p. 98.
38 *SIPRI Yearbook 1978*, Taylor and Francis, London, 1978, p. 320.
39 From an interview with Nick Carrera, ACDA, Washington DC, 9 December 1986.
40 From an interview with Warren Heckrotte, Livermore Laboratory, California, 3 December 1986.
41 *UN Disarmament Yearbook 1977*, p. 102.
42 CCD/536.
43 Ibid.
44 *UN Disarmament Yearbook 1977*, p. 104.
45 *US Perspectives on Peaceful Nuclear Explosions*, unpublished ms, ACDA, Washington DC, 7 February 1973. A condensed version was later published by ACDA under the title *The American Experience With Peaceful Nuclear Explosions*, US Arms Control and Disarmament Agency, US Government Printing Office, Washington DC, 1978.
46 *US Perspectives on Peaceful Nuclear Explosions*, p. 1.
47 Ibid., p. 2.
48 Ibid., p. 3.
49 From an interview with Nick Carrera, ACDA, Washington DC, 9 December 1986.
50 R. L. Garthoff, *Détente and Confrontation*, The Brookings Institution, Washington DC, 1985, p. 756.
51 From an interview with Warren Heckrotte, Livermore Laboratory, California, 3 December 1986.
52 From an interview with Herbert York, La Jolla, California, 1 December 1986.
53 *UN Disarmament Yearbook 1977*, p. 104.
54 *UN Disarmament Yearbook 1978*, United Nations, New York, 1979, p. 199.
55 From interviews with Warren Heckrotte, Livermore Laboratory, California, 3 December

1986 and Nick Carrera, ACDA, Washington DC, 9 December 1986.

56 *Final Document of the Special Session of the General Assembly on Disarmament, 1978*, UN Department of Public Information, New York, 1978, p. 11.

57 Ibid., p. 10.

58 *UN Disarmament Yearbook 1978*, p. 202.

59 CCD/PV. 801.

60 *UN Disarmament Yearbook 1979*, United Nations, New York, 1980, p. 125.

61 Ibid., p. 128.

62 Ibid., p. 134.

63 *SIPRI Yearbook 1980*, Taylor and Francis, London, 1980, p. 429.

64 Ibid., p. 360.

65 Committee on Disarmament or the Conference on Disarmament, Document CD/130.

66 *UN Disarmament Yearbook 1980*, United Nations, New York, 1981, p. 154.

67 This was announced at the same time as the announcement that the United States would not ratify the TTBT or PNET due to alleged deficiencies in the TTBT's verification provisions (*UN Disarmament Yearbook 1982*, United Nations, New York, 1983, p. 257).

68 CD/346.

69 Instead of 'without delay' as in the Trilateral Report, the Soviets used 'promptly', and instead of a vague undertaking to 'keep under consideration' the PNE issue, they used the stronger expression 'consideration shall be given'.

70 The text read as follows:

3. A moratorium shall be declared on nuclear explosions for peaceful purposes, under which the parties to this treaty shall refrain from causing, encouraging, or in any way participating in the carrying out of such explosions until an appropriate procedure has been elaborated.

4. Promptly after the entry into force of this Treaty, consideration shall be given to the question of the procedure for the carrying out of nuclear explosions for peaceful purposes. Such a procedure, to be agreed upon, may take the form of a special agreement or agreements constituting an integral part of this Treaty (CD/346).

71 CD/381.

72 Ibid.

73 CD/405.

74 CD/383.

75 *UN Disarmament Yearbook 1983*, United Nations, New York, 1984, p. 202.

76 Ibid., p. 202.

77 J. Goldblat, 'Multilateral Arms Control Efforts' in *SIPRI Yearbook 1986*, Oxford University Press, Oxford, 1986, pp. 451–52.

# Epilogue: The Future of PNEs

*The prevention of nuclear war is so vital to the survival of the human race that it would be a world tragedy if inconsequential peaceful applications of nuclear explosives were allowed to interfere with that goal.*[1]

Herbert Scoville, 1975

SINCE the NPT entered into force in 1970, the 'potential benefits' of PNEs have not eventuated, no negotiations on a PNE regime have commenced and there has been a dramatic decline in interest in nuclear dynamite. The industrialised states, even those with large land areas, such as Australia and Canada, have rejected the use of PNEs. Most of the developing countries, including those like Egypt and Thailand which were once genuinely interested, have also quietly shelved the idea. Since the Indian example, none of the developing nations has attempted to detonate a so-called 'peaceful nuclear explosion'. Of the nuclear weapon states, the United Kingdom, France and China have all declined to experiment with PNEs, while the United States has closed Plowshare down.

The decline in international interest in PNEs is in part attributable to the complex web of international legal restrictions that has enveloped the use of nuclear dynamite, including the PTBT, the Treaty of Tlatelolco, the NPT, the (still unratified) PNET, a series of safeguards agreements between the IAEA and its members, and bilateral safeguards agreements between states which regulate the purposes to which nuclear material may be put. The newest constraint is the South Pacific Nuclear-Free Zone (SPNFZ) Treaty or Treaty of Raratonga.

It is partly a reflection of the strong anti-PNE sentiment of Australia and New Zealand and partly a response to the ambiguities of the Treaty of Tlatelolco that this treaty specifically bans PNEs. In Article 3 of the

Treaty, each party undertakes *inter alia* 'not to manufacture or otherwise acquire, possess or have control over any nuclear explosive device by any means anywhere inside or outside the South Pacific Nuclear-Free Zone'.[2] Rather more clearly than Tlatelolco, Article I of the treaty defines a nuclear explosive device as 'any nuclear weapon or other explosive device capable of releasing nuclear energy, irrespective of the purpose for which it could be used'.[3] According to David Sadleir, Australian chairman of the South Pacific Forum working group which drafted the treaty:[4]

> The term nuclear explosive device was used so as to exclude so-called 'peaceful' nuclear explosive devices as well as nuclear weapons since technically and in non-proliferation terms it was impossible to distinguish between them.

This definition, combined with the Treaty's prohibition on the 'emplantation' or 'emplacement' of nuclear explosive devices on territories of states parties, rules out the use of PNEs even if provided by a nuclear weapon state in accordance with Article V of the NPT. On this reading, parties to the SPNFZ are the only countries in the world to have totally and unambiguously renounced the use of PNEs. In acceding to the treaty's protocols, the nuclear weapon states also renounce the use of nuclear dynamite in the South Pacific.[5]

The existing international legal constraints on PNEs, such as the SPNFZ, have, however, arisen as a by-product of attempts to achieve other arms control goals, such as a nuclear weapon test ban and a nonproliferation treaty, rather than as a result of attempts to control PNEs *per se*. Edward Teller and his collaborators have lamented this situation, although for different reasons than those of the arms controllers:[6]

> ... international agreements or treaties which are designed to limit the development of nuclear weapons may, *unfortunately* [emphasis added], apply to the peaceful uses of nuclear explosives as well.

The PNE problem has in fact rarely been confronted directly. It has been skirted around, handled ambiguously, placed in the 'too hard' basket or simply ignored. The resulting patchwork of constraints on PNEs restricts their development and use but has not killed them off altogether. The nuclear weapon states are still legally free to detonate PNEs. Only France and China, not having acceded to the PTBT, are permitted to do so in any environment, while the two superpowers have informally restricted themselves to detonations under 150 kilotons. As for the non-nuclear weapon states, those which are not party to the PTBT, the NPT or a nuclear-free zone agreement, are still legally permitted to detonate PNEs in any environment. Gaps such as these in the international regulation of PNEs could be filled either by a universal comprehensive test ban treaty or by a separate treaty specifically banning PNEs.

The political obstacles to a total ban on PNEs are today fewer than they have ever been. There is almost total international agreement that PNEs are not a viable technology and should be proscribed for the sake of arms control. Even the annual non-aligned CTB resolution at the United Nations General Assembly, co-sponsored by Sweden and Mexico, which had always been ambiguously worded in regard to the PNE issue — in order to attract the support of the threshold states — was changed in 1985. Until then the resolution, while calling for 'the prohibition of *all* nuclear test explosions' in its preamble, requested the CD to begin negotiations on a treaty only on the prohibition of nuclear *weapon* tests. The co-sponsors of the Western CTB resolution, Australia and New Zealand, made repeated unsuccessful approaches to both Mexico and Sweden to remove this ambiguity. This would have facilitated either amalgamation of the competing CTB resolutions, or permitted the Western pro-CTB states to vote in favour of the non-aligned text. In 1985, for reasons unknown, the Mexican/Swedish resolution finally acknowledged the PNE problem by calling on the CD to negotiate a 'nuclear test ban' rather than a 'nuclear weapon test ban'. With this obstacle removed, the remaining obstacles to a complete ban on PNEs are the Soviet Union's continuing PNE program and the attitude of the so-called 'threshold' states, Argentina, Brazil and India.

The Soviet Union is currently the only country with a PNE program. Although the program was suspended during the unilateral Soviet testing moratorium from August 1985 to January 1987, General Secretary Gorbachev noted with regret that this had involved the Soviet Union foregoing 'economic benefits'.[7] After ending the moratorium, the Soviets resumed their PNE program on 19 April 1987 with two detonations in the Urals region of Perm. The detonations, according to *Tass,* had yields of up to 20 kilotons and were staged 'in the interests of the national economy',[8] to perfect 'blasting techniques for civil construction projects'.[9] On 6 May and 6 June 1987 another two tests below 20 kilotons were conducted, which *Tass* claimed were designed to check the findings of research into the physics of nuclear detonations.[10] Another three explosions in Siberia were detected on 7 July, 24 July and 12 August.[11] The Swedish Hagfors Observatory also reported a 'presumed underground nuclear explosion' on 3 October 1987 north-east of the Caspian Sea and south of Aktyubinsk, the first ever detected in the area.[12] As there is no advantage, in terms of weapons development, for the Soviets to detonate nuclear devices outside their normal weapon testing grounds, it can only be assumed that all these explosions were PNEs. On this assumption the Soviet PNE program is clearly still vigorous.

Whether the Soviets view their continuing program as essential is, however, difficult to discern. In August 1986 it was reported that the proposed Pechora/Kama river diversion project, with or without PNEs, had finally been cancelled because of opposition by General Secretary Gorbachev.[13] This removed, once and for all, a major incentive for the development of nuclear excavation technology. The Director of the Research Coordination Centre of the Soviet Foreign Ministry, Mr V.V. Shoustov, has confirmed that the Soviet PNE excavation program was shelved principally because of environmental concerns. The Soviets, he said, would only conduct such PNEs after careful consideration.[14] The Soviet Union however has other uses for PNEs which it apparently still regards as effective, economic and environmentally sound, principally gas storage and deep seismic sounding. It remains to be seen whether it would sacrifice such uses for the sake of a CTB, although this is now highly likely.

At the 1985 session of the United Nations General Assembly the Soviet bloc's annual nuclear test ban resolution, tabled by Hungary, heralded a shift in Soviet policy back towards the spirit of the trilateral agreement. While the resolution still referred, as in previous years, to a 'nuclear *weapon* test ban', it also called for 'provisions, acceptable to all, preventing the circumvention of this ban by means of nuclear explosions for peaceful purposes'. For the first time, a Soviet nuclear test ban resolution specifically recognised the PNE problem. In a letter to the *Bulletin of the Atomic Scientists* in May 1986, the Hungarian Mission to the United Nations in New York described this as 'an absolutely new operative paragraph'.[15] The paragraph has been retained in subsequent annual Soviet resolutions. It was also reproduced in the new draft 'Basic Provisions of a Treaty on the Complete and General Prohibition of Nuclear Weapon Tests', tabled by the Soviet Union in the CD on 17 June 1987.

However the most convincing evidence of recent Soviet mellowing on the PNE issue was the tabling in the CD on 24 April 1987 of a Socialist group working paper which contained the following paragraph:[16]

> ... ways and means will have to be found, *preventing the circumvention of this ban* [a CTB] *by means of nuclear explosions for peaceful purposes* [PNE]. There is no clear-cut solution to this issue yet. All alternative proposals to this end as well as future ones should be considered in order to agree on a provision acceptable to all. A solution could provide, e.g. for:
> — a complete prohibition of such explosions from the outset;
> — the carrying out of such explosions under certain conditions.
> An *interim solution* could envisage that States parties to the NTB [Nuclear Test Ban] would not carry out PNEs until international arrangements for conducting them are worked out. Such arrangements will have to be consistent with the treaty and the obligations of each party under other relevant agreements.

This new policy, designed to cover all contingencies, represents the ultimate in flexibility. The Soviets know well the difficulties in devising a foolproof method to prevent weapons-related information being derived from PNEs under a CTB. Should negotiations on a CTB commence, Soviet policy is likely to prove more or less identical to that of the West, although the need for new face-saving devices on the issue cannot be ruled out.

The major difficulty facing a CTB over the PNE issue probably derives not from the Soviet attitude, but from that of India, Argentina and Brazil. These three continue to assert, with varying degrees of emphasis and persistence, that PNEs can be distinguished from nuclear weapon tests by the intentions of the state involved; that attempts to ban PNEs are another plot by the developed countries to stifle economic development in the Third World; and that the PNE issue is peripheral to the negotiation of a CTB. Quite apart from allowing them to preserve their nuclear weapon option, PNEs have become a symbol for such states, albeit a tarnished one, of their right to achieve industrial and economic maturity.

India's attitude is currently the most uncertain of the three. In 1987 the Indian representative on the Palme Commission registered the only reservation to its report, *Toward a Comprehensive Test Ban*, by stating that:[17]

> The core of the CTB is to prevent the testing and development of nuclear weapons through nuclear explosions. It is not the purpose to stop the utilisation of nuclear technology, including explosions technology, for peaceful purposes. A problem arises, however, because the technology of explosions is the same for both peaceful and military purposes. The need to prevent explosions for military purposes has to be reconciled with the possibility of not blocking off a whole range of technology to the non-nuclear weapon countries. It should be possible to negotiate an international regime for the purpose...

The Indians now at least recognise the problem, even though they propose an unworkable solution. On 22 May 1987 Prime Minister Rajiv Gandhi joined five other non-aligned heads of state, the orginators of the so-called Five Continent Initiative, in calling for a 'halting of all nuclear testing'.[18] India has also voted in favour of the new Mexican/Swedish CTB resolution with its reference to the prohibition of all nuclear test explosions.

Argentina's policy is more strident. While it also joined the Five Continent Initiative statement and votes for the Mexican/Swedish resolution, its representative told the CD in 1986 during a speech on the CTB issue that:[19]

> Again, as is true in the consideration of other disarmament measures, the final purpose criterion should be applied to resolve questions which might arise from the dual use, peaceful and military, that can be made of any technology. The use of technology to promote economic and social development according to the needs, priorities and interests of each state should not be hindered.

Brazil's PNE policy remains the most vociferous of the three. In explaining its decision to abstain on the Mexican/Swedish resolution at the 1985 session of the UN General Assembly, Brazil said that the draft 'prejudged' the scope of a future CTB and did not make 'the necessary distinction between nuclear explosions for peaceful purposes and nuclear weapon explosions'.[20]

Despite the almost unanimous sentiment in the UN General Assembly in favour of a CTB, there is currently no prospect that negotiations on such a treaty will resume at an early date. In the absence of such negotiations, one way to encouarge a resolution of the PNE problem would be a multilateral PNE Treaty (a draft of which may be found at Annex V). Such an agreement would commit states to renounce the use of PNEs for all time. Among the nuclear weapon states, the United Kingdom has virtually taken such a position already, while the United States is in a *de facto* position of having renounced PNEs. Neither China nor France has shown any interest in taking up the PNE option. This leaves only the Soviet Union to be convinced. Such a treaty would place pressure on India, Brazil and Argentina to give up the PNE option. As Jozef Goldblat points out, 'the arguments against peaceful nuclear explosions would be more convincing if the nuclear weapon powers decided to forego them altogether'.[21] It would also clarify a confused legal situation, especially in regard to the Treaty of Tlatelolco. Being open to signature by all states on equal terms, a multilateral PNE treaty would permit non-NPT states to renounce the use of PNEs without acceding to the allegedly discriminatory NPT. Finally, it would supersede the 1976 Peaceful Nuclear Explosions Treaty, which lends credence to the PNE concept, and obviate the need for the superpowers to proceed with ratification.

As in the PNET, a multilateral PNE treaty could define PNEs as being those explosions conducted outside designated nuclear weapon test sites. While the Indians, Argentinians and Brazilians are likely to argue that this would discriminate in favour of the nuclear weapon states, such an objection could be countered by permitting any state wishing to sign the treaty to designate a nuclear weapon test site. India, Brazil and Argentina are unlikely to have the nerve to do so, while most other states, even if so inclined, would already be prevented by their adherence to other international argeements, such as the NPT. Verification of such a PNE Treaty could be carried out by so-called National Technical Means, which includes satellite surveillance, signals intelligence and human intelligence, and the international seismic monitoring network planned by the CD for the verification of a CTB. The monitoring of a PNE treaty would in fact be a useful first task for the network, in preparation for the more difficult task of monitoring a CTB.

With or without a PNE treaty, the future of nuclear dynamite looks bleak. Circumscribed by a web of international treaties, environmentally, technically and economically discredited (except apparently in the Soviet Union), and widely recognised as a hazard to arms control, there appear to be few supporters of PNEs left. In September 1987 the Executive Director of the New York-based Scientists and Engineers for Secure Energy, Miro Todorovich, said he felt that the Soviet Union had 'outsmarted' the United States by pursuing PNEs. He admitted, however, that the US public had a 'low threshold' for anything nuclear. 'This formalism', he said, 'has really stifled many projects'.[22]

Only one man appears to believe in a rosy future for PNEs. Despite the damage done to arms control, the obvious practical drawbacks and the waste of human and material resources involved, Edward Teller remains a true believer. When asked in late 1986 how he viewed Project Plowshare in retrospect he retorted, 'Not in retrospect, in prospect!'[23] Clearly he regards the abandonment of Plowshare after 1973 as a temporary setback. He blames 'prejudices' and 'politics', 'environmentalists' and 'Communists' for Plowshare's fate. When pressed as to whether he truly believed nuclear dynamite had a future, he replied unhesitatingly that he wasn't sure if the United States or Australia had a future, but he knew that Plowshare did. After all, Teller added, it was sixty years after the telephone was invented that the White House finally had one installed. 'Sometimes progress takes a little while...'

## *Notes*

1 H. Scoville Jr., 'Peaceful Nuclear Explosions — An Invitation to Proliferation' in A.W. Marks (ed.), *NPT: Paradoxes and Problems*, Arms Control Association, Washington DC, 1975, p. 53.
2 *Report of the Chairman of the Working Group on a South Pacific Nuclear Free Zone to the South Pacific Forum, Raratonga, 4–6 August 1985*, Australian Department of Foreign Affairs, Canberra, 1985, p. 48.
3 Ibid., p. 46.
4 Ibid., p. 19.
5 To date only the Soviet Union and China have done so.
6 E. Teller *et al.*, *The Constructive Uses of Nuclear Explosives*, McGraw-Hill, New York, 1968, p.v.
7 I.Y. Borg, *Nuclear Explosions for Peaceful Purposes*, Lawrence Livermore National Laboratory Paper No. UCRL–95026, Rev. II, November 1986, p. 17.
8 The *Canberra Times*, 20 April 1987, p. 6.
9 *Arms Control Chronicle*, 20, June 1987, p. 20.
10 *Arms Control Chronicle*, 21, August 1987, p. 16.
11 Australian Seismological Centre telexes of 8 July, 24 July and 12 August 1987.
12 Hagfors Observatory, Sweden, telex to Australian Seismological Centre, Canberra, 6 October 1987.
13 The *Canberra Times*, 27 August 1986.
14 From a conversation with Mr Shoustov at Sochi, USSR, 25 March 1988.
15 *Bulletin of the Atomic Scientists*, May 1986, p. 48.
16 Committee (later Conference) on Disarmament, Geneva, Document CD/746.
17 *Towards A Comprehensive Test Ban*, The Palme Commission on Disarmament and Security Issues, Stockholm, 1987, p. 31.
18 CD/758.
19 CD/PV. 334.
20 UN General Assembly, First Committee Document A/C. 1/40/PV. 45, 26 November 1985.
21 J. Goldblat, 'The Indian Nuclear Test and the NPT' in A.W. Marks (ed.), p. 33.
22 *Washington Times*, 2 September 1987, p. 5.
23 T.C. Findlay, 'Twenty Minutes with Edward Teller', *Newsletter*, Peace Research Centre, Australia National University, Canberra, 2, 1, 1987, p. 2.

# Annex I

## Plowshare Experiments (1957–73)

| *Purpose* | *Type of Explosive* | | | |
|---|---|---|---|---|
| | *Nuclear* | | *Conventional High Explosive** | |
| Phenomenology | 1957 | Rainier** | | |
| | | | 1959 | Pre-Gnome |
| | 1961 | Gnome | | |
| | 1962 | Hardhat** | | |
| | 1963 | Shoal** | | |
| | 1964 | Handcar+ | | |
| | 1965 | Piledriver** | | |
| Excavation | | | 1960 | Toboggan |
| | | | 1960 | Stagecoach |
| | | | 1960 | Scooter |
| | | | 1960 | Buckboard |
| | | | 1961 | Rowboat |
| | 1962 | Danny Boy** | | |
| | 1962 | Sedan | | |
| | | | 1963 | Pre-Buggy I |
| | | | 1963 | Pre-Buggy II |
| | | | 1964 | Pre-Schooner |
| | | | 1964 | Dugout |
| | 1964 | Sulky | | |
| | 1964 | Palanquin | | |
| | | | 1965 | Pre-Schooner II |
| | 1968 | Cabriolet | | |
| | 1968 | Buggy | | |
| | | | 1968 | Pre-Gondola III |
| | 1968 | Schooner | | |
| Underground Storage | 1964 | Salmon** | | |
| Oil Shale Development | | | 1960 | Pinot |
| Gas Stimulation | 1967 | Gasbuggy | | |
| | 1969 | Rulison | | |
| | 1973 | Rio Blanco | | |
| PNE Explosives Development | 1963 | Tornillo | | |
| | 1964 | Klickitat | | |

| | |
|---|---|
| | 1964 Ace |
| | 1966 Templar |
| | 1966 Saxon |
| | 1966 Simms |
| | 1967 Switch |
| | 1968 Stoddard |
| | 1970 Flask |
| | 1971 Miniata (for gas stimulation) |
| Emplacement Technique Experiments | 1964 Dub |
| | 1967 Marvel |
| Heavy Element Production and Neutron Physics Experiments | 1962 Anacostia |
| | 1963 Kaweah |
| | 1963 Gerbil** |
| | 1963 Kennebec# |
| | 1963 Anchovy** |
| | 1963 Greys** |
| | 1964 Oconto** |
| | 1964 Bye** |
| | 1964 Par |
| | 1964 Barbel** |
| | 1964 Parrot** |
| | 1965 Scaup** |
| | 1965 Tweed** |
| | 1965 Petrel** |
| | 1966 Duryea** |
| | 1966 Cyclamen** |
| | 1966 Kankakee* |
| | 1966 Vulcan++ |
| | 1967 Persimmon** |
| | 1967 Hutch** |

* pre-1960 high explosive experiments have been omitted

** non-Plowshare experiments with Plowshare participation or contribution to Plowshare technology

\# not announced as Plowshare or listed by Norris *et al.* as such

\+ described in the *GURC Report* as a dedicated Plowshare shot, but by Borg as a multipurpose shot

++ listed by Norris *et al.* as Plowshare

(*Sources*: I. Y. Borg, *Nuclear Explosions for Peaceful Purposes*, Lawrence Livermore National Laboratory, Paper No. UCRL–95026, Rev. II, November 1986, p. 28; R. S. Norris, T. B. Cochran and W. Arkin, *Nuclear Weapons Databook* Working Paper NWD 86-2, 'Known US Nuclear Tests July 1945 to 31 December 1985', February 1986; and the *GURC Report*, Vol. II, pp. 206–11.)

# Annex II

## Western Draft Treaty on the Discontinuance of Nuclear Weapon Tests, 18 April 1961

Article 13
Detonations for Peaceful Purposes

1. Each of the Parties to this Treaty undertakes to detonate, or assist others in the detonation of, nuclear devices for peaceful purposes only in accordance with the provisions of this Article. The detonations carried out pursuant to the provisions of this Article shall not be regarded as a violation of Article I.

2. A Party intending to carry out or assist in such a detonation shall provide the Commission, at least four months in advance of the proposed detonation date, with a plan containing the following information.

A. The date, site and purpose of the proposed detonation;
B. The procedure it will follow to comply with paragraph 4 of this Article;
C. The expected yield of the device;
D. The measures to be taken to ensure that there will be no substantial fallout oustide the immediate vicinity; and
E. The measurements to be taken and any experimentation to be conducted therewith.

3. Within two months after the receipt of the plan, the Commission shall authorise the Party to proceed with, or assist in, the proposed detonation, unless the Commission shall find that such detonation would not be carried out in accordance with paragraph 4 of this Article. If, as a result of observations at the proposed site, the Commission determines that there is a lack of compliance with paragraph 4, it shall immediately so notify the Party planning to conduct or assist in the detonation. The Party shall thereupon refrain from carrying out or assisting in the detonation until notified by the Commission that it has determined that the detonation will be carried out in accordance with paragraph 4.

4. Each of the original Parties shall be given an adequate opportunity at a designated inspection site to inspect externally and internally any nuclear device to be detonated pursuant to this Article and to examine detailed drawings of the device, provided that such detailed drawings may not be reproduced or taken away from the inspection site. The device to be detonated shall, after inspection and reassembly, be under the continual surveillance of members of the Organisation staff until detonation.

5. Members of the Organisation staff shall, in addition to maintaining surveillance of the device to be detonated, observe all preparation for, and the actual firing of, the device and shall at all times have unrestricted access to the vicinity of the detonation to ensure that the device employed is the one provided in accordance with paragraph 4 of this Article.

6. Representatives of the original Parties shall be given adequate opportunity to accompany and to participate with members of the Organisation staff in the exercise of their functions under paragraphs 4 and 5 of this Article.

7. The Commission may, with the concurring votes of the original Parties, provide for any other system of safeguards to ensure that nuclear detonations for peaceful purposes are carried out in accordance with the objectives of this Treaty.

(*Source*: US Disarmament Administration, *Geneva Conference on the Discontinuance of Nuclear Weapon Tests: History and Analysis of Negotiations*, US Department of State, Washington DC, 1961, pp. 492–93.)

# Annex III

## Treaty for the Prohibition of Nuclear Weapons in Latin America

### Explosions for Peaceful Purposes

Article 18

1. The Contracting Parties may carry out explosions of nuclear devices for peaceful purposes — including explosions which involve devices similar to those used in nuclear weapons — or collaborate with other parties for the same purpose, provided that they do so in accordance with the provisions of this article and other articles of the Treaty, particularly articles 1 and 5.

2. Contracting Parties intending to carry out, or to co-operate in carrying out, such an explosion shall notify the Agency and the International Atomic Energy Agency, as far in advance as the circumstances require, of the date of the explosion and shall at the same time provide the following information:

   (a) The nature of the nuclear device and the source from which it was obtained;
   (b) The place and purpose of the planned explosion;
   (c) The procedures which will be followed in order to comply with paragraph 3 of this article;
   (d) The expected force of the device; and
   (e) The fullest possible information on any possible radioactive fallout that may result from the explosion or explosions, and measures which will be taken to avoid danger to the population, flora, fauna and territories of any other Party or Parties.

3. The General Secretary and the technical personnel designated by the Council and the International Atomic Energy Agency may observe all the preparations, including the explosion of the device, and shall have unrestricted access to any area in the vicinity of the site of the explosion in order to ascertain whether the device and the procedures followed during the explosion are in conformity with the information supplied under paragraph 2 of this article and with other provisions of the Treaty.

4. The Contracting Parties may accept the collaboration of third parties for the purpose set forth in paragraph 1 of the present article, in accordance with paragraphs 2 and 3 thereof.

(*Source: Status of Multilateral Arms Regulation and Disarmament Agreements*, United Nations Department for Disarmament Affairs, New York, 1982, pp. 55–56.)

# Annex IV

## Treaty Between the United States of America and the Union of Soviet Socialist Republics on Underground Nuclear Explosions for Peaceful Purposes

Signed at Washington and Moscow May 28, 1976

The United States of America and the Union of Soviet Socialist Republics, hereinafter referred to as the Parties,

Proceeding from a desire to implement Article III of the Treaty between the United States of America and the Union of Soviet Socialist Republics on the Limitation of Underground Nuclear Weapon Tests, which calls for the earliest possible conclusion of an agreement on underground nuclear explosions for peaceful purposes,

Reaffirming their adherence to the objectives and principles of the Treaty Banning Nuclear Weapon Tests in the Atmosphere, in Outer Space and Under Water, the Treaty on Non-Proliferation of Nuclear Weapons, and the Treaty on the Limitation of Underground Nuclear Weapon Tests, and their determination to observe strictly the provisions of these international agreements,

Desiring to assure that underground nuclear explosions for peaceful purposes shall not be used for purposes related to nuclear weapons,

Desiring that utilization of nuclear energy be directed only toward peaceful purposes,

Desiring to develop appropriately cooperation in the field of underground nuclear explosions for peaceful purposes,

Have agreed as follows:

Article I

1. The Parties enter into this Treaty to satisfy the obligations in Article III of the Treaty on the Limitation of Underground Nuclear Weapon Tests, and assume additional obligations in accordance with the provisions of this Treaty.

2. This Treaty shall govern all underground nuclear explosions for peaceful purposes conducted by the Parties after March 31, 1976.

Article II

For the purposes of this Treaty:

(a) "explosion" means any individual or group underground nuclear explosion for peaceful purposes;

(b) "explosive" means any device, mechanism or system for producing an individual explosion;

(c) "group explosion" means two or more individual explosions for which the time interval between successive individual explosions does not exceed five seconds and for which the emplacement points of all explosives can be interconnected by straight line segments, each of which joins two emplacement points and each of which does not exceed 40 kilometers.

Article III

1. Each Party, subject to the obligations assumed under this Treaty and other international agreements, reserves the right to:

(a) carry out explosions at any place under its jurisdiction or control outside the geographical boundaries of test sites specified under the provisions of the Treaty on the Limitation of Underground Nuclear Weapon Tests; and

(b) carry out, participate or assist in carrying out explosions in the territory of another State at the request of such other State.

2. Each Party undertakes to prohibit, to prevent and not to carry out at any place under its jurisdiction or control, and further undertakes not to carry out, participate or assist in carrying out anywhere:

(a) any individual explosion having a yield exceeding 150 kilotons;

(b) any group explosion:

(1) having an aggregate yield exceeding 150 kilotons except in ways that will permit identification of each individual explosion and determination of the yield of each individual explosion in the group in accordance with the provisions of Article IV of and the Protocol to this Treaty;

(2) having an aggregate yield exceeding one and one-half megatons;

(c) any explosion which does not carry out a peaceful application;

(d) any explosion except in compliance with the provisions of the Treaty Banning Nuclear Weapon Tests in the Atmosphere, in Outer Space and Under Water, the Treaty on the Non-Proliferation of Nuclear Weapons, and other international agreements entered into by that Party.

3. The question of carrying out any individual explosion having a yield exceeding the yield specified in paragraph 2(a) of this article will be considered by the Parties at an appropriate time to be agreed.

Article IV

1. For the purpose of providing assurance of compliance with the provisions of this Treaty, each Party shall;

(a) use national technical means of verification at its disposal in a manner consistent with generally recognized principles of international law; and

(b) provide to the other Party information and access to sites of explosions and furnish assistance in accordance with the provisions set forth in the Protocol to this Treaty.

2. Each Party undertakes not to interfere with the national technical means of verification of the other Party operating in accordance with paragraph 1(b) of this article.

Article V

1. To promote the objectives and implementation of the provisions of this Treaty, the Parties shall establish promptly a Joint Consultative Commission within the framework of which they will:

(a) consult with each other, make inquiries and furnish information in response to such inquiries, to assure confidence in compliance with the obligations assumed;

(b) consider questions concerning compliance with the obligations assumed and related situations which may be considered ambiguous;

(c) consider questions involving unintended interference with the means for assuring compliance with the provisions of this Treaty;

(d) consider changes in technology or other new circumstances which have a bearing on the provisions of this Treaty; and

(e) consider possible amendments to provisions governing underground nuclear explosions for peaceful purposes.

2. The Parties through consultation shall establish, and may amend as appropriate, Regulations for the Joint Consultative Commission governing procedures, composition and other relevant matters.

Article VI

1. The Parties will develop cooperation on the basis of mutual benefit, equality, and reciprocity in various areas related to carrying out underground nuclear explosions for peaceful purposes.

2. The Joint Consultative Commission will facilitate this cooperation by considering specific areas and forms of cooperation which shall be determined by agreement between the Parites in accordance with their constitutional procedures.

3. The Parties will appropriately inform the International Atomic Energy Agency of results of their cooperation in the field of underground nuclear explosions for peaceful purposes.

Article VII

1. Each Party shall continue to promote the development of the international agreement or agreements and procedures provided for in Article V of the Treaty on the Non-Proliferation of Nuclear Weapons, and shall provide appropriate assistance to the International Atomic Energy Agency in this regard.

2. Each Party undertakes not to carry out, participate or assist in the carrying out of any explosion in the territory of another State unless that State agrees to the implementation in its territory of the international observation and procedures contemplated by Article V of the Treaty on the Non-Proliferation of Nuclear Weapons and the provisions of Article IV of and the Protocol to this Treaty, including the provision by that State of the assistance necessary for such implementation and of the privileges and immunities specified in the Protocol.

Article VIII

1. This Treaty shall remain in force for a period of five years, and it shall be extended for successive five-year periods unless either Party notifies the other of its termination no later than six months prior to its expiration. Before the expiration of this period the Parties may, as necessary, hold consultations to consider the situation relevant to the substance of this Treaty. However, under no circumstances shall either Party be entitled to terminate this Treaty while the Treaty on the Limitation of Underground Nuclear Weapon Tests remains in force.

2. Termination of the Treaty on the Limitation of Underground Nuclear Weapon Tests shall entitle either Party to withdraw from this Treaty at any time.

3. Each Party may propose amendments to this Treaty. Amendments shall enter into force on the day of the exchange of instruments of ratification of such amendments.

Article IX

1. This Treaty including the Protocol which forms an integral part hereof, shall be subject to ratification in accordance with the constitutional procedures of each Party. This Treaty shall enter into force on the day of the exchange of instruments of ratification which exchange shall take place simultaneously with the exchange of instruments of ratification of the Treaty on the Limitation of Underground Nuclear Weapon Tests.

2. This Treaty shall be registered pursuant to Article 102 of the Charter of the United Nations.

**DONE** at Washington and Moscow, on May 28, 1976, in duplicate, in the English and Russian languages, both texts being equally authentic.

For the United States of America:

**GERALD R. FORD,**

*The President of the United States of America.*

For the Union of Soviet Socialist Republics:

**L. BREZHNEV,**

*General Secretary of the Central Committee of the CPSU.*

## Protocol to the Treaty Between the United States of America and the Union of Soviet Socialist Republics on Underground Nuclear Explosions for Peaceful Purposes

The United States of America and the Union of Soviet Socialist Republics, hereinafter referred to as the Parties,

Having agreed to the provisions in the Treaty on Underground Nuclear Explosions for Peaceful Purposes, hereinafter referred to as the Treaty,

Have agreed as follows:

## Article I

1. No individual explosion shall take place at a distance, in meters, from the ground surface which is less than 30 times the 3.4 root of its planned yield in kilotons.

2. Any group explosion with a planned aggregate yield exceeding 500 kilotons shall not include more than five individual explosions, each of which has a planned yield not exceeding 50 kilotons.

## Article II

1. For each explosion, the Party carrying out the explosion shall provide the other Party:

(a) not later than 90 days before the beginning of emplacement of the explosives when the planned aggregate yield of the explosion does not exceed 100 kilotons, or not later than 180 days before the beginning of emplacement of the explosives when the planned aggregate yield of the explosion exceeds 100 kilotons, with the following information to the extent and degree of precision available when it is conveyed:

(1) the purpose of the planned explosion;

(2) the location of the explosion expressed in geographical coordinates with a precision of four or less kilometers, planned date and aggregate yield of the explosion;

(3) the type or types of rock in which the explosion will be carried out, including the degree of liquid saturation of the rock at the point of emplacement of each explosive; and

(4) a description of specific technological features of the project, of which the explosion is a part, that could influence the determination of its yield and confirmation of purpose; and

(b) not later than 60 days before the beginning of emplacement of the explosives the information specified in subparagraph 1(a) of this article to the full extent and with the precision indicated in that subparagraph.

2. For each explosion with a planned aggregate yield exceeding 50 kilotons, the Party carrying out the explosion shall provide the other Party, not later than 60 days before the beginning of emplacement of the explosives, with the following information:

(a) the number of explosives, the planned yield of each explosive, the location of each explosive to be used in a group explosion relative to all other explosives in the group with a precision of 100 or less meters, the depth of emplacement of each explosive with a precision of one meter and the time intervals between individual explosions in any group explosion with a precision of one-tenth second; and

(b) a description of specific features of geological structure or other local conditions that could influence the determination of the yield.

3. For each explosion with a planned aggregate yield exceeding 75 kilotons, the Party carrying out the explosion shall provide the other Party, not later than 60 days before the beginning of emplacement of the explosives,

with a description of the geological and geophysical characteristics of the site of each explosion which could influence determination of the yield, which shall include; the depth of the water table; a stratigraphic column above each emplacement point; the position of each emplacement point relative to nearby geological and other features which influenced the design of the project of which the explosion is a part; and the physical parameters of the rock, including density, seismic velocity, porosity, degree of liquid saturation, and rock strength, within the sphere centered on each emplacement point and having a radius, in meters, equal to 30 times the cube root of the planned yield in kilotons of the explosive emplaced at that point.

4. For each explosion with a planned aggregate yield exceeding 100 kilotons, the Party carrying out the explosion shall provide the other Party, not later than 60 days before the beginning of emplacement of the explosives, with:

(a) information on locations and purposes of facilities and installations which are associated with the conduct of the explosion;

(b) information regarding the planned date of the beginning of emplacement of each explosive; and

(c) a topographic plan in local coordinates of the areas specified in paragraph 7 of Article IV, at a scale of 1:24,000 or 1:25,000 with a contour interval of 10 meters or less.

5. For application of an explosion to alleviate the consequences of an emergency situation involving an unforeseen combination of circumstances which calls for immediate action for which it would not be practicable to observe the timing requirements of paragraphs 1, 2 and 3 of this article, the following conditions shall be met:

(a) the Party deciding to carry out an explosion for such purposes shall inform the other Party of the decision immediately after it has been made and describe such circumstances;

(b) the planned aggregate yield of an explosion for such purpose shall not exceed 100 kilotons; and

(c) the Party carrying out an explosion for such purpose shall provide to the other Party the information specified in paragraph 1 of this article, and the information specified in paragraph 2 and 3 of this article if applicable, after the decision to conduct the explosion is taken, but not later than 30 days before the beginning of emplacement of the explosives.

6. For each explosion, the Party carrying out the explosion shall inform the other Party, not later than two days before the explosion, of the planned time of detonation of each explosive with a precision of one second.

7. Prior to the explosion, the Party carrying out the explosion shall provide the other Party with timely notification of changes in the information provided in accordance with this article.

8. The explosion shall not be carried out earlier than 90 days after notification of any change in the information provided in accordance with

this article which requires more extensive verification procedures than those required on the basis of the original information, unless an earlier time for carrying out the explosion is agreed between the Parties.

9. Not later than 90 days after each explosion the Party carrying out the explosion shall provide the other Party with the following information:

(a) the actual time of the explosion with a precision of one-tenth second and its aggregate yield;

(b) when the planned aggregate yield of a group explosion exceeds 50 kilotons, the actual time of the first individual explosion with a precision of one-tenth second, the time interval between individual explosions with a precision of one milli-second and the yield of each individual explosion; and

(c) confirmation of other information provided in accordance with paragraphs 1, 2, 3 and 4 of this article and explanation of any changes or corrections based on the results of the explosion.

10. At any time, but not later than one year after the explosion, the other Party may request the Party carrying out the explosion to clarify any item of the information provided in accordance with this article. Such clarification shall be provided as soon as practicable, but not later than 30 days after the request is made.

### Article III

1. For the purposes of this Protocol:

(a) "designated personnel" means those nationals of the other Party identified to the Party carrying out an explosion as the persons who will exercise the rights and functions provided for in the Treaty and this Protocol; and

(b) "emplacement hole" means the entire interior of any drill-hole, shaft, adit or tunnel in which an explosive and associated cables and other equipment are to be installed.

2. For any explosion with a planned aggregate yield exceeding 100 kilotons but not exceeding 150 kilotons if the Parties, in consultation based on information provided in accordance with Article II and other information that may be introduced by either Party, deem it appropriate for the confirmation of the yield of the explosion, and for any explosion with a planned aggregate yield exceeding 150 kilotons, the Party carrying out the explosion shall allow designated personnel within the areas and at the locations described in Article V to exercise the following rights and functions:

(a) confirmation that the local circumstances, including facilities and installations associated with the project, are consistent with the stated peaceful purposes;

(b) confirmation of the validity of the geological and geophysical information provided in accordance with Article II through the following procedures:

(1) examination by designated personnel of research and measure-

ment data of the Party carrying out the explosion and of rock core or rock fragments removed from each emplacement hole, and of any logs and drill core from existing exploratory holes which shall be provided to designated personnel upon their arrival at the site of the explosion;

(2) examination by designated personnel of rock core or rock fragments as they become available in accordance with the procedures specified in subparagraph 2(b)(3) of this article; and

(3) observation by designated personnel of implementation by the Party carrying out the explosion of one of the following four procedures, unless this right is waived by the other Party:

(i) construction of that portion of each emplacement hole starting from a point nearest the entrance of the emplacement hole which is at a distance, in meters, from the nearest emplacement point equal to 30 times the cube root of the planned yield in kilotons of the explosive to be emplaced at that point and continuing to the completion of the emplacement hole; or

(ii) construction of that portion of each emplacement hole starting from a point nearest the entrance of the emplacement hole which is at a distance, in meters, from the nearest emplacement point equal to six times the cube root of the planned yield in kilotons of the explosive to be emplaced at that point and continuing to the completion of the emplacement hole as well as the removal of rock core or rock fragments from the wall of an existing exploratory hole, which is substantially parallel with and at no point more than 100 meters from the emplacement hole, at locations specified by designated personnel which lie within a distance, in meters, from the same horizon as each emplacement point of 30 times the cube root of the planned yield in kilotons of the explosive to be emplaced at that point; or

(iii) removal of rock core or rock fragments from the wall of each emplacement hole at locations specified by designated personnel which lie within a distance, in meters, from each emplacement point of 30 times the cube root of the planned yield in kilotons of the explosive to be emplaced at each such point; or

(iv) construction of one or more new exploratory holes so that for each emplacement hole there will be a new exploratory hole to the same depth as that of the emplacement of the explosive, substantially parallel with and at no point more than 100 meters from each emplacement hole, from which rock cores would be removed at locations specified by designated personnel which lie within a distance, in meters, from the same horizon as each emplacement point of 30 times the cube root of the planned yield in kilotons of the explosive to be emplaced at each such point:

(c) observation of the emplacement of each explosive, confirmation of the depth of its emplacement and observation of the stemming of each emplacement hole;

(d) unobstructed visual observation of the area of the entrance to each emplacement hole at any time from the time of emplacement of each explosive until all personnel have been withdrawn from the site for the detonation of the explosion; and

(e) observation of each explosion.

3. Designated personnel, using equipment provided in accordance with paragraph 1 of Article IV, shall have the right, for any explosion with planned aggregate yield exceeding 150 kilotons, to determine the yield of each individual explosion in a group explosion in accordance with the provisions of Article VI.

4. Designated personnel, when using their equipment in accordance with paragraph 1 of Article IV, shall have the right, for any explosion with planned aggregate yield exceeding 500 kilotons, to emplace, install and operate under the observation and with the assistance of personnel of the Party carrying out the explosion, if such assistance is requested by designated personnel, a local seismic network in accordance with the provisions of paragraph 7 of Article IV. Radio links may be used for the transmission of data and control signals between the seismic stations and the control center. Frequencies, maximum power output of radio transmitters, directivity of antennas and times of operation of the local seismic network radio transmitters before the explosion shall be agreed between the Parties in accordance with Article X and time of operation after the explosion shall conform to the time specified in paragraph 7 of Article IV.

5. Designated personnel shall have the right to:

(a) acquire photographs under the following conditions:

(1) the Party carrying out the explosion shall identify to the other Party those personnel of the Party carrying out the explosion who shall take photographs as requested by designated personnel;

(2) photographs shall be taken by personnel of the Party carrying out the explosion in the presence of designated personnel and at the time requested by designated personnel for taking such photographs. Designated personnel shall determine whether these photographs are in conformity with their requests and, if not, additional photographs shall be taken immediately;

(3) photographs shall be taken with cameras provided by the other Party having built-in, rapid developing capability and a copy of each photograph shall be provided at the completion of the development process to both Parties;

(4) cameras provided by designated personnel shall be kept in agreed secure storage when not in use; and

(5) the request for photographs can be made, at any time, of the following:

(i) exterior views of facilities and installations associated with the conduct of the explosion as described in subparagraph 4(a) of Article II;

(ii) geological samples used for confirmation of geological and geophysical information, as provided for in subparagraph 2(b) of this article and the equipment utilized in the acquistion of such samples;

(iii) emplacement and installation of equipment and associated cables used by designated personnel for yield determination;

(iv) emplacement and installation of the local seismic network used by designated personnel;

(v) emplacement of the explosives and the stemming of the emplacement hole; and

(vi) containers, facilities and installations for storage and operation of equipment used by designated personnel;

(b) photographs of visual displays and records produced by the equipment used by designated personnel and photographs within the control centers taken by cameras which are component parts of such equipment; and

(c) receive at the request of designated personnel and with the agreement of the Party carrying out the explosion supplementary photographs taken by the Party carrying out the explosion.

## Article IV

1. Designated personnel in exercising their rights and functions may choose to use the following equipment of either Party, of which choice the Party carrying out the explosion shall be informed not later than 150 days before the beginning of emplacement of the explosives:

(a) electrical equipment for yield determination and equipment for a local seismic network as described in paragraphs 3, 4 and 7 of this article; and

(b) geologist's field tools and kits and equipment for recording of field notes.

2. Designated personnel shall have the right in exercising their rights and functions to utilize the following additional equipment which shall be provided by the Party carrying out the explosion, under procedures to be established in accordance with Article X to ensure that the equipment meets the specifications of the other Party: portable short-range communication equipment, field glasses, optical equipment for surveying and other items which may be specified by the other Party. A description of such equipment and operating instructions shall be provided to the other Party not later than 90 days before the beginning of emplacement of the explosives in connection with which such equipment is to be used.

3. A complete set of electrical equipment for yield determination shall consist of:

(a) sensing elements and associated cables for transmission of electrical power, control signals and data;

(b) equipment of the control center, electrical power supplies and cables for transmission of electrical power, control signals and data; and

(c) measuring and calibration instruments, maintenance equipment and spare parts necessary for ensuring the functioning of sensing elements, cables and equipment of the control center.

4. A complete set of equipment for the local seismic network shall consist of:

(a) seismic stations each of which contains a seismic instrument, electrical power supply and associated cables and radio equipment for receiving and transmission of control signals and data or equipment for recording control signals and data;

(b) equipment of the control center and electrical power supplies; and

(c) measuring and calibration instruments, maintenance equipment and spare parts necessary for ensuring the functioning of the complete network.

5. In case designated personnel, in accordance with paragraph 1 of this article, choose to use equipment of the Party carrying out the explosion for yield determination or for a local seismic network, a description of such equipment and installation and operating instructions shall be provided to the other Party not later than 90 days before the beginning of emplacement of the explosives in connection with which such equipment is to be used. Personnel of the Party carrying out the explosion shall emplace, install and operate the equipment in the presence of designated personnel. After the explosion, designated personnel shall receive duplicate copies of the recorded data. Equipment for yield determination shall be emplaced in accordance with Article VI. Equipment for a local seismic network shall be emplaced in accordance with paragraph 7 of this article.

6. In case designated personnel, in accordance with paragraph 1 of this article, choose to use their own equipment for yield determination and their own equipment for a local seismic network, the following procedures shall apply:

(a) the Party carrying out the explosion shall be provided by the other Party with the equipment and information specified in subparagraphs (a)(1) and (a)(2) of this paragraph not later than 150 days prior to the beginning of emplacement of the explosives in connection with which such equipment is to be used in order to permit the Party carrying out the explosion to familarize itself with such equipment, if such equipment and information has not been previously provided, which equipment shall be returned to the other Party not later than 90 days before the beginning of emplacement of the explosives. The equipment and information to be provided are:

(1) one complete set of electrical equipment for yield determination as described in paragraph 3 of this article, electrical and mechanical design information, specifications and installation and operating instructions concerning this equipment; and

(2) one complete set of equipment for the local seismic network described in paragraph 4 of this article, including one seismic station, electrical and mechanical design information, specifications and installation and operating instructions concerning this equipment.

8. The Party carrying out the explosion shall have the right to examine

in the presence of designated personnel all equipment, instruments and tools of designated personnel specified in subparagraph 1(b) of this article.

9. The Joint Consultative Commission will consider proposals that either Party may put forward for the joint development of standardized equipment for verification purposes.

Article V

1. Except as limited by the provisions of paragraph 5 of this article, designated personnel in the exercise of their rights and functions shall have access along agreed routes:

(a) for an explosion with a planned aggregate yield exceeding 100 kilotons in accordance with paragraph 2 of Article III:

(1) to the locations of facilities and installations associated with the conduct of the explosion provided in accordance with subparagraph 4(a) of Article II; and

(2) to the locations of activities described in paragraph 2 of Article III; and

(b) for any explosion with a planned aggregate yield exceeding 150 kilotons, in addition to the access described in subparagraph 1(a) of this article:

(1) to other locations within the area circumscribed by circles of 10 kilometers in radius centered on points on the surface of the earth above the points of emplacement of the explosives in order to confirm that the local circumstances are consistent with the stated peaceful purposes;

(2) to the locations of the components of the electrical equipment for yield determination to be used for recording data when, by agreement between the Parties, such equipment is located outside the area described in subparagraph 1(b)(1) of this article; and

(3) to the sites of emplacement of the equipment of the local seismic network provided for in paragraph 7 of Article IV.

2. The Party carrying out the explosion shall notify the other Party of the procedure it has chosen from among those specified in subparagraph 2(b)(3) of Article III not later than 30 days before beginning the implementation of such procedure. Designated personnel shall have the right to be present at the site of the explosion to exercise their rights and functions in the areas and at the locations described in paragraph 1 of this article for a period of time beginning two days before the beginning of the implementation of the procedure and continuing for a period of three days after the completion of this procedure.

3. Except as specified in paragraph 5 of this article, designated personnel shall have the right to be present in the areas and at the locations described in paragraph 1 of this article:

(a) for an explosion with a planned aggregate yield exceeding 100 kilotons but not exceeding 150 kilotons, in accordance with paragraph 2 of Article III, at any time beginning five days before the beginning of emplacement of the explosives and continuing after the explosion and after safe access to evacuated areas has been established according to standards determined by the Party carrying out the explosion for a period of two days; and

(b) for any explosion with a planned aggregate yield exceeding 150 kilotons, at any time beginning 20 days before the beginning of emplacement of the explosives and continuing after the explosion and after safe access to evacuated areas has been established according to standards determined by the Party carrying out the explosion for a period of:

(1) five days in the case of an explosion with a planned aggregate yield exceeding 150 kilotons but not exceeding 500 kilotons; or

(2) eight days in the case of an explosion with a planned aggregate yield exceeding 500 kilotons.

4. Designated personnel shall not have the right to be present in those areas from which all personnel have been evacuated in connection with carrying out an explosion, but shall have the right to re-enter those areas at the same time as personnel of the Party carrying out the explosion.

5. Designated personnel shall not have or seek access by physical, visual or technical means to the interior of the canister containing an explosive, to documentary or other information descriptive of the design of an explosive nor to equipment for control and firing of explosives. The Party carrying out the explosion shall not locate documentary or other information descriptive of the design of an explosive in such ways as to impede the designated personnel in the exercise of their rights and functions.

6. The number of designated personnel present at the site of an explosion shall not exceed:

(a) for the exercise of their rights and functions in connection with the confirmation of the geological and geophysical information in accordance with the provisions of subparagraph 2(b) and applicable provisions of paragraph 5 of Article III—the number of emplacement holes plus three;

(b) for the exercise of their rights and functions in connection with confirming that the local circumstances are consistent with the information provided and with the stated peaceful purposes in accordance with the provisions in subparagraphs 2(a), 2(c), 2(d) and 2(e) and applicable provisions of paragraph 5 of Article III—the number of explosives plus two;

(c) for the exercise of their rights and functions in connection with confirming that the local circumstances are consistent with the information provided and with the stated peaceful purposes in accordance with the provisions in subparagraphs 2(a), 2(c), 2(d) and 2(e) and applicable provisions of paragraph 5 of Article III and in connection with the use of electrical equipment for determination of the yield in accordance with paragraph 3 of Article III—the number of explosives plus seven; and

(d) for the exercise of their rights and functions in connection with confirming that the local circumstances are consistent with the information provided and with the stated peaceful purposes in accordance with the provisions in subparagraph 2(a), 2(c), 2(d) and 2(e) and applicable provisions of paragraph 5 of Article III and in connection with the use of electrical equipment for determination of the yield in accordance with paragraph 3 of Article III and with the use of the local seismic network in accordance with paragraph 4 of Article III—the number of explosives plus 10.

7. The Party carrying out the explosion shall have the right to assign its personnel to accompany designated personnel while the latter exercise their rights and functions.

8. The Party carrying out an explosion shall assure for designated personnel telecommunications with their authorities, transportation and other services appropriate to their presence and to the exercise of their rights and functions at the site of the explosion.

9. The expenses incurred for the transportation of designated personnel and their equipment to and from the site of the explosion, telecommunications provided for in paragraph 8 of this article, their living and working quarters, subsistence and all other personal expenses shall be the responsibility of the Party other than the Party carrying out the explosion.

10. Designated personnel shall consult with the Party carrying out the explosion in order to coordinate the planned program and schedule of activities of designated personnel with the program of the Party carrying out the explosion for the conduct of the project so as to ensure that designated personnel are able to conduct their activities in an orderly and timely way that is compatible with the implementation of the project. Procedures for such consultations shall be established in accordance with Article X.

## Article VI

For any explosion with a planned aggregate yield exceeding 150 kilotons, determination of the yield of each explosive used shall be carried out in accordance with the following provisions:

1. Determination of the yield of each individual explosion in the group shall be based on measurements of the velocity of propagation, as a function of time, of the hydrodynamic shock wave generated by the explosion, taken by means of electrical equipment described in paragraph 3 of Article IV.

2. The Party carrying out the explosion shall provide the other Party with the following information:

(a) not later than 60 days before the beginning of emplacement of the explosives, the length of each canister in which the explosive will be contained in the corresponding emplacement hole, the dimensions of the tube or other device used to emplace the canister and the cross-sectional dimensions of the emplacement hole to a distance, in meters, from the emplacement point of 10 times the cube root of its yield in kilotons;

(b) not later than 60 days before the beginning of emplacement of the explosives, a description of materials, including their densities, to be used to stem each emplacement hole; and

(c) not later than 30 days before the beginning of emplacement of the explosives, for each emplacement hole of a group explosion, the local coordinates of the point of emplacement of the explosive, the entrance of the emplacement hole, the point of the emplacement hole most distant from the entrance, the location of the emplacement hole at each 200 meters distance from the entrance and the configuration of any known voids larger than one cubic meter located within the distance, in meters, of 10 times the cube root of the planned yield in kilotons measured from the bottom of the canister containing the explosive. The error in these coordinates shall not exceed one percent of the distance between the emplacement hole and the nearest other emplacement hole or one percent of the distance between the point of measurement and the entrance of the emplacement hole, whichever is smaller, but in no case shall the error be required to be less than one meter.

3. The Party carrying out the explosion shall emplace for each explosive that portion of the electrical equipment for yield determination described in subparagraph 3(a) of Article IV, supplied in accordance with paragraph 1 of Article IV, in the same emplacement hole as the explosive in accordance with the installation instructions supplied under the provisions of paragraph 5 or 6 of Article IV. Such emplacement shall be carried out under the observation of designated personnel. Other equipment specified in subparagraph 3(b) of Article IV shall be emplaced and installed:

(a) by designated personnel under the observation and with the assistance of personnel of the Party carrying out the explosion, if such assistance is requested by designated personnel; or

(b) in accordance with paragraph 5 of Article IV.

4. That portion of the electrical equipment for yield determination described in subparagraph 3(a) of Article IV that is to be emplaced in each emplacement hole shall be located so that the end of the electrical equipment which is farthest from the entrance to the emplacement hole is at a distance, in meters, from the bottom of the canister containing the explosive equal to 3.5 times the cube root of the planned yield in kilotons of the explosive when the planned yield is less than 20 kilotons and three times the cube root of the planned yield in kilotons of the explosive when the planned yield is 20 kilotons or more. Canisters longer than 10 meters containing the explosive shall only be utilized if there is prior agreement between the Parties establishing provisions for their use. The Party carrying out the explosion shall provide the other Party with data on the distribution of density inside any other canister in the emplacement hole with a transverse cross-sectional area exceeding 10 square centimetres located within a distance, in meters, of 10 times the cube root of the planned yield in kilotons of the explosion from the bottom of the canister containing the explosive. The Party carrying out the explosion shall provide the other Party with access to confirm such data on density distribution within any such canister.

5. The Party carrying out an explosion shall fill each emplacement hole, including all pipes and tubes contained therein which have at any transverse

section an aggregate cross-sectional area exceeding 10 square centimetres in the region containing the electrical equipment for yield determination and to a distance, in meters, of six times the cube root of the planned yield in kilotons of the explosive from the explosive emplacement point, with material having a density not less than seven-tenths of the average density of the surrounding rock, and from that point to a distance of not less than 60 meters from the explosive emplacement point with material having a density greater than one gram per cubic centimeter.

6. Designated personnel shall have the right to:

(a) confirm information provided in accordance with subparagraph 2(a) of this article;

(b) confirm information provided in accordance with subparagraph 2(b) of this article and be provided, upon request, with a sample of each batch of stemming material as that material is put into the emplacement hole; and

(c) confirm the information provide in accordance with subparagraph 2(c) of this article by having access to the data acquired and by observing, upon their request, the making of measurements.

7. For those explosives which are emplaced in separate holes, the emplacement shall be such that the distance D, in meters, between any explosive and any portion of the electrical equipment for determination of the yield of any other explosive in the group shall be not less than 10 times the cube root of the planned yield in kilotons of the larger explosive of such a pair of explosives. Individual explosions shall be separated by time intervals, in milliseconds, not greater than one-sixth the amount by which the distance D, in meters, exceeds 10 times the cube root of the planned yield in kilotons of the larger explosive of such a pair of explosives.

8. For those explosives in a group which are emplaced in a common emplacement hole, the distance, in meters, between each explosive and any other explosive in that emplacement hole shall be not less than 10 times the cube root of the planned yield in kilotons of the larger explosive of such a pair of explosives, and the explosives shall be detonated in sequential order, beginning with the explosive farthest from the entrance to the emplacement hole, with the individual detonations separated by time intervals, in milliseconds, of not less than one times the cube root of the planned yield in kilotons of the largest explosive in this emplacement hole.

## Article VII

1. Designated personnel with their personal baggage and their equipment as provided in Article IV shall be permitted to enter the territory of the Party carrying out the explosion at an entry port to be agreed upon by the Parties, to remain in the territory of the Party carrying out the explosion for the purpose of fulfilling their rights and functions provided for in the Treaty and this Protocol, and to depart from an exit port to be agreed upon by the Parties.

2. At all times while designated personnel are in the territory of the Party carrying out the explosion, their persons, property, personal baggage, archives and documents as well as their temporary official and living quarters shall be accorded the same privileges and immunities as provided in Articles 22, 23, 24, 29, 30, 31, 34 and 36 of the Vienna Convention on Diplomatic Relations of 1961 to the persons, property, personal baggage, archives and documents of diplomatic agents as well as to the premises of diplomatic missions and private residences of diplomatic agents.

3. Without prejudice to their privileges and immunities it shall be the duty of designated personnel to respect the laws and regulations of the State in whose territory the explosion is to be carried out insofar as they do not impede in any way whatsoever the proper exercising of their rights and functions provided for by the Treaty and this Protocol.

## Article VIII

The Party carrying out an explosion shall have sole and exclusive control over and full responsibility for the conduct of the explosion.

## Article IX

1. Nothing in the Treaty and this Protocol shall affect proprietary rights in information made available under the Treaty and this Protocol and in information which may be disclosed in preparation for and carrying out of explosions; however, claims to such proprietary rights shall not impede implementation of the provisions of the Treaty and this Protocol.

2. Public release of the information provided in accordance with Article II or publication of material using such information, as well as public release of the results of observation and measurements obtained by designated personnel, may take place only by agreement with the Party carrying out an explosion; however, the other Party shall have the right to issue statements after the explosion that do not divulge information in which the Party carrying out the explosion has rights which are referred to in paragraph 1 of this article.

## Article X

The Joint Consultative Commission shall establish procedures through which the Parties will, as appropriate, consult with each other for the purpose of ensuring efficient implementation of this Protocol.

**DONE** at Washington and Moscow, on May 28, 1976.

For the United States of America:

**GERALD R. FORD,**

*The President of the United States of America.*

For the Union of Soviet Socialist Republics:

**L. BREZHNEV,**
*General Secretary of the Central Committee of the CPSU.*

## AGREED STATEMENT

*May 13, 1976*

The Parties to the Treaty Between the United States of America and the Union of Soviet Socialist Republics on Underground Nuclear Explosions for Peaceful Purposes, hereinafter referred to as the Treaty, agree that under subparagraph 2(c) of Article III of the Treaty:

(a) Development testing of nuclear explosives does not constitute a "peaceful application" and any such development tests shall be carried out only within the boundaries of nuclear weapon test sites specified in accordance with the Treaty between the United States of America and the Union of Soviet Socialist Republics on the Limitation of Underground Nuclear Weapon Tests;

(b) Associating test facilities, instrumentation or procedures related only to testing of nuclear weapons or their effects with any explosion carried out in accordance with the Treaty does not constitute a "peaceful application".

# Annex V

## Model Multilateral Peaceful Nuclear Explosions Treaty

The States Parties to this Treaty,

Seeking to achieve the discontinuance of all test explosions of nuclear devices for all times,[1]

Aware that the technology for producing nuclear weapons is identical to that for producing other nuclear explosive devices, including those for peaceful purposes,

Mindful of the possibility that military benefits may be derived from the conduct of nuclear explosions for peaceful purposes,

Convinced that the prevention of nuclear war is so vital to the survival of the human race that it would be a world tragedy if inconsequential peaceful applications of nuclear explosives were allowed to interfere with that goal,

Determined to forego any potential benefits from any peaceful applications of nuclear explosions in order to achieve that goal,

Declaring their intention to achieve at the earliest possible date the cessation of the nuclear arms race and to take effective measures toward reductions in strategic arms, nuclear disarmament, and general and complete disarmament,[2]

Have agreed as follows:

### Article I[3]

1. Each State Party to the Treaty undertakes to prohibit, prevent and not to carry out at any place under its jurisdiction or control any nuclear explosion for peaceful purposes.
2. Each State Party further undertakes not to carry out, participate or assist in carrying out anywhere any nuclear explosion for peaceful purposes.

### Article II

For the purposes of this Treaty:

A 'nuclear explosion for peaceful purposes' is any nuclear explosion conducted outside designated nuclear weapon test sites in accordance with Articles III and IV of this treaty.

### Article III

For the purposes of this Treaty each State Party may designate one only nuclear weapon test site at any place under its jurisdiction or anywhere else.

### Article IV[4]

For the purpose of ensuring verification of compliance with the obligations of the Parties under the Treaty, each Party shall provide the following data for the Consultative Committee:

(a) the geographic co-ordinates of the boundaries of any nuclear weapon test site at any place under its jurisdiction and of the boundaries of the geophysically distinct testing areas therein;

(b) information on the geology of the testing areas of the sites (the rock characteristics of geological formations and the basic physical properties of the rock, i.e., density, seismic velocity, water saturation porosity and the depth of the water table);

(c) the geographic co-ordinates of underground nuclear weapon tests, after they have been conducted;

(d) yield, date, time, depth and co-ordinates for two nuclear weapon tests for calibration purposes from each geographically distinct testing area where underground nuclear weapon tests have been and are to be conducted.

Article V[5]

1. For the purpose of providing assurance of compliance with the provisions of this Treaty, each Party may use national technical means at its disposal in a manner consistent with the generally recognised principles of international law and undertakes not to interfere with such means of verification.

2. Each Party to this Treaty undertakes to co-operate in good faith in an effective international exchange of seismological data in order to facilitate the monitoring of this Treaty. The arrangements for this international exchange are laid down in Protocol I annexed to this Treaty.

3. The Parties to this Treaty undertake to consult one another and to co-operate in good faith for the clarification of all events pertaining to the subject matter of this Treaty. In accordance with this provision, each Party to the Treaty is entitled to:
(a) request and receive information from any other Party;
(b) request an on-site inspection in the territory of any other Party for the purpose of ascertaining whether or not a specified event was a nuclear explosion for peaceful purposes. Procedures for such inspections and the manner of their conduct, including the rights and functions of the inspecting personnel, are laid out in Protocol II to this Treaty.

4. In order to avoid unfounded accusations or misinterpretations of large non-nuclear explosions the Party conducting such an explosion may invite an inspection at the site of the explosion. The rules and procedures for such inspections are laid down in Protocol II to this Treaty.

5. For the purposes set forth in this Article a Consultative Commmittee shall be established to oversee the implementation of the Treaty and of the international verification arrangements. A Technical Expert Group and a permanent Secretariat shall be established to assist the Consultative Committee. The functions and rules of procedure of these bodies are set out in Protocol III of this Treaty.

## Article VI[6]

This Treaty does not affect obligations which have been assumed by Parties under other international agreements, including the Treaty Banning Nuclear Weapon Tests in the Atmosphere, in Outer Space and Under Water.

## Article VII[7]

Any Party may propose amendments to this Treaty. Such proposals shall be submitted to the Depositary, who shall, in consultation with States Parties, take appropriate action. Amendments shall enter into force for each Party accepting them upon their acceptance by a majority of the Parties to the Treaty and thereafter for each remaining Party on the date of acceptance by it.

## Article VIII[8]

Five years after the entry into force of this Treaty, a conference of Parties to the Treaty shall be held in Geneva, Switzerland, in order to review the operation of this Treaty with a view to ensuring that the purposes of the preamble and the provisions of the Treaty are being realised. At intervals of five years thereafter, a majority of the Parties to the Treaty may obtain, by submitting a proposal to this effect to the Depositary, the convening of a further conference with the same objective of reviewing the operation of the Treaty.

## Article IX[9]

1. This Treaty shall be open to all States for signature. Any state which does not sign the Treaty before its entry into force in accordance with paragraph 3 of this article may accede to it at any time.

2. This Treaty shall be subject to ratification by signatory States. Instruments of ratification and instruments of accession shall be deposited with the Secretary-General of the United Nations, who shall be the Depositary of this Treaty.

3. This Treaty shall enter into force upon the deposit with the Depositary of instruments of ratification by twenty Governments, including the Governments of the United States of America and the Union of Soviet Socialist Republics.

4. For those States whose instruments of ratification or accession are deposited after the entry into force of this Treaty it shall enter into force on the date of their deposit of their instruments of ratification or accession.

5. The Depositary shall promptly inform all signatory and acceding States of the date of each signature, the date of deposit of each instrument of ratification or of accession and the date of the entry into force of this Treaty and of any amendments thereto, any notice of withdrawal, as well as of the receipt of any notice. He shall also inform the Security Council of the United Nations of any notice of withdrawal.

6. This Treaty shall be registered by the Depositary in accordance with Article 102 of the Charter of the United Nations.

## *Notes*

1 Adapted from the preamble of the 1963 Partial Test Ban Treaty.
2 Taken from the Preamble of the 1972 ABM Treaty.
3 Adapted from Article III of the 1976 Peaceful Nuclear Explosions Treaty.
4 Adapted from Article I of the Protocol to the 1974 Threshold Test Ban Treaty.
5 Adapted from Article IV of the Swedish Draft Comprehensive Test Ban Treaty, CD/381, of 14 June 1983.
6 Taken from Article III of CD/381.
7 Taken from Article VI of CD/381.
8 Adapted from Article VII of CD/381.
9 Adapted from Article VIII of CD/381.

# Bibliography

## *Monographs*


Beaton, L., *Must the Bomb Spread?*, Penguin, Harmondsworth, 1966

Beckman, R. L., *Nuclear Non-Proliferation: Congress and the Control of Peaceful Nuclear Activities*, Westview Press, Boulder, Colorado, 1985

Bellany, I., *Australia in the Nuclear Age*, Sydney University Press, Sydney, 1972

Blumberg, S. A. and Owens, G., *Energy and Conflict: The Life and Times of Edward Teller*, G. P. Putnam's Sons, New York, 1976

Bolt, B. A., *Nuclear Explosions and Earthquakes: The Parted Veil*, W. H. Freeman and Co., San Francisco, 1976

Boskey, B. and Willrich, M. (eds), *Nuclear Proliferation: Prospects for Control*, Dunellen Publishing Co., New York, 1970

Brenner, M. J., *Nuclear Power and Non-Proliferation: The Remaking of US Policy*, Cambridge University Press, Cambridge, 1981

Broad, W. J., *The Star Warriors*, Simon and Schuster, New York, 1985

Brooks, D. B. and Krutilla, J. V., *Peaceful Use of Nuclear Explosives: Some Economic Aspects*, Resources for the Future, Inc., Washington DC, 1969

Carnesale, A. and Haass, R. (eds), *Learning From Experience With Arms Control*, Office of Sponsored Research, Presidents and Fellows of Harvard College, Cambridge, Mass., 1986

Chidambaram, R. and Ramanna, R., *Some Studies on India's Peaceful Nuclear Explosion Experiment*, Bhabha Atomic Research Centre, Trombay, 1974

Clarke, D. L., *Politics of Arms Control: The Role and Effectiveness of the US Arms Control and Disarmament Agency*, The Free Press, New York, 1979

Dahlman, O. and Israelson, H., *Monitoring Underground Nuclear Explosions*, Elsevier Scientific Publishing Co., Amsterdam, 1977

Danielsson, B. and Danielsson, M., *Poisoned Reign: French Nuclear Colonialism in the Pacific*, Penguin, Harmondsworth, 1986

Dyson, F., *Disturbing the Universe*, Pan, London, 1981

Epstein, W., *The Last Chance*, The Free Press, New York, 1976

Farnsworth, D. N. and McKenney, J. W., *US/Panama Relations 1903–1978: A Study in Linkage Politics*, Westview Press, Boulder, Colorado, 1983

Fischer, D. and Szasz P., *Safeguarding the Atom*, Stockholm International Peace Research Institute, Stockholm, 1985

Fischer, G., *The Non-Proliferation of Nuclear Weapons*, Europa Publications, London, 1971

Ford, D., *The Cult of the Atom*, Simon and Schuster, New York, 1982

Freeman, L. J., *Nuclear Witnesses: Insiders Speak Out*, W. W. Norton and Co., New York, 1981

Garthoff, R. L., *Detente and Confrontation*, The Brookings Institution, Washington DC, 1985
Goldblat, J. (ed.), *Non-proliferation: The Why and Wherefore*, SIPRI, Taylor and Francis, London, 1985
Goldblat, J. and Cox, D. (eds), *Nuclear Weapon Tests: Prohibition or Limitation?*, SIPRI, Oxford University Press, Oxford, 1988
Green, H. P. and Rosenthal, A., *Government of the Atom*, Atherton Press, New York, 1963
Greenberg, D. S., *The Politics of American Science*, Penguin, Harmondsworth, 1969
Greenwood, T., Feiveson, H. A. and Taylor, T. B., *Nuclear Proliferation: Motivations, Capabilities and Strategies for Control*, McGraw-Hill, New York, 1977
Guhin, M. A., *Nuclear Paradox: Security Risks of the Peaceful Atom*, American Enterprise Institute for Public Policy Research, Washington DC, 1976
Izrael, Y. A., *Peaceful Nuclear Explosions and The Environment*, US Joint Publications Research Service (JPRS 63502), Washington DC, 22 November 1974
Jacobsen, H. K. and Stein, E., *Diplomats, Scientists and Politicians: The US and the Nuclear Test Ban Negotiations*, University of Michigan Press, Ann Arbor, Michigan, 1976
Jain, J. P., *Nuclear India*, Radiant Publishers, New Delhi, 1974
Jungk, R., *Brighter Than A Thousand Suns*, Penguin, Harmondsworth, 1960
Kalkstein, M., *International Arrangements for the Peaceful Applications of Nuclear Explosives*, Stockholm International Peace Research Institute, Stockholm, 1970
Kincade, W. H. and Bertram, C. (eds), *Nuclear Proliferation in the 1980s*, Macmillan, London, 1982
Kissinger, H., *Years of Upheaval*, Little, Brown and Co., Boston, 1982
Kramish, A., *The Peaceful Atom in Foreign Policy*, Harper and Row, New York, 1963
Kreith, F. and Wrenn, C. B., *The Nuclear Impact*, Westview Press, Boulder, Colorado, 1976
Lamm, V., *The Utilisation of Nuclear Energy and International Law*, Budapest, 1984
Lawrence, R. M. and Larus, J., *Nuclear Proliferation Phase II*, University of Kansas Press, Witchita, 1974
Marks, A. W. (ed.), *NPT: Paradoxes And Problems*, Arms Control Association, Washington DC, 1975
Mastny, V. (ed.), *Disarmament and Nuclear Tests 1964–69*, Facts on File Inc., New York, 1970
Marwah, O. and Shulz, A. (eds), *Nuclear Proliferation and the Near-Nuclear Countries*, Ballinger Publishing Co., Cambridge, Mass., 1975
McBride, J. H., *The Test Ban Treaty: Military, Technological and Political Implications*, Henry Regency Co., Chicago, 1967
McNaught, K., *The Pelican History of Canada*, Penguin, Harmondsworth, 1975
Metzger, H. P., *The Atomic Establishment*, Simon and Schuster, New York, 1972
Milliken, R., *No Conceivable Injury*, Penguin, Ringwood, 1986
Mirchandani, G. G. and Namboodiri, P. K. S., *Nuclear India: A Technological Assessment*, Vision Books, New Delhi, 1981
Novik, S., *The Careless Atom*, Delta, New York, 1969
Pringle, P. and Spigelman, J., *The Nuclear Barons*, Avon Books, New York, 1981

Quester, G. H., *The Politics of Nuclear Proliferation*, Johns Hopkins University Press, Baltimore, Maryland, 1973

Rhodes, R., *The Making of the Atomic Bomb*, Simon & Schuster, New York, 1986

Sanders, R., *Project Plowshare*, Public Affairs Press, Washington DC, 1962

Scheinman, L., *The IAEA & World Nuclear Order*, Resources for the Future, Washington DC, 1987

Seaborg, G. T., *Nuclear Milestones*, W. H. Freeman and Co., San Francisco, 1972

Seaborg, G. T. and Corliss, W. R., *Man and Atom: Building a New World Through Nuclear Technology*, E. P. Dutton and Co., New York, 1971

Seaborg, G. T. and Loeb, B. S., *Kennedy, Khrushchev and the Test Ban*, University of California Press, Berkeley, 1981

Seaborg, G. T. and Loeb, B. S., *Stemming the Tide: Arms Control in the Johnson Years*, Lexington Books, Lexington, Mass., 1987

Shaker, M. I., *The Nuclear Non-Proliferation Treaty: Origin and Implementation 1959–1979*, Vols 1 and 2, Oceana Publications, London, 1980

Simpson, J. and McGrew, A. G. (eds), *The International Nuclear Non-Proliferation System*, Macmillan, London, 1984

Sobel, L. A. (ed.), *Disarmament and Nuclear Tests 1960–1963*, Facts on File Inc., New York, 1964

Smith, J., *Clouds of Deceit: The Deadly Legacy of Britain's Bomb Tests*, Faber and Faber, London, 1985

Spector, L. S., *Nuclear Proliferation Today*, Vintage Books, New York, 1984

Subrahmanyam, K., *Nuclear Myths and Realities: India's Dilemma*, ABC Publishing House, New Delhi, 1981

Teller, E., *Energy From Heaven and Earth*, W. H. Freeman and Co., San Francisco, 1979

Teller, E. and Latter, A. L., *Our Nuclear Future*, Criterion Books, New York, 1958

Teller, E., *The Legacy of Hiroshima*, Doubleday and Co., New York, 1982

Teller, E. *et al.*, *The Constructive Uses of Nuclear Explosives*, McGraw-Hill, New York, 1968

Voss, E. H., *Nuclear Ambush: The Test Ban Trap*, Henry Regency Co., Chicago, 1963

Weaver, L. E. (ed.), *Education for Peaceful Uses of Nuclear Explosives*, University of Arizona Press, Tucson, Arizona, 1970

Williams, R. C. and Cantelon, P. L. (eds), *The American Atom: A Documentary History of Nuclear Policies From the Discovery of Fission to the Present, 1939–1984*, University of Pennsylvania Press, Philadelphia, 1984

Willrich, M., *Global Politics of Nuclear Energy*, Praeger, New York, 1971

Willrich, M., *The Non-Proliferation Treaty: Framework for Nuclear Arms Control*, The Mitchie Company, Charlottesville, Virginia, 1969

Wilson, A. R. W., *Technical Aspects of Peaceful Nuclear Explosions*, SIPRI, Taylor and Francis, London, 1979

Yager, J. A. (ed.), *Nonproliferation and US Foreign Policy*, Brookings Institution Press, Washington DC, 1980

York, H. F. (ed.), *Arms Control*, W. H. Freeman and Co., San Francisco, 1973

York, H. F., *The Advisors: Oppenheimer, Teller and the Superbomb*, W. H. Freeman and Co., San Francisco, 1976

Young, E., *A Farewell to Arms Control?*, Penguin, Harmondsworth, 1972

## *Articles*

'AEC Plans Peaceful Explosions But Raises Anger', *Nature*, Vol. 241, 23 February 1973

Aja Espil, J. A., 'Argentina' in Goldblat, J. (ed.), *Non-proliferation: The Why and Wherefore*, SIPRI, Taylor and Francis, London, 1985

Bechhoefer, B. G., 'Plowshare Control' in Boskey, B. and Willrich, M. (eds), *Nuclear Proliferation: Prospects for Control*, Dunellen Publishing Co., New York, 1970

Bethe, H., 'Comments on the History of the H-Bomb, 1954' in Williams, R. C. and Cantelon, P. L. (eds), *The American Atom: A Documentary History of Nuclear Policies From the Discovery of Fission to the Present, 1939–1984*, University of Pennsylvania Press, Philadelphia, 1984

Betts, R. K., 'India, Pakistan and Iran' in Yager, J. A. (ed.), *Nonproliferation and US Foreign Policy*, Brookings Institution Press, Washington DC, 1980

Borg, I. Y., 'Nuclear Explosives — The Peaceful Side', *New Scientist*, 8 March 1984

Borg, I. Y., 'Peaceful Nuclear Explosions in Soviet Gas Condensate Fields', *Energy and Technology Review*, Lawrence Livermore National Laboratory, May 1983

Briggs, J. C. Jr., 'The Sea-Level Panama Canal: Potential Biological Catastrophe', *Bioscience*, January 1969

Brooks, D. B. and Myers, H. R., 'Plowshare Evaluation' in Boskey, B. and Willrich, M. (eds), *Nuclear Proliferation: Prospects for Control*, Dunellen Publishing Co., New York, 1970

Carr, F., 'The View From the Third World' in Kincade, W. H. and Bertram, C. (eds), *Nuclear Proliferation in the 1980s*, Macmillan, London, 1982

Clark, F. J. Jr., 'Plowshare Program Development Activities' in Weaver, L. E. (ed.), *Education for Peaceful Uses of Nuclear Explosives*, University of Arizona Press, Tucson, Arizona, 1970

Coffer, H. F. and Higgins, G. H., 'Future Contained Nuclear Explosives Experiments' in Weaver, L. E. (ed.), *Education for Peaceful Uses of Nuclear Explosives*, University of Arizona Press, Tucson, Arizona, 1970

Courtney, W. H., 'Brazil and Argentina: Nuclear Choices for Friendly Rivals' in Yager, J. A. (ed.), *Nonproliferation and US Foreign Policy*, Brookings Institution Press, Washington DC, 1980

Ehrlich, T., 'Legal Problems (of PNEs)' in Weaver, L. E. (ed.), *Education for Peaceful Uses of Nuclear Explosives*, University of Arizona Press, Tucson, Arizona, 1970

Emelyanov, V. A., 'On the Peaceful Use of Nuclear Explosions' in *Nuclear Proliferation Problems*, SIPRI, Almqvist and Wiksell, Stockholm, 1974

Ferber, G. J. and List, R. J., 'Prediction of External Gamma Dose From Nuclear Excavation of a Sea-level Canal', *Bioscience* , March 1969

Findlay, T. C., 'Twenty Minutes with Edward Teller', *Newsletter*, Peace Research Centre, Australian National University, Canberra, 2, 1, 1987

Fischer, D., 'The Role of the IAEA' in Simpson, J. and McGrew, A. G. (eds), *The International Nuclear Non-Proliferation System*, Macmillan, London, 1984

Garcia Robles, A., 'The Tlatelolco Treaty' in *SIPRI Yearbook 1969–1970*, SIPRI, Taylor and Francis, London, 1970

Gofman, J. W., 'Hazards To Man From Radioactivity', paper presented to the Third Plowshare Symposium, reproduced in US Congress, JCAE, Hearings on Peaceful Applications of Nuclear Explosives — Plowshare, 5 January 1965, US Government Printing Office, Washington DC, 1965.

Goldblat, J., 'Multilateral Arms Control Efforts' in *SIPRI Yearbook 1986*, Oxford University Press, Oxford, 1986

Goldblat, J., 'The Indian Nuclear Test and the NPT' in Marks, A. W. (ed.), *NPT: Paradoxes and Problems*, Arms Control Association, Washington DC, 1975

Goldblat, J., 'The Nuclear Explosion Limitation Treaties' in Goldblat, J. and Cox, D. (eds), *Nuclear Weapon Tests: Prohibition or Limitation?*, SIPRI, Oxford University Press, Oxford, 1988

Heckrotte, W., 'Negotiating with the Soviets', *Energy and Technology Review*, Lawrence Livermore National Laboratory, May 1983

Jain, G., 'India' in Goldblat, J., (ed.), *Nonproliferation: The Why and Wherefore*, SIPRI, Taylor and Francis, London, 1985

Jaipal, R., 'The Indian Nuclear Situation', *International Security*, Vol. I, No. 4, Spring 1977

Johnson, G. W., 'The Plowshare Program — History and Goals', US Congress, JCAE, Hearings on Peaceful Applications of Nuclear Explosives — Plowshare, 5 January 1965, US Government Printing Office, Washington DC, 1965

Johnson, G. W., 'Plowshare at the Crossroads', *Bulletin of the Atomic Scientists*, June 1970

Johnson, G. W., 'Technological Development of Nuclear Explosives Engineering' in Weaver, L. E. (ed.), *Education for Peaceful Uses of Nuclear Explosives*, University of Arizona Press, Tucson, Arizona, 1970

Johnson, G. W., 'The Soviet Program for Industrial Applications of Explosions', *Bulletin of the Atomic Scientists*, November 1960

Johnson, G. W. and Brown, H., 'Non-military Uses of Nuclear Explosives' in York, H. F. (ed.), *Arms Control*, W. H. Freeman and Co., San Francisco, 1973

Jones, R. W., 'India' in Goldblat, J. (ed.), *Nonproliferation: The Why and Wherefore*, SIPRI, Taylor and Francis, London, 1985

Kedrov, O., *Capabilities for Verification of a Nuclear Test Ban*, paper presented at the International Symposium on Seismic Monitoring, Australian National University, Canberra, August 1986

Koop, J., 'Plowshare and the Nonproliferation Treaty', *Orbis*, XII, No. 3, Fall 1968

Kruger, P., 'A Survey of University Courses in Nuclear Civil Engineering' in Weaver, L. E. (ed.), *Education for Peaceful Uses of Nuclear Explosives*, University of Arizona Press, Tucson, Arizona, 1970

Leggett, J., Clark, R. and Westaway, R., *The Possible Role of Confidence-Building Measures in Arriving at a Ban on Nuclear Weapons Testing*, paper presented at the Soviet Academy of Sciences' Forum for Peace, Moscow, 15 February 1987

Lomas, P., 'Attitudes of the Nuclear Threshold Countries' in Goldblat, J. and Cox, D. (eds), *Nuclear Weapon Tests: Prohibition or Limitation?*, SIPRI, Oxford University Press, Oxford, 1988

Long, F. A., 'Peaceful Nuclear Explosions', *Bulletin of the Atomic Scientists*, October 1976

Lough, T. S., 'Peaceful Nuclear Explosions and Disarmament', *Peace Research Reviews*, Canadian Peace Research Institute, Ottawa, 1968

Marwah, O., 'Epilogue: The NPT Review Conference, Geneva, 1975' in Marwah, O. and Shulz, A. (eds), *Nuclear Proliferation and the Near-Nuclear Countries*, Ballinger Publishing Co., Cambridge, Mass., 1975

McEwan, A. C., 'Environmental Effects of Underground Nuclear Explosions' in

Goldblat, J. and Cox, D. (eds), *Nuclear Weapon Tests: Prohibition or Limitation?*, SIPRI, Oxford University Press, Oxford, 1988
Millar, T. B., 'Australia: A Recent Ratification' in Lawrence, R. M. and Larus, J. (eds.), *Nuclear Proliferation Phase II*, University of Kansas Press, Witchita, 1974
Morrison, P., 'Constructive Uses of Nuclear Explosives' in York, H. (ed.), *Arms Control*, W. H. Freeman and Co., San Francisco, 1973
Reines, F., 'Are There Peaceful Engineering Uses of Atomic Explosives?', *Bulletin of the Atomic Scientists*, June 1950
Ridenour, L., 'The Hydrogen Bomb' in York H. F. (ed.), *Arms Control*, W. H. Freeman and Co., San Francisco, 1973
Rosenbaum, H. J., 'Brazil's Nuclear Aspirations' in Marwah, O. and Shulz, A. (eds), *Nuclear Proliferation and the Near-Nuclear Countries*, Ballinger Publishing Co., Cambridge, Mass., 1975
Rowen, H. S. and Brody, R., 'Nuclear Potential and Possible Contingencies' in Yager, J. A. (ed.), *Nonproliferation and US Foreign Policy*, Brookings Institution Press, Washington, DC, 1980
Scheimer, J. F. and Borg, I. Y., 'Deep Seismic Sounding with Nuclear Explosives in the Soviet Union', *Science*, Vol. 226, No. 4676, 16 November 1984
Scoville, H. Jr., 'Peaceful Nuclear Explosions — An Invitation to Proliferation' in Marks, A. W. (ed.), *NPT: Paradoxes and Problems*, Arms Control Association, Washington DC, 1975
Smart, I., 'Nonproliferation Treaty: Status and Prospects' in Marks, A. W. (ed.), *NPT: Paradoxes and Problems*, Arms Control Association, Washington DC, 1975
Smith, C., 'A Policy of Ambiguity: Nuclear Proliferation in South Asia', *ADIU Report*, Armament and Disarmament Information Unit, University of Sussex, Vol. 9, No. 4, July-August 1987
Subrahmanyam, K., 'India's Nuclear Policy' in Marwah, O. and Shulz, A. (eds), *Nuclear Proliferation and the Near-Nuclear Countries*, Ballinger Publishing Co., Cambridge, Mass., 1975
Tape, G. F., 'National Policy on Peaceful Uses of Nuclear Explosives' in Weaver, L. E. (ed.), *Education for Peaceful Uses of Nuclear Explosives*, University of Arizona Press, Tucson, Arizona, 1970
Taylor, T. B., 'Endless Generations of Nuclear Weapons', *Bulletin of the Atomic Scientists*, November 1986
Teller, E., 'The University's Role in Nuclear Explosives Engineering Research' in Weaver, L. E. (ed.), *Education for Peaceful Uses of Nuclear Explosives*, University of Arizona Press, Tucson, Arizona, 1970
'The Livermore Lab's Nuclear Test Detectives', *The Washington Post*, Washington DC, 10 June 1986
Trivedi, V. C., 'India's Approach Towards Nuclear Energy and Non-Proliferation of Nuclear Weapons' in Marks, A. W. (ed.), *NPT: Paradoxes and Problems*, Arms Control Association, Washington DC, 1975
Vogt, J. R., 'Radionuclide Production for the Nuclear Excavation of an Isthmian Canal', *Bioscience*, February 1969
Vortman, L. J., 'Nuclear Excavation' in Weaver L. E. (ed), *Education for Peaceful Uses of Nuclear Explosives*, University of Arizona Press, Tucson, Arizona, 1970
Wiesner, J. B. and York, H. F., 'National Security and the Nuclear Test Ban' in York, H. F. (ed.), *Arms Control*, W. H. Freeman and Co., San Francisco, 1973
Wilson, A. R. W., 'Current Status of Civil Engineering and Mineral Resources

Development Applications of Peaceful Nuclear Explosions' in Proceedings of the Fourth International Conference on the Peaceful Uses of Nuclear Energy, 6–16 September 1971, IAEA, New York and Vienna, 1972

Wilson, A. R. W., 'Educational Needs of Industry' in Weaver, L. E. (ed.), *Education for Peaceful Uses of Nuclear Explosives*, University of Arizona Press, Tucson, Arizona, 1970

Wilson, A. R. W., 'Technical Aspects of Peaceful Nuclear Explosions Relevant to their Possible Role in the Future Proliferation of Weapon-Usable Nuclear Materials' in *Nuclear Energy and Nuclear Weapon Proliferation*, SIPRI, Taylor and Francis, London, 1979

York, H. F., 'The Origins of the Lawrence Livermore Laboratory', *Bulletin of the Atomic Scientists*, September 1975

## Yearbooks

*SIPRI Yearbook 1968–69*, Almqvist and Wiksell, Stockholm, 1969

*SIPRI Yearbook 1969–70*, Taylor and Francis, London, 1970

*SIPRI Yearbook 1973*, Taylor and Francis, London, 1973

*SIPRI Yearbook 1974*, Almqvist and Wiksell, Stockholm, 1974

*SIPRI Yearbook 1975*, Taylor and Francis, London, 1975

*SIPRI Yearbook 1976*, Almqvist and Wiksell, Stockholm, 1976

*SIPRI Yearbook 1977*, Almqvist and Wiskell, Stockholm, 1977

*SIPRI Yearbook 1978*, Taylor and Francis, London, 1978

*SIPRI Yearbook 1980*, Taylor and Francis, London, 1980

*SIPRI Yearbook 1981*, Taylor and Francis, London, 1981

*SIPRI Yearbook 1983*, Taylor and Francis, London, 1983

*SIPRI Yearbook 1985*, Taylor and Francis, London, 1985

*SIPRI Yearbook 1986*, Oxford University Press, Oxford, 1986

*SIPRI Yearbook 1987*, Oxford University Press, Oxford, 1987

*SIPRI Yearbook 1988*, Oxford University Press, Oxford, 1988

*The United Nations and Disarmament 1945–1970*, Department of Political and Security Affairs, United Nations, New York, 1970

*UN Disarmament Yearbook 1976*, United Nations, New York, 1977

*UN Disarmament Yearbook 1977*, United Nations, New York, 1978

*UN Disarmament Yearbook 1978*, United Nations, New York, 1979

*UN Disarmament Yearbook 1979*, United Nations, New York, 1980

*UN Disarmament Yearbook 1980*, United Nations, New York, 1981

*UN Disarmament Yearbook 1982*, United Nations, New York, 1983

*UN Disarmament Yearbook 1983*, United Nations, New York, 1984

*UN Disarmament Yearbook 1986*, United Nations, New York, 1987

## Reports

*A Legal Framework for Industrial Plowshare Applications*, Atomic Industrial Forum Inc., New York, 1969

*An Analysis of the Economic Feasibility, Technical Significance and Time Scale for Application of Peaceful Nuclear Explosions in the US, with Special Reference to the GURC Report Thereon (The Long Report)*, Program on Science, Technology and Society, Cornell University, US Arms Control and Disarmament Agency,

Washington DC, 1975

Arnett, J. T., *Industry and Nuclear Explosives: Facts About An Emerging Technology and Its Immense Potential for Benefit*, Atomic Industrial Forum Inc., New York, July 1973

Australian Atomic Energy Commission (AAEC), *Annual Report 1968–69*, Australian Government Publishing Service, Canberra, 1969

Australian Atomic Energy Commission, *Peaceful Uses of Nuclear Explosions — An Evaluation, for Australian Purposes, of Proposed Civil Engineering and Mining Applications*, AAEC, Sydney, 1964

Borg, I. Y., *Nuclear Explosions for Peaceful Purposes*, Lawrence Livermore National Laboratory Paper No. UCRL–95026, Rev. II, November 1986 [reprinted in Goldblat, J. and Cox, D. (eds), *Nuclear Weapon Tests: Prohibition or Limitation?*, SIPRI, Oxford University Press, Oxford, 1988]

*Environment of the Cape Thompson Region, Alaska*, USAEC, US Government Printing Office, Washington DC, 1966

Ericsson, U., *The Question of Nuclear Explosions for Peaceful Purposes by Non-Nuclear Weapon States and the Possibility of Misuse of Such Technology for the Production of Nuclear Weapons*, Document A/CONF. 35/DOC. 3, Conference of Non-Nuclear Weapon States, Geneva, 1968

*Fourth Plowshare Symposium Summaries*, American Nuclear Society, Hinsdale, Illinois, 1970

Ginsburg, T., *The Question of Peaceful Explosions for the Benefit of Non-Nuclear Weapon States*, Document A/CONF. 35/DOC. 2, Conference of Non-Nuclear Weapon States, Geneva, July 1968

*Interdisciplinary Seminar on Nuclear Technology, Society and Health*, University of California, Berkeley, Transcript of seventh meeting (unpublished), 30 October 1986

International Atomic Energy Agency (IAEA), *Annual Report 1975*, GC(XX)/565, IAEA, Vienna, 1976

International Court of Justice, *Application Instituting Proceedings on behalf of Australia against the French Republic*, Australian Department of Foreign Affairs, Canberra, 1972

Long, F. A., *Peaceful Nuclear Explosions*, unpublished manuscript for ACDA, Washington DC, 1 September 1975

*Major Activities in the Atomic Energy Programs, January-December 1959*, US Atomic Energy Commission, US Government Printing Office, Washington DC, January, 1960

*Negotiations on A Comprehensive Test Ban 1963–71*, US Arms Control and Disarmament Agency, Unclassified Document Series, DDS Ref. #587, Washington DC, 22 December 1971

*PNE Activity Projections for Arms Control Planning* (the *GURC Report*), Gulf Universities Research Consortium, Galveston, Texas, US Arms Control and Disarmament Agency Document, ACDA/PAB-253, Washington DC, 1975

*Report of the Ad Hoc Advisory Group on Nuclear Explosions for Peaceful Purposes*, IAEA Document GOV/1854, 26 August 1977, IAEA, Vienna, 1977

*Report of the Chairman of the Working Group on a South Pacific Nuclear Free Zone to the South Pacific Forum, Raratonga, 4–6 August 1985*, Australian Department of Foreign Affairs, Canberra, 1985

*Report of the US Delegation to the International Conference on the Peaceful Uses*

*of Atomic Energy, 1-13 September 1958*, USAEC, US Government Printing Office, Washington DC, 1958
*Review of the Peaceful Uses of Nuclear Explosives (PNE)*, Aldermaston Weapons Research Establishment (AWRE), Document 10/74, June 1974, HM Stationary Office, London
Seaborg, G. T. *et al.*, *Plowshare*, Division of Technical Information, US Atomic Energy Commission, Oak Ridge, Tennessee, 1966
*The American Experience with Peaceful Nuclear Explosions*, US Arms Control and Disarmament Agency, US Government Printing Office, Washington DC, 1978
*Towards a Comprehensive Test Ban*, The Palme Commission on Disarmament and Security Issues, Stockholm, 1987
US Atomic Energy Commission, *Annual Report to Congress, 1970*, US Government Printing Office, Washington DC, 1970
*US Perspectives on Peaceful Nuclear Explosions*, unpublished ms, ACDA, Washington DC, 7 February 1978

## Documents

*Arms Control and Disarmament Agreements*, United States Arms Control and Disarmament Agency, Washington DC, 1982
Australian Parliamentary Hansard, 15 October 1968, 25 February 1969, 25 March 1969, 13 May 1969, 19 September 1969
Australian Parliamentary Hansard, Senate Estimates Committee, 15 October 1968
Australian Senate Record, 25 February 1969, 19 September 1969
Conference of the Committee on Disarmament, Geneva, Documents CCD/130, CCD/346, CCD/348, CCD/523, CCD/536, CCD/562/Rev. 1, CCD/PV. 512, CCD/PV.576, CCD/PV.643, CCD/PV.647, CCD/PV.801
Committee (later Conference) on Disarmament, Geneva, Documents CD/130, CD/346, CD/381, CD/383, CD/405, CD/746, CD/758, CD/PV. 344.
Conference of Non-Nuclear Weapon States, Geneva, 1968, Document A/CONF. 35/DOC. 15
Eighteen-Nation Disarmament Committee (ENDC) Documents, ENDC/PV.192, ENDC/PV.193/Rev. 1, ENDC/PV.234, ENDC/PV.242, ENDC/PV.280, ENDC/PV.284, ENDC/PV.285, ENDC/PV.287, ENDC/PV.288, ENDC/PV.293, ENDC/PV.294, ENDC/PV.295, ENDC/PV.298, ENDC/PV.299, ENDC/PV.302, ENDC/PV.310, ENDC/PV.327, ENDC/PV.329, ENDC/PV.330, ENDC/PV.331, ENDC/PV.334, ENDC/PV.335, ENDC/PV.337, ENDC/PV.343, ENDC/PV.344, ENDC/PV.345, ENDC/PV.373, ENDC/PV.387, ENDC/PV.404, ENDC/PV.409, ENDC/PV.414, ENDC/PV.415
*Final Document of the Special Session of the General Assembly on Disarmament, 1978*, UN Department of Public Information, New York, 1978
International Atomic Energy Agency, Vienna, Documents GC(XXX)/544, GOV/COM.23/2, GOV/COM.23/15, GOV/COM.1756
NPT Review Conference Documents NPT/CONF. II/CII/36, NPT/CONF. III/II, NPT/CONF./12
*Status of Multilateral Arms Regulation and Disarmament Agreements*, United Nations, Second Edition, UN Department for Disarmament Affairs, New York, 1983
'Text of the Agreement of 22 July 1977 Between Argentina and the Agency for

the Application of Safeguards in Connection with a Cooperation Agreement Between Argentina and Canada', INFCIRC/251, 25 November 1977

UN General Assembly, First Committee Document A/C.1/40/PV.45

UN General Assembly, Resolution 3261D (XXIX), 9 December 1974

US Congress, Joint Committee on Atomic Energy (JCAE), Hearings on AEC Authorizing Legislation for Fiscal Year 1961, US Government Printing Office, Washington DC, 1960

US Congress, JCAE, Hearings on AEC Authorizing Legislation for Fiscal Year 1965, US Government Printing Office, Washington DC, 1964

US Congress, JCAE, Hearings on AEC Authorizing Legislation for Fiscal Year 1966, US Government Printing Office, Washington DC, 1965

US Congress, JCAE, Hearings on AEC Authorizing Legislation for Fiscal Year 1967, US Government Printing Office, Washington DC, 1966

US Congress, JCAE, Hearings on AEC Authorizing Legislation for Fiscal Year 1970, US Government Printing Office, Washington DC, 1969

US Congress, JCAE, Hearings on AEC Authorizing Legislation for Fiscal Year 1971, US Government Printing Office, Washington DC, 1970

US Congress, JCAE, Hearings on AEC Authorizing Legislation for Fiscal Year 1972, US Government Printing Office, Washington DC, 1971

US Congress, JCAE, Hearings on AEC Authorizing Legislation for Fiscal Year 1973, US Government Printing Office, Washington DC, 1972

US Congress, JCAE, Hearings on Frontiers in Atomic Energy Research, 22–26 March 1960, US Government Printing Office, Washington DC, 1960

US Congress, JCAE, Hearings on Peaceful Application of Nuclear Explosives — Plowshare, 5 January 1965, US Government Printing Office, Washington DC, 1965

US Congress, JCAE, Hearings on PNE Services for Domestic and Foreign Users, 8–9 May and 17 July 1969, US Government Printing Office, Washington DC, 1969

US Congress, *Nuclear Non-Proliferation Fact Book*, US Government Printing Office, Washington DC, 1977

US Congress, Senate Foreign Relations Committee, Hearings on the Nuclear Test Ban Treaty, 12–15 August, 19–23 August, 26–27 August 1963, US Government Printing Office, Washington DC, 1963

US Congress, Senate Foreign Relations Committee, Hearings on the Threshold Test Ban Treaty and Peaceful Nuclear Explosions Treaty, 28 July, 3 August, 8 and 15 September 1977, US Government Printing Office, Washington DC, 1977

US Congress, Senate Foreign Relations Committee, Hearings on the Treaty of Tlatelolco, 22 September 1970 and 23 February 1971, US Government Printing Office, Washington DC, 1971

US Congress, Senate Foreign Relations Committee, Hearings on the NPT, 18 and 20 February 1969, US Government Printing Office, Washington DC, 1969

US Congress, Senate Foreign Relations Committee, Hearings on the TTBT/PNET, 13 and 15 January 1987, US Government Printing Office, Washington DC, 1987

US Congress, Subcommittee on Public Lands of the Senate Committee on Interior and Insular Affairs, Hearings, 11 May 1973 (*The Haskell Hearings*), US Government Printing Office, Washington DC, 1973

US Congressional Record, 28 January and 14 April 1969, US Government Printing Office, Washington DC, 1969

US Disarmament Administration, *Geneva Conference on the Discontinuance of Nuclear Weapon Tests: History and Analysis of Negotiations*, US Department of

State, Washington DC, 1961
Western Australian Parliamentary Hansard, 2 January 1969, Thursday 27 March 1969

## *Journals*

*ADIU Report*
*Arms Control Chronicle*
*Arms Control Reporter*
*Arms Control Update*
*Bioscience*
*Bulletin of the Atomic Scientists*
*Current Notes*
*Disarmament Newsletter*, Department of Foreign Affairs, Canberra
*Energy and Technology Review*, Lawrence Livermore National Laboratory
*International Security*
*Nature*
*New Scientist*
*Newsletter*, Peace Research Centre, Australian National University, Canberra
*Orbis*

## *Newspapers*

The *Age* (Melbourne)
The *Canberra Times*
The *Herald* (Melbourne)
The *International Herald Tribune*
The *New York Times*
The *Sun-Herald* (Sydney)
The *Sun* (Melbourne)
The *Sydney Morning Herald*
*Washington Post*
*Washington Times*

# Index